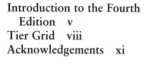CONTENTS

M O D U L A R
Mathematics
for GCSE

Brian Gaulter and Leslye Buchanan

OXFORD
UNIVERSITY PRESS

OXFORD

Great Clarendon Street, Oxford OX2 6DP

Oxford University Press is a department of the University
of Oxford. It furthers the University's objective of excellence in
research, scholarship, and education by publishing worldwide in

Oxford New York
Auckland Bangkok Buenos Aires Cape Town Chennai
Dar es Salaam Delhi Hong Kong Istanbul Karachi Kolkata
Kuala Lumpur Madrid Melbourne Mexico City Mumbai Nairobi
São Paulo Shanghai Taipei Tokyo Toronto

Oxford is a registered trade mark of Oxford University Press
in the UK and in certain other countries

First published 1991
National Curriculum edition 1993
Third edition 1997
Fourth edition 1999

British Library Cataloguing in Publication Data

Data available

ISBN 0 19 914762 0 (School and College Edition)
ISBN 0 19 914763 9 (Bookshop edition)

10   9   8   7   6   5

Typeset and illustrated by Tech-Set Limited

Printed and bound in Great Britain

Modular syllabuses for GCSE Mathematics

This book is designed for the 'mature' student who hopes to obtain a GCSE certificate in Mathematics. A mature student is anyone over the age of 16.

This edition is specifically designed for the revised GCSE Modular Mathematics Syllabus as examined by the Southern Examining Group (SEG) from 2000 (i.e. examinations on or after Winter 1999). In this syllabus, three different sections of mathematics are studied in turn, with an examination at the end of each module or section. The Modular Mathematics course has been created to offer a new opportunity in mathematics for mature students, and the modules contain mathematics relevant to everyday life, particularly the Money Management and Number module. Mature students usually have considerable skills in mathematics, of which they are often not aware. The authors' intention is for the students to build upon these skills, while gaining an appreciation of their use in GCSE mathematics.

This book is divided into three modules, each corresponding to a module of the SEG exam. Some basic numeracy has been incorporated at the beginning of the Money Management and Number module because the National Curriculum Attainment Target 2 is partly assessed through the module. The coverage of the attainment targets is as follows.

Module	National Curriculum Attainment Target
Money Management and Number	The Number part of 2 (Number and Algebra)
Statistics and Probability	4 (Handling data)
Terminal Module	The Algebra part of 2 (Number and Algebra), and 3 (Shape and space)

There are three differentiated levels of entry for the examination in each module: Higher tier (Grades A*, A, B C), Intermediate tier (Grades B, C, D, E), and Foundation tier (Grades D, E, F, G).

Material which is only relevant at the Higher Level, i.e. Grades A* and A, is indicated with an asterisk (*). Some material which is not on the National Curriculum is included to enable mature students to see how Mathematics is used in real life (e.g. Personal Savings). Such work is indicated with a dagger (†) to show that these topics need not be covered for the examination.

The book contains all the topics in other GCSE Mathematics syllabuses.

Examination Technique

In the new (2000) syllabus one half of the marks in each of the modules, including the Money Management and Number module, is allocated to questions that must be solved without the use of a calculator. For this reason an Appendix has been added on pages 501–509 to cover non-calculator techniques. (This Appendix replaces the Aural Tests which are no longer needed.)

The accuracy required in the examination questions will be as specific in the National Curriculum. Answers involving money should be given to the nearest penny and exact answers should always be given where possible. Where appropriate, other levels of accuracy can be required, and in Probability, fractional answers are usually expected.

As a guide, students at the Higher levels should use at least four figures in their working and give answers to three significant figures. The answer must be rounded to the third figure.

Students working at lower levels of the National Curriculum will often be giving answers to two significant figures. These students must always use at least three figures in their calculations. Answers such as 142 apples should always be left as 142 and *not* rounded to 140. Exact answers should always be given wherever possible.

There are many occasions (e.g. when using trigonometry or using iteration) when it is more convenient to use the memory function of the calculator and retain all figures until the end, rather than keep rounding off at each stage.

Students who show working and show step-by-step solutions gain credit in examination even when the final answer is incorrect, whereas students who obtain an incorrect answer and have not written down any working are unable to gain any credit.

It is essential that students consider the sense of their answers. Decimal points are frequently misplaced in error, and, for instance, electricity bills of £17,411 should be checked to see whether 17 411 pence (or £174.11) was the actual answer!

Applications paper

In Mathematics examinations set to test the Mathematics National Curriculum, Attainment Target 1 (Using and Applying Mathematics) is assessed separately and weighted at 20% of the whole course. This attainment target includes skills which cannot be assessed by means of a normal written examination paper containing a number of relatively short questions. Among these skills are the ability to produce imaginative and creative work arising from mathematical ideas, and the conducting of individual enquiry and experiment.

Sometimes, this Attainment Target is assessed by coursework. In the SEG Modular Examination, this Attainment Target is assessed by the Applications Paper, which forms part of the Terminal Module. The two-hour Applications Paper (two-and-a-half-hour Paper at the Higher Tier) consists of two tasks. Students may attempt both tasks but are advised to attempt *one* task only in depth rather than attempt two tasks in less depth.

In 'Using and Applying Mathematics' the work produced will be assessed under the three headings:

- Making and monitoring decisions to solve problems
- Communicating mathematically
- Developing skills of mathematical reasoning

It is essential that the work you produce is able to be assessed in each of these three areas. To produce work excelling in just one of these three areas is not profitable; you should aim to reach a creditable standard in **all three** areas.

The mathematical techniques used must be appropriate and relevant to the problem. To obtain a high-level result students must show an appropriately high level of competence and understanding. Coursework submitted as part of a higher-tier entry would gain benefit from including techniques which only occur on the higher-tier

syllabus. For example, in Statistics, standard deviation and the mean of grouped data could be included where appropriate.

Additional work, which simply shows the same skills as have previously been shown, does not gain credit. Extra work on a topic to produce more material which shows *only* the same skills would gain no more credit than the original piece of work. Extra work to show *new* skills by 'following new lines of enquiry' **would** gain more credit. **Quality, not quantity**, is the important criterion.

Presentation of work

All work needs a suitable writing up. This should satisfy the following guidelines.

There are three parts to any work, which require significant allocations of time.

These are the planning, the work itself, and the writing up.

The following is only a guide on what to include in your work. Although your work is unlikely to fit this format exactly, try to include as many suggestions as possible, as this is *vital* to your grade.

Remember that a good investigation can be let down by a poor write-up.

Planning

Once you have chosen which task to attempt, or even both tasks, you should spend some time thinking about:

- the starting point for the work
- what information you will need
- what you expect this information to tell you, your prediction of the result you expect to find and how it will relate back to your starting point.

Having a plan like this will mean that you have a good idea of how much work and time the coursework will involve before you even start. This should allow you to organise your time efficiently and not waste it on unnecessary tasks. As the work progresses, you may have to change your plan in the light of your increasing knowledge of the topic. If you keep notes of the plan and any changes made as you proceed, you will have the information to look back on when you write up your project.

Carrying out your work

Try to fit your plan of action into the time available for your work.

Keep notes on what you do and whether you feel it has been successful. This will help when you come to write up your summary.

You should include algebra in your working on the Applications Paper if at all possible. If you consider that you can find an equation which represents a 'solution' to the task, you should clearly:

- show how you have obtained the equation
- show that you have tested the equation which you have found, to try and prove that it is correct.

Writing up

Making and monitoring decisions to solve problems. This could include why you chose to use a particular method and not others that you may have considered. It should state what you expected to find (e.g. a hypothesis or general result). Include any problems you expected to encounter in carrying out your investigation. State what problems, if any, you had in obtaining the result. State how each part of your investigation relates to the task.

Communicating mathematically. Give details of how you obtained your information, emphasising why you chose to use your method rather than another method. Each table or diagram should have a number and a title. The title should make it obvious what the figures refer to. This section should be totally factual and consist of a commentary linking each table or diagram. The conclusion should briefly outline your investigation and what you found, using mathematical symbols where appropriate. You need to make interpretations and judgements about your results and how you have developed your original work further. It should include any generalisations or conclusions that can reasonably be made from your results.

Mathematical reasoning. The work produced should involve explaining the reasoning behind your work, the way you made a generalisation and how you tested out this theory before moving onto the next stage. If appropriate, give details of how you checked the accuracy of your measurements.

You will need to justify any of your results or conclusions.

That's all the advice we can offer you for the moment. We hope you will find our book useful, and wish you every success in the examination.

Brian Gaulter and Leslye Buchanan
Hampshire, May 1999

Tier Grid

This grid shows which sections(s) are appropriate to each tier (H = Higher, I = Intermediate, F = Foundation) of the SEG Examination (+ topics not required for examination).

Section	H	I	F	Section	H	I	F
1.1	•	•	•	8.2	•	•	•
1.2	•	•	•	8.3	•	•	
1.3	•	•	•	8.4	•	•	
1.4	•			8.5	•	•	•
1.5	•	•	•	8.6	•		
1.6	•	•	•	9.1	•	•	•
2.1	•	•		9.2	•	•	
2.2	•	•		10.1	•	•	•
2.3	•	•		10.2	•	•	•
2.4	•	•		11.1	•	•	•
3.1	•	•	•	11.2	•	•	
3.2	•	•	•	11.3	+		
3.3	•	•	•	12.1	•	•	•
3.4	•	•	•	12.2	+		
4.1	•	•	•	12.3	+		
4.2	•	•	•	13.1	•		
4.3	•	•	•	13.2	•	•	•
4.4	•	•		13.3	•	•	
4.5	•	•	•	13.4	•	•	
4.6	•	•		14.1	•	•	•
5.1	•	•	•	14.2	•	•	•
5.2	•	•	•	14.3	•	•	•
6.1	•	•	•	15.1	•	•	
6.2	•	•	•	15.2	•	•	•
6.3	•	•	•	15.3	•		
6.4	•	•	•	15.4	•	•	•
6.5	•	•	•	16.1	•	•	•
7.1	•	•	•	16.2	•		
7.2	•			17.1	•	•	
7.3	•			17.2	•		
8.1	•	•	•	17.3	•	•	•

Section	H	I	F	Section	H	I	F
18	●		●	27.2	●	●	●
19.1	●	●	●	27.3	●		
19.2	●			27.4	●		
20	●	●	●	28.1	●	●	●
21.1	●	●	●	28.2	●	●	
21.2	●	●	●	28.3	●	●	
21.3	●	●	●	29.1	●	●	●
21.4	●			29.2	●	●	●
21.5	●	●		29.3	●	●	●
21.6	●	●	●	29.4	●	●	●
21.7	●	●	●	29.5	●	●	●
21.8	●	●		29.6	●	●	
22.1	●	●	●	30.1	●	●	●
22.2	●	●	●	30.2	●	●	●
22.3	●	●	●	30.3	●	●	●
23.1	●	●	●	30.4	●	●	
23.2	●	●	●	30.5	●	●	●
23.3	●	●	●	30.6	●	●	●
23.4	●	●	●	31.1	●	●	●
23.5	●			31.2	●	●	●
23.6	●			31.3	●	●	●
23.7	●	●	●	31.4	●	●	●
23.8	●			32.1	●	●	●
23.9	+			32.2	●	●	●
24.1	●	●	●	32.3	●	●	●
24.2	●	●	●	32.4	●	●	●
24.3	●	●		32.5	●		
24.4	●	●		33.1	●	●	●
25.1	●	●	●	33.2	●	●	
25.2	●	●	●	33.3	●	●	
25.3	●	●	●	33.4	●		
25.4	●	●	●	33.5	●	●	
25.5	●	●		33.6	●		
26.1	●	●		34.1	●	●	●
26.2	●	●		34.2	●		
26.3	●	●		34.3	●		
26.4	●	●		34.4	●		
27.1	●	●	●	35.1	●	●	

Section	H	I	F	Section	H	I	F
35.2	●	●		42.3	●	●	+
35.3	●	●		42.4	●	●	●
35.4	●	●		42.5	●	●	
35.5	●	●		42.6	●	●	
36.1	●	●		43.1	●	●	●
36.2	●	●		43.2	●	●	●
37.1	●	●	●	44.1	●	●	●
37.2	●	●	●	44.2	●	●	●
37.3	●	●	●	44.3	●	●	●
37.4	●	●	●	44.4	●	●	●
37.5	●	●	●	44.5	●	●	●
37.6	●	●	●	44.6	●	●	
37.7	●	●	●	45.1	●	●	
37.8	●	●	●	45.2	●	●	
38.1	●	●	●	45.3	●	●	
38.2	●	●	●	45.4	●		
38.3	●	●	●	46.1	●		
39	●	●	●	46.2	●	●	
40.1	●	●	●	46.3	●		
40.2	●	●	●	46.4	●		
40.3	●	●		47.1	●		
40.4	●	●	●	47.2	●		
41.1	●	●		47.3	●		
41.2	●	●		47.4	●		
41.3	●			47.5	●		
41.4	●			47.6	●	●	
42.1	●	●		47.7	●	●	
42.2	●	●	●	47.8	●		

Acknowledgements

The publisher and authors are grateful to the following organisations for permission to reproduce previously published material:

Abbeylink, Debenhams, Marks and Spencer, and Principles for credit and store cards, and the Association for Payment Clearing Services, for the Cheque Guarantee logo. (p. 95)
Daily Mail, for the InterCity headline (p. 50)
HMSO, for extracts from *Social Trends* and the *Annual Abstract of Statistics* (pp. 258–259)
Jacobs Suchard Ltd, for the use of Toblerone (p. 448)
Lloyds Bank, for the Access card and cheques and statement forms (pp. 83, 86, 95)
National Westminister Bank for cheque forms (p. 84)
The Office of Population Censuses and Surveys, for a table (p. 150)
Pedigree Pet Foods, for allowing a mention of Pedigree Chum (p. 53)
Service Publications Ltd, for the map of York, here much simplified (p. 384)
Southern Examining Group, for questions from past examination papers (Sp = Spring, S = Summer, W = Winter, followed by year, P = part question) (pp. 5, 12, 19, 23, 35, 40, 47, 53, 61, 67, 78, 81, 93, 101, 123, 129, 130, 136, 141, 143, 147, 174, 179, 180, 181, 195, 201, 210, 211, 219, 220, 221, 224, 227, 232, 233, 271, 282, 288, 292, 305, 308, 312, 323, 346, 353, 356, 357, 362, 369, 375, 377, 379, 387, 404, 419, 427, 443, 445, 450, 452, 453, 466, 468, 469, 472, 483, 490, 491)
Which?, for information on insurance (p. 121)

The cover image ('Conjugate attractors') was created by Michael Field and Martin Golubitsley and is based on a picture in their book *Symmetry in Chaos* (published by OUP in 1992).

Every reasonable effort has been made to contact copyright owners, but we apologise for any unknown errors or omissions. The list will be corrected, if necessary, in the next reprint.

1 Basic Numeracy

Even in prehistoric times the basis of arithmetic was being developed. The earliest number system was one, two, many. Different number systems were adopted as different cultures evolved, but, in time, the **decimal system** became the most widely used.

This system, which we all use today, is based on the number 10, probably because human beings have ten fingers which are convenient for counting.

The numbers below 10 are called **digits**.

1.1 Numbers

- **Natural numbers** are the counting numbers 1, 2, 3, ...
  When zero is added to the set we have the
- **positive integers**: 0, 1, 2, 3, ...
  The positive integers and negative integers gives the set of
- **integers** or **directed numbers**: $-4, -3, -2, -1, 0, 1, 2, \ldots$.
  When we add **fractions**, e.g. $\frac{7}{10}$, $3\frac{1}{2}$, -1.25, we obtain the set of
- *****rational numbers** which are numbers that can be written in the form:

$$\frac{a}{b} \text{ where } a \text{ and } b \text{ are both integers.}$$

- *****Irrational numbers** are numbers which cannot be written as fractions, or as exact decimals, e.g. π, $\sqrt{2}$, $3^{\frac{1}{4}}$.

A **terminating decimal** is a decimal that can be written exactly with a fixed number of digits.
For example, $2\frac{1}{4} = 2.25$ is a terminating decimal whereas
$2\frac{1}{7} = 2.142\,857\,14\ldots$ is not. It is a **recurring decimal**.

Even number: integer (whole number) divisible by 2,
  e.g. 2, 4, 6, . . . or $-2, -4, -6, \ldots$

Odd number: integer not divisible by 2,
  e.g. 1, 3, 5, . . . or $-1, -3, -5, \ldots$

Factor: a number which divides exactly into a given number leaving no remainder, e.g. 4 is a factor of 12.

Multiple: a number which is formed by multiplying a given number by an integer is a multiple of the given number, e.g. 20 is a multiple of 2, 4, 5 and 10.

Prime number: a number which has precisely two factors, the number itself and 1, e.g. 2, 3, 5, 7, 11, . . .

Square number: a number which is the result of multiplying a number by itself, e.g. 1, 4, 9, 16, ...

MONEY MANAGEMENT AND NUMBER

The **four rules of arithmetic** are:

Addition: the answer to which is called the **sum**

Subtraction: the answer to which is called the **difference**

Multiplication: the answer to which is called the **product**

Division: the answer to which is called the **quotient**.

Example 1

Divide 30 by numbers below 30.

Those which give an integer result (i.e. a whole number) are factors.

Factors of 30 are: 1, 2, 3, 5, 6, 10, 15, 30.

Example 2

Write down which of the numbers 6, 7, 15, 19, 28, 29 are prime.

If any number has a factor, other than itself and 1, it is not prime.

$\dfrac{28}{2} = 14$, means that 28 has a factor (of 2) and 28 is not prime.

[Note that all the factors of 28 are 1, 2, 4, 7, 14, 28).

The prime numbers are: 7, 19, 29.

Exercise 1A

1 From the set of numbers, 4, 17, 24, 41, 49, write down:

  a a multiple of 8

  b two square numbers

  c a prime number

  d a factor of 8

  e a number which is the sum of two numbers in the set.

2 Find the sum of the first four odd numbers.

3 **a** Find the product of 100 and 6013.

  b Write your answer to **a** in words.

4 What is the remainder when 1538 is divided by 39?

5 **a** Which number from 1 to 25 has the largest number of factors?

  b Which number from 1 to 25 has the smallest number of factors?

6 What is the difference between five thousand and twenty two and eight thousand, two hundred and five?
Give your answer in words.

7 Write down the nearest integer to each of the following numbers:

  a -3.75 **b** $5\frac{1}{3}$ **c** $\sqrt{17}$

  d π **e** $(1.92)^2$

8 From the set of numbers, $\sqrt{16}$, $\sqrt{19}$, $\sqrt{7}$, $\sqrt{49}$, $\sqrt{81}$, write down any which are:

  ***a** irrational numbers

  b square numbers

  c even numbers

  d odd numbers

  e prime numbers.

A calculator should *not* be used in the following questions:

9 Add 17, 33 and 12.

10 Write down the seventh prime number.

11 Subtract 583 from 1000.

12 From the set of numbers, -8, 4, $\sqrt{5}$, π, 19, write down those which are:

   a integers

   b natural numbers

  *****c** irrational numbers

   d prime numbers

   e multiples of 2.

13 The attendance at a football match was eighteen thousand and forty three. Write this number in figures.

14 Write the amount twenty thousand, one hundred and six pounds in figures.

15 Write the number 130 402 in words.

16 List, in ascending order, all the factors of 24.

17 From the set of numbers, 2, $\sqrt{16}$, 9, $\sqrt{101}$, 15, 16, 19, 45, write down:

   a two prime numbers

   b two square numbers

  *****c** an irrational number

   d a factor of 30

   e a multiple of 9.

18 List the numbers $(2.9)^2$, 9.34, 3^2, $\sqrt{63}$, $\sqrt{65}$, in descending order of size.

19 Find:

a 1.05×10		**f** 0.0356×0
b $1.05 \div 10$		**g** 45.7×100
c 0.003×10		**h** $45.7 \div 0$
d $0.12 \div 20$		**i** 0.6×200
e 67.03×1000		

1.2 Directed numbers

The negative sign has two distinct uses in mathematics:

  (i) as a **subtraction** operation, e.g. $6 - 4 = 2$,
 (ii) as a **direction** symbol, e.g. $-7\,°C$.

If we wish to show a temperature which is $7\,°C$ *below* zero, we can write $^-7\,°C$ *or* $-7\,°C$.

If a car travels 20 miles in one direction and then 15 miles in the reverse direction, we can write the distances travelled as $^+$**20 miles** and $^-$**15 miles.**

The numbers $^-7$, $^+20$ and $^-15$ are called **directed numbers**.

On your calculator you will see that there are two keys with negative symbols:

$\boxed{-}$ for subtraction

$\boxed{+/-}$ for direction.

Directed numbers can be represented on a horizontal or a vertical number line.

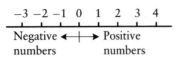

$$-3 \ -2 \ -1 \ \ 0 \ \ 1 \ \ 2 \ \ 3 \ \ 4$$

Negative $\longleftrightarrow$ Positive
numbers   numbers

3
2
1   Positive numbers ▲
0
−1   Negative numbers ▼
−2
−3

Addition and subtraction

Think of additions and subtractions of numbers as movements *up* and *down* the vertical number line.

Start at zero, move up 7 and then down 3.
Your position is now 4 *above* zero.

i.e. $\qquad\qquad\qquad\qquad 7 + (-3) = +4$

Subtracting two numbers is the same as finding the difference between them:

$$5 - (-3)$$

is the difference between being 3 *below* zero and being 5 *above* zero on the number line.

From 3 below to 5 above you must move up 8. Therefore

$$5 - (-3) = +8 \ or \ 5 - (-3) = 5 + 3 = 8$$

(Note that $-(-3) = +3$.)

$$-4 - (-2)$$

is the difference between being 2 below zero and being 4 below zero.

From 2 *below* to 4 *below* you must move DOWN 2, i.e.

$$-4 - (-2) = -2$$

or use the rule $- \times - = +$ to give

and $\qquad\qquad -4 - (-2) = -4 + 2 = -2$

7
6
5
4
3
2
1
0
−1
−2
−3
−4
−5
−6
−7

Exercise 1B

Find the answers to the following, without the aid of a calculator:

1   a $-5 + 7$ c $-2 + -8$ e $-21 + 27$ g $-19 + 19$

    b $3 + (-7)$ d $16 + (-12)$ f $35 + 5$ h $-35 + (-14)$

2   a $7 - 8$ c $-9 - 4$ e $-10 - (-7)$ g $25 - (-25)$

    b $12 - (-5)$ d $11 - (-9)$ f $-15 - (-15)$ h $100 - 64$

For the following question you may use a calculator:

3   a $306 - (-219) + (-513)$ d $-697 + (-518) - (-29)$

    b $1427 + (-472) - 318$ e $-6724 + 3410 + (-148)$

    c $5061 - (-725) + (-491)$ f $5803 - 4190 - (-153)$

Negative numbers in context
(e.g. bank accounts)

Bank statements (see page 86) describe an account as **overdrawn** when the total of money in the account is a negative amount.

If your account is overdrawn you 'owe' the bank money, because you have taken out more money from your bank account than you had in it.

The total in your account is called the **balance** on your account.

┌─ Example 1 ────────────────────────

The bank pays a cheque for £30 from your account. You were £72 overdrawn. What is the new balance on your account?

This means that you had −£72 and then spent another £30.

∴   Amount overdrawn = −£72 − £30 = −£102

or   You owed the bank £72 and then spent another £30.

∴   You now owe the bank £102.

Hence the new balance on your account is £102 overdrawn.

┌─ Example 2 ────────────────────────

Alexia's bank account is £34.25 overdrawn. She pays into her bank account a cheque for £72.47. What is the new balance on Alexia's account?

The amount in Alexia's account was −£34.25.

She pays in the cheque for £72.47

∴   Amount in the account = −£34.25 + £72.47
                            = £38.22

i.e.   The new balance on Alexia's account is £38.22.

▇ Exercise 1C ▇▇▇▇▇▇▇▇▇▇▇▇▇▇▇▇▇▇▇▇▇▇▇

1   Maeve's monthly bank statement shows that she has an overdraft of £74.32.

   She pays £100 into her bank account immediately.

   After this payment, what is the new balance on Maeve's account?   (SEG W95)

2   In Gdansk, Poland, the temperature on one day in January rose from −7 °F to +22 °F.
   By how many degrees did the temperature rise?   (SEG S95)

3   Dawn has £30.25 in her bank account. She buys a coat costing £64 and pays for it by a cheque which is accepted by her bank. What is the new balance of her bank account?   (SEG W93)

4   Poppy's bank account is overdrawn by £79. She pays in a cheque and then has a balance of £112.

   How much has she paid into her bank account?   (SEG S96)

Multiplication and division

Starting at zero and going *up* 2 three times would mean you were now 6 *above* zero. So

$$(+2) \times 3 = +6$$

or $\quad\quad\quad 2 \times 3 = 6 \quad\quad\quad$ also $\quad\quad 3 \times 2 = 6$

Starting at zero and going *down* 2 three times would mean you were now 6 *below* zero. So

$$(-2) \times 3 = -6$$

or $\quad\quad\quad -2 \times 3 = -6 \quad\quad\quad$ also $\quad\quad 3 \times -2 = -6$

But what is $(-2) \times (-3)$?

Remember that $-(-6) = +6$

and $\quad\quad\quad (-2) \times 3 = -6 = -(2 \times 3)$

$\therefore \quad\quad\quad (-2) \times (-3) = -(2 \times -3) = -(-6) = +6$

i.e. the reverse of moving 6 *down* is moving 6 *up*.

The rules for multiplication and division are similar:

Two numbers with like signs give a positive answer.
Two numbers with unlike signs give a negative answer.

Example 1

Find the product of $-4 \times -7 \times -3$

$$(-4 \times -7) \times -3 = +28 \times -3$$
$$= -84$$

Example 2

Evaluate $(-12 \div 4) \times -6$

$$(-12 \div 4) \times -6 = -3 \times -6$$
$$= +18$$

Exercise 1D

Do *not* use a calculator in the following questions:

1
a -3×6

b $15 \div -5$

c -7×-4

d $-7 \div -4$

e $10 \times -3 \div 2$

f $-3 \times -3 \times -3$

g $8 \div -\frac{1}{2} \div -2$

h $(-3 + -6) \times -4$

i $-7 \div (3 - (-4))$

j $\dfrac{20 \times -6}{-4 \times -3}$

You *may* use a calculator for the following questions:

2
a $42 \div (-8)$

b $(-47) \times (-73)$

c $(-26) \times 15$

d $\dfrac{252 \times (-20)}{(-30)}$

e $\dfrac{(-16) \times (-27)}{(-36) \times (-10)}$

f $\dfrac{(-19) + (-17)}{46 - (-26)}$

3 Given that $X = \dfrac{-2Y + 3Z}{3Y - 2Z}$, find X when:

a $Y = 6, Z = 3$

b $Y = -3, Z = \frac{1}{2}$

c $Y = 6, Z = -1$

Order of operations

The order in which arithmetic operations are performed is as follows:

>Brackets
>Of
>Division
>Multiplication
>Addition
>Subtraction

(Taking the first letter of each word gives the **BODMAS** rule.)

Example 1

Evaluate $[5 - (-3)] \times [4 + (-1)]$

$$[5 - (-3)] \times [4 + (-1)] = 8 \times 3$$
$$= 24$$

Example 2

Evaluate $5 - (-3) \times 4 + (-1)$

$$5 - (-3) \times 4 + (-1) = 5 - (-12) + (-1)$$
$$= 5 + 12 - 1$$
$$= 16$$

Example 3

Evaluate $\frac{1}{2}$ of $18 - 7$

$$\tfrac{1}{2} \text{ of } 18 - 7 = 9 - 7$$
$$= 2$$

Example 4

Evaluate $\frac{1}{2}$ of $(18 - 7)$

$$\tfrac{1}{2} \text{ of } (18 - 7) = \tfrac{1}{2} \text{ of } 11$$
$$= 5\tfrac{1}{2}$$

Exercise 1E

1 Use the BODMAS rule to evaluate the following:

  a $-4(2 - 6) + (-6) - (-3)$

  b $-3(16 - 12) - (-1)(12 - 16)$

  c $\frac{1}{2}$ of $(-35 - 21) - (-72)$

  d $12 - (7 - 3) - 4 \times 6 + 2$

2 The formula for converting degrees Fahrenheit (°F) to degrees Celsius (°C) is:

$$°C = \frac{5}{9}(°F - 32)$$

Calculate the number of degrees Celsius (to 1 decimal place) which is equivalent to:

  a 70 °F   **b** 28 °F   **c** ⁻4 °F   **d** ⁻10 °F

3 Given that $P = -Q(R - 2S)$, find P when:

  a $Q = 4, R = 3, S = 6$

  b $Q = -10, R = -7.5, S = 2.5$

  c $Q = 8.3, R = -4.6, S = -2.9$

1.3 Powers, roots and reciprocals

Powers

Patterns of dots like this

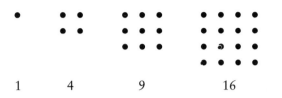

1   4   9   16

form squares, and the numbers 1, 4, 9, 16... are called the **square numbers**.

The square numbers are also formed by finding the product of an integer with itself:

$$1 \times 1 = 1$$
$$2 \times 2 = 4$$
$$3 \times 3 = 9$$
$$4 \times 4 = 16, \text{ etc.}$$

The sums of odd numbers also produce the square numbers:

$$1 \qquad\qquad = 1$$
$$1 + 3 \qquad\quad = 4$$
$$1 + 3 + 5 \qquad = 9$$
$$1 + 3 + 5 + 7 \quad = 16$$
$$1 + 3 + 5 + 7 + 9 = 25, \text{ etc.}$$

If any number is multiplied by itself, the product is called the square of the number. For example:

the square of 1.2 is 1.44
the square of $\sqrt{2}$ is 2.

The square of 3 or 3 squared $= 3 \times 3 = 3^2$, and the 2 is called the **index**.

Similarly:

the cube of 4 = 4 cubed $= 4 \times 4 \times 4 = 4^3$

and

the fourth power of $2 = 2 \times 2 \times 2 \times 2 = 2^4$

Roots

The square of $4 = 4 \times 4 = 16$
The number 4, in this example, is called the **square root** of 16, i.e. the number which was squared to give 16.

16 has another square root because $-4 \times -4 = 16$.

Hence the square roots of 16 are $+4$ and -4.

The symbol $\sqrt{}$ indicates the positive square root, and hence $\sqrt{16} = +4$.

Similarly, a fourth root of a given number is the number which has a fourth power equal to the given number. For example:

the positive fourth root of 16 or $\sqrt[4]{16} = 2$
$(2^4 = 2 \times 2 \times 2 \times 2 = 16)$

Reciprocals

The **reciprocal** of a number is equal to the number one divided by that number. To find the reciprocal of a number we invert it. For example:

the reciprocal of 3 or $\frac{3}{1}$ is $\frac{1}{3}$
the reciprocal of $\frac{1}{5}$ is 5
the reciprocal of $\frac{2}{3}$ is $\frac{3}{2}$ or $1\frac{1}{2}$.

To find the reciprocal of a mixed number (e.g. $2\frac{1}{4}$) it must first be converted to an improper fraction:

$$2\frac{1}{4} = \frac{9}{4}$$

Therefore the reciprocal of $2\frac{1}{4} = \frac{4}{9}$.

Exercise 1F

1 Write down the first 10 square numbers.

2 Write down the first 6 cube numbers.

3 Find the value of:
a 2^5 **b** 3^6 **c** $(-5)^2$ **d** $(\frac{1}{2})^3$ **e** $(-4)^3$

4 Find the value of:

a $2^3 \times 2^2$ **c** $2^4 \div 2^2$

b $3^4 \times 3^2 \times 3$ **d** $\dfrac{3^4 \times 3^2}{3 \times 3^3}$

5 Find the positive square roots of the following numbers:
a 25 **b** 121 **c** 625 **d** 225 **e** 196

6 **a** Find the cube root of (i) 27 (ii) −64 (iii) 729

b Find the fifth root of 32.

c Find the fourth root of (i) 81 (ii) 625

7 Find the reciprocals of:
a $\frac{1}{6}$ **b** $\frac{3}{4}$ **c** 0.125 **d** $3\frac{3}{4}$ **e** 9.5

8 **a** What is the reciprocal of $\dfrac{1}{\frac{3}{4}}$?

b What is the value of $\dfrac{1}{\frac{3}{4}}$?

9 Write 25 as the sum of a sequence of odd numbers.

10 What is the reciprocal of:
a 11^2 **b** $\sqrt{144}$ **c** $\dfrac{2^3}{3^2}$ **d** $\dfrac{1}{(-3)^3}$ **e** 1?

A calculator should *not* be used for the following questions.

11 What is the third power of 2?

12 What is 3 cubed?

13 Write down the ninth square number.

14 Write down the fifth cube number.

15 Find the value of:
a 2^5 **b** 3^4 **c** $(\frac{1}{4})^2$ **d** $(-2)^3$ **e** $(0.5)^2$

16 Write down the cube root of:
a 8 **b** 27 **c** 216 **d** $\frac{1}{8}$ **e** −64

17 What is the fourth root of 81?

18 Find the reciprocals of the following numbers, as whole numbers or fractions:
a 4 **b** $\frac{1}{9}$ **c** 0.5 **d** $3\frac{1}{2}$ **e** 8.5

19 Evaluate $5^2 \times \sqrt{16}$.

20 Show that $\sqrt{(16 \times 9)} = \sqrt{16} \times \sqrt{9}$.

*1.4   Surds

Numbers written as square roots, such as $\sqrt{12}$, are called **surds**.

Most surds are irrational numbers (see page 1) since they cannot be given as a terminating decimal or as a recurring decimal or as a fraction.
For example, $\sqrt{16} = 4$, a rational number, but $\sqrt{12} = 3.4641\ldots$, an irrational number.

Simplifying surds

A surd which has been 'simplified' will be expressed in terms of the smallest possible surd. For example, $\sqrt{12}$ and $2\sqrt{3}$ have the same value, but $\sqrt{3}$ is less than $\sqrt{12}$, so

$$2\sqrt{3} \text{ is the simplified form of } \sqrt{12}.$$

Such surds are said to be simplified because they enable you to see whether you can combine two surds. For example, you will notice that

$$2\sqrt{3} + 5\sqrt{3} = 7\sqrt{3}$$

but it is more difficult to notice that you can add $\sqrt{12}$ and $\sqrt{75}$, which is the same addition!

For example,
$$\begin{array}{ll}
\sqrt{48} = \sqrt{(16 \times 3)} & \text{or} \quad \sqrt{48} = \sqrt{(4 \times 4 \times 3)} \\
\quad = \sqrt{16} \times \sqrt{3} & \qquad = \sqrt{(4^2 \times 3)} \\
\quad = 4 \times \sqrt{3} & \qquad = 4 \times \sqrt{3} \\
\quad = 4\sqrt{3} & \qquad = 4\sqrt{3}
\end{array}$$

If at first, you only found the square factor 4, you could still obtain the same result by repeating the method:

$$\begin{aligned}
\sqrt{48} &= \sqrt{(4 \times 12)} \\
&= \sqrt{4} \times \sqrt{12} \\
&= 2 \times \sqrt{12} \\
&= 2 \times \sqrt{4} \times \sqrt{3} \\
&= 2 \times 2 \times \sqrt{3} \\
&= 4\sqrt{3}
\end{aligned}$$

Example

Simplify $\sqrt{180}$

$$\begin{array}{ll}
\sqrt{180} = \sqrt{(36 \times 5)} & \text{or} \quad \sqrt{180} = \sqrt{(10 \times 18)} \\
\quad = \sqrt{36} \times \sqrt{5} & \qquad = \sqrt{(2 \times 5 \times 2 \times 9)} \\
\quad = 6\sqrt{5} & \qquad = \sqrt{[(2 \times 2) \times 5 \times (3 \times 3)]} \\
& \qquad = 2 \times 3 \times \sqrt{5} \\
& \qquad = 6\sqrt{5}
\end{array}$$

Multiplying and dividing surds

Any two or more surds may be multiplied or divided by combining them
under the same root sign and then simplifying as above.

Example 1

Simplify $\sqrt{75} \times \sqrt{21}$

$$
\begin{aligned}
\sqrt{75} \times \sqrt{21} &= \sqrt{(75 \times 21)} \\
&= \sqrt{(3 \times 25 \times 3 \times 7)} \\
&= \sqrt{(3 \times 3)} \times \sqrt{25} \times \sqrt{7} \\
&= 3 \times 5 \times \sqrt{7} \\
&= 15\sqrt{7}
\end{aligned}
$$

Example 2

Multiply $2\sqrt{5}$ by $4\sqrt{15}$, leaving your answer
in surd form.

$$
\begin{aligned}
2\sqrt{5} \times 4\sqrt{15} &= 2 \times 4 \times \sqrt{(5 \times 15)} \\
&= 8 \times \sqrt{(5 \times 3 \times 5)} \\
&= 8 \times 5 \times \sqrt{3} \\
&= 40\sqrt{3}
\end{aligned}
$$

Example 3

Simplify: **a** $\dfrac{\sqrt{180}}{\sqrt{60}}$ **b** $\dfrac{\sqrt{56}}{\sqrt{63}}$

a $\dfrac{\sqrt{180}}{\sqrt{60}} = \sqrt{\dfrac{180}{60}} = \sqrt{3}$

b $\dfrac{\sqrt{56}}{\sqrt{63}} = \sqrt{\dfrac{56}{63}} = \sqrt{\dfrac{8}{9}} = \dfrac{\sqrt{4 \times 2}}{\sqrt{9}} = \dfrac{2\sqrt{2}}{3}$

Adding and subtracting surds

As with fractions, only terms *of the same type* may be added or subtracted,
i.e. $2\sqrt{2}$ may be added to $3\sqrt{2}$ but *not* to $3\sqrt{3}$.

Example 1

Simplify $2\sqrt{2} + 5\sqrt{2} - 3\sqrt{2}$

$$
\begin{aligned}
2\sqrt{2} + 5\sqrt{2} - 3\sqrt{2} &= (2 + 5 - 3)\sqrt{2} \\
&= 4\sqrt{2}
\end{aligned}
$$

Example 2

Find the sum of $\sqrt{50}$ and $\sqrt{242}$ in surd form.

$$
\begin{aligned}
\sqrt{50} + \sqrt{242} &= \sqrt{(25 \times 2)} + \sqrt{(2 \times 121)} \\
&= 5\sqrt{2} + 11\sqrt{2} \\
&= 16\sqrt{2}
\end{aligned}
$$

Exercise 1G

Simplify the following, leaving your answers in surd form.

1 $\sqrt{18}$	4 $\sqrt{147}$	7 $\sqrt{50} - \sqrt{8}$	10 $\sqrt{45} \times \sqrt{80}$	13 $\dfrac{\sqrt{72}}{\sqrt{720}}$
2 $\sqrt{216}$	5 $\sqrt{27} + \sqrt{108}$	8 $3\sqrt{75} - 5\sqrt{12}$	11 $\dfrac{\sqrt{180}}{\sqrt{45}}$	14 $\sqrt{112} \div \sqrt{108}$
3 $\sqrt{20}$	6 $\sqrt{45} + \sqrt{180}$	9 $\sqrt{18} \times \sqrt{200}$	12 $\dfrac{\sqrt{90}}{\sqrt{450}}$	15 $\sqrt{84} \div \sqrt{28}$

Irrational numbers

On page 1, an irrational number was defined as a number which cannot be written as a fraction or an exact decimal.

Using surds you can simplify some numbers which appear to be irrational, in order to determine whether they are rational or not.

For example $\dfrac{\sqrt{12}}{\sqrt{3}} = \dfrac{2\sqrt{3}}{\sqrt{3}} = 2$, which is rational.

Example

Find pairs of irrational numbers whose:

  (i) sum is rational
 (ii) difference is rational
(iii) product is rational.

  (i) Since $\sqrt{2} + (2 - \sqrt{2}) = 2$ (*or*, the sum of $\sqrt{2}$ and $2 - \sqrt{2}$ is 2), $\sqrt{2}$ and $(2 - \sqrt{2})$ are two irrational numbers whose sum is rational.

 (ii) $(2 + \sqrt{2}) - \sqrt{2} = 2$
     $\therefore$ $(2 + \sqrt{2})$ and $\sqrt{2}$ are two irrational numbers whose difference is rational.

(iii) $\sqrt{8} \times \sqrt{50} = 2\sqrt{2} \times 5\sqrt{2} = 10\sqrt{4} = 20$
     $\therefore$ $\sqrt{8}$ and $\sqrt{20}$ are two irrational numbers whose product is rational.

Exercise 1H

1 State which of the following are rational:

    $\sqrt{2}, \quad (\sqrt{3})^2, \quad (\sqrt{5})^3, \quad 2 + \sqrt{2}, \quad 2 - \sqrt{2}, \quad \pi.$

2 **a** Write down:
       (i) a rational number between 4 and 5
      (ii) an irrational number between 4 and 5.

    b (i) Is $\dfrac{\sqrt{100}}{\sqrt{75}}$ rational or irrational?
      (ii) Give a reason for your answer.   (SEG S96)

3 **a** When r and s are both rational numbers, $r + s$ is also rational. Write down one example to illustrate this rule.

    b When r and s are both irrational numbers, $r + s$ can be either rational or irrational.
      (i) Write down one example to show that $r + s$ can be irrational.
     (ii) Write down one example to show that $r + s$ can be rational.
                                        (SEG W95)

4 **a** State which of these numbers is irrational:

      $\sqrt{5}, \quad \pi, \quad \sqrt{2.56}, \quad (\sqrt{3})^3, \quad 2\tfrac{1}{7}, \quad \sqrt[3]{8}.$

    b Write down two different irrational numbers whose product is a rational number.   (SEG S95)

1.5   Factors and multiples

A **factor** is an integer which divides into a given number exactly, i.e. there is no remainder.

For example, 3 and 4 are factors of 12.

A **multiple** of a number is the number multiplied by any integer.

For example, multiples of 6 are 12, 18, 24, and so on.

12 is a multiple of 3 and 4, since 3 and 4 are factors of 12.

Example 1

Find all the factors of 42 whose product is 42 and list them in pairs.

The factors of 42 are 1, 2, 3, 6, 7, 14, 21, 42.

Listed in pairs they are 1×42, 2×21, 3×14, 6×7.

Example 2

Which of the following numbers are multiples of 9?

a 63   **b** 732   **c** 1944

a $63 \div 9 = 7$ means that 63 is a multiple of 9.

b $732 \div 9 = 81.3$ means that 732 is not a multiple of 9.

c $1944 \div 9 = 216$ means that 1944 is a multiple of 9.

Note that if the sum of the digits of a number is a multiple of 9, the number is divisible by 9.

Exercise 1I

1   Find all the factors of each of the following whose product is the number itself and list the factors in pairs:

   a 15   **b** 27   **c** 36   **d** 64   **e** 100

2   **a** Write down the factors of   (i) 48   (ii) 72.

   b List the factors which are common to 48 and 72.

3   Of which of the following numbers is 4 a factor?
   34,   64,   144,   642,   116,   2620.

4   State which of 5, 10, 20, 25 are factors of the following:

   a 300   **b** 1050   **c** 470   **d** 875   **e** 2360

5   Which of the following numbers are divisible by   **a** 3   **b** 6   **c** 15?

   125,   240,   87,   96,   146,   255,   1073.

Prime factors

A **prime factor** is a **factor** which is also a **prime number**.

Example 1

a What are the prime factors of 12?

b Write 12 as a product of prime factors.

a Factors of 12 are   1, 2, 3, 4, 6, 12
   Prime factors of 12 are   1, 2, 3

b As a product of prime factors,
$$12 = 1 \times 2 \times 2 \times 3$$
$$= 2^2 \times 3 \text{ (1 is not usually included).}$$

Example 2

Find the prime factors of 360.

Method

Begin dividing by the lowest prime factor of the given number and continue dividing until the quotient is no longer divisible by this prime number.

Find the next lowest prime factor and continue dividing.

When the quotient is 1, all prime factors have been found.

Write your answer using index notation.

2	360
2	180
2	90
3	45
3	15
5	5
	1

$$360 = 2 \times 2 \times 2 \times 3 \times 3 \times 5$$
$$= 2^3 \times 3^2 \times 5$$

Example 3

a Find three common multiples for the set of numbers 3, 4, 6.

b Describe the set of common multiples.

a Three common multiples are 12, 24 and 36. (There are, of course, many other possible answers.)

b The common multiples are all multiples of 12.

Exercise 1J

1   Write the following as products of prime numbers:

   a   60   b   315   c   225   d   8820   e   1694

2   (i) List the prime factors of the following pairs of numbers.
     (ii) List the factors which each pair have in common.

   a   36 and 90   c   312 and 260   e   918 and 468

   b   168 and 72   d   154 and 220   f   735 and 1050.

3   a   Find three common multiples for each of the following sets of
         numbers:

      (i) 7, 14   (ii) 6, 15   (iii) 4, 8, 12   (iv) 2, 3, 5

   b   Describe the set of common multiples in each case.

1.6   Highest common factor and lowest common multiple

The **highest common factor (HCF)** is the largest number which will divide exactly into each of the given numbers.

The **lowest common multiple (LCM)** is the smallest number into which each of the given numbers will divide exactly.

Example 1

Find the highest common factor of 120 and 252.

2	120
2	60
2	30
3	15
5	5
	1

2	252
2	126
3	63
3	21
7	7
	1

$$120 = 2^3 \times 3 \times 5 \qquad 252 = 2^2 \times 3^2 \times 7$$

The factors which are common to both numbers are 2^2 and 3.

The HCF of 120 and 252 $= 2^2 \times 3$
                        $= 12$

Example 2

Find the lowest common multiple of 24 and 54.

2	24
2	12
2	6
3	3
	1

$24 = 2^3 \times 3$

2	54
3	27
3	9
3	3
	1

$54 = 2 \times 3^3$

A common multiple must include the highest power of each prime factor.

The LCM of 24 and 54 $= 2^3 \times 3^3$
$= 216$

Exercise 1K

1 Find the HCF of:

  a 64, 96   **d** 84, 56, 98

  b 105, 75   **e** 345, 920

  c 196, 252   **f** 154, 330, 242

2 Find the LCM of:

  a 15, 25, 10   **d** 900, 4200

  b 16, 18   **e** 9, 21, 28

  c 52, 78   **f** 24, 16, 36

3 **a** Find the HCF of 156 and 216.

  b Hence write $\dfrac{156}{216}$ in its lowest terms.

4 **a** Find the HCF of 900 and 2925.

  b Hence write $\dfrac{900}{2925}$ in its lowest terms.

5 Fractions $\dfrac{5}{8}$, $\dfrac{7}{12}$, $\dfrac{4}{9}$ are to be added.

  Find the LCM of their denominators. (See page 26 for a definition of 'denominator'.)

6 A yeti and its mate are walking through the snow. The yeti's footprints are 5 metres apart, its mate's footprints are 4 metres apart.
If its mate steps in the yeti's first footprint:

  a how many steps will its mate take before the prints match again?

  b how many steps will the yeti take before the prints match again?

  c what is the distance between the first and third matching prints?

7 Two buoys have lights attached which flash at regular intervals. The light on the first buoy flashes every 8 minutes and that on the second buoy, every 12 minutes. At 1.00 am the lights flash simultaneously.
At what time will they next flash simultaneously?

8 The planet Athyr has three moons: Artemis, Khons and Soma. Artemis takes 18 days to orbit the planet, Khons takes 28 days and Soma takes 36 days.

  a If the three moons can be viewed in line on one particular evening, how many days will elapse before the moons are again in line?

  b How many orbits of the planet Athyr has each moon made in this time?

9 **a** A manufacturer sells plastic drinking cups which are packed in boxes.
One customer has a regular order for 3750 cups and another has a regular order for 7000 cups.
How many cups should the manufacturer pack in each box so that he can fulfil both orders with complete boxes?

  b How many boxes will each customer receive?

10 A farmer delivers boxes of six eggs which are packed into cartons. To prevent damage, each carton is completely filled with boxes of eggs. His customers require 108 boxes, 90 boxes or 144 boxes.
How many boxes should each carton be designed to hold so that he can deliver full cartons to all his customers?

2 Using a Calculator

2.1 Approximations

Most calculators display answers of up to 10 digits.

In most cases, this is too many digits because this level of accuracy is not usually required. Therefore, when using a calculator, it is necessary to give an *approximate* answer which contains fewer digits (but which has a sensible degree of accuracy).

There are two methods in common usage:
rounding to a number of **decimal places** (e.g. 2 decimal places or 2 d.p.) and rounding to a number of **significant figures** (e.g. 3 significant figures or 3 s.f.).

The rule for rounding to, for example, 2 decimal places is:
If the digit in the third decimal place is **5 or more, round up**, i.e. increase the digit in the second decimal place by 1.

If the digit in the third decimal place is **less than 5, round down**, i.e. the digit in the second place remains the same.

Example 1

Give the following numbers, from a calculator display, to 2 d.p.
7.92341,   25.675231,   0.06666,   0.9999999

Calculator display	Degree of accuracy required (2 d.p.)	Answer correct to 2 d.p.
7.92341	7.92|341	7.92
25.675231	25.67|5231	25.68
0.066666	0.06|6666	0.07
0.9999999	0.99|99999	1.00

The rules for significant figures are similar, but you need to take care with zeros.

Zeros at the beginning of a decimal number or at the end of an integer are not counted as significant figures, but must be included in the final result. All other zeros are significant.

For example,
70 631.9 given correct to 3 significant figures is 70 600.

The three significant figures are 7, 0 and 6. The last two zeros are not significant (i.e. do not count as fourth and fifth figures), but are essential so that 7 retains its value of 70 thousand and 6 its value of 6 hundred.

---**Example 2**---

Give the following numbers, from a calculator display, to 3 s.f.

7.92341, 25.675231, 0.066666, 24380, 0.999999

Calculator display	Degree of accuracy required (3 s.f.)	Answer correct to 3 s.f.
7.92341	7.92|341	7.92
25.675231	25.6|75231	25.7
0.066666	0.0666|66	0.0667
24380.	243|80.	24400
0.999999	0.999|999	1.00

In the last example, the digit 1 becomes the first significant figure, and the 2 zeros are the second and third figures.

Exercise 2A

Give each of the following numbers to the degree of accuracy requested in brackets:

1   9.736   (3 s.f.)
2   0.362 18   (2 d.p.)
3   147.49   (1 d.p.)
4   28.613   (2 s.f.)
5   0.5252   (2 s.f.)

6   4.1983   (2 d.p.)
7   1245.4   (3 s.f.)
8   0.004 25   (3 d.p.)
9   273.6   (2 s.f.)
10 459.973 14 (1 d.p.)

2.2   Estimation

The answer displayed on a calculator will be correct for the values you have entered, but a calculator cannot tell you if you have pressed the wrong key or entered your numbers in the wrong order.

Each number you enter into the calculator should be checked for accuracy and the final answer should be checked by comparing it with an *estimated* answer.

As in the example below, to find an estimated answer to a calculation, first round all the numbers to one significant figure. Then do the calculation using these approximations. The value you obtain is an estimate of the true value.

---**Example**---

Estimate the value of 31.41×79.6.

31.41 is approximately 30
79.6   is approximately 80

An estimated value is therefore $30 \times 80 = 2400$.

If, when you do the calculation with the accurate values your calculator displays 25002.36 you will realise that a mistake has been made, as the answer should be near 2400. The correct answer is 2500.236.

Exercise 2B

1 By rounding all numbers to 1 significant figure, find an estimated value of each calculation:

a 52.2×67.4

d $607 \div 1.86$

g $\dfrac{520.4 \times 8.065}{99.53}$

b 6143×0.0381

e $48.2 \div 0.203$

c 607×1.86

f $3784 \div 412$

h $\dfrac{807}{391.2 \times 0.38}$

2 Find an estimate for each of the following calculations, by choosing an appropriate approximation for each number:

a $82.3 \div 9.1$

b $0.364 \div 6.29$

c $\dfrac{31.73 \times 6.282}{7.918}$

3 By finding an estimate of the answer, state which of the following calculations are obviously incorrect. (Do not use a calculator.)

a $8.14 \times 49.6 = 403.74$

f $\dfrac{42.3 \times 3.97}{1635} = 10.27$

b $23.79 \div 5.57 = 4.27$

g $\sqrt{1640} = 40.5$

c $324 \div 196 \times 0.5 = 226$

h $\sqrt{650} = 80.6$

d $3.14 \times 9.46^2 = 882.35$

i $(0.038)^2 = 0.001\,44$

e $23.79 \div 0.213 = 11.169$

j $(0.205)^3 = 0.0862$

4 A Students' Union pays £3980 to hire a boat for a leavers' cruise, and 249 students wish to purchase a ticket for the cruise. In order to cover the hire charge, Derek estimates the cost of each ticket to be £16.

a Which two numbers would you divide to find an estimate of the cost of each ticket?

b Is Derek's estimate correct? (SEG S95)

5 Five students celebrate their examination successes by taking their friends out to dinner. The dinner costs £20.85 each and there are a total of 14 in the dinner party. Wendy is horrified when she thinks that it will cost each of the five of them £583.80

a Which two numbers would you multiply to find a quick estimate of the total cost of the meal?

b Hence estimate how much each of the five students has to pay. (SEG S94)

2.3 Standard form

Standard form is used when dealing with very large or very small numbers.

In standard form the number is always written as a number between 1 and 10 multiplied by a power of 10, i.e. as

$$A \times 10^n \quad \text{where} \quad 1 \leqslant A < 10 \quad \text{and } n \text{ is an integer.}$$

This form is also known as **scientific notation.**

Example 1

The velocity of light is 300 000 000 metres per second.

Write this number in standard form.

(i) Write down A, the number between 1 and 10:   3.0

(ii) Count the number of places to the *right* which the decimal point must be moved to give the velocity of light: 8 places.

(iii) Write the number in standard form: 3.0×10^8

or
$$300\,000\,000 = 3.0 \times 100\,000\,000$$
$$= 3.0 \times 10^8$$

Example 2

0.000 000 000 000 000 000 000 000 001 67 kg is the mass of a hydrogen atom.

Write this number in standard form.

(i) $A = 1.67$

(ii) In this case the decimal point must be moved 27 places to the *left*: $n = -27$

(iii) In standard form, the mass $= 1.67 \times 10^{-27}$ kg

Example 3

Convert 3.15×10^4 to a 'normal' number.

$n = 4$ so the decimal point must be moved 4 places to the right:
$$3.15 \times 10^4 = 31\,500$$

or
$$3.15 \times 10^4 = 3.15 \times 10\,000$$
$$= 31\,500$$

Example 4

Multiply (2.7×10^4) by (5×10^{-2})

$$(2.7 \times 10^4) \times (5 \times 10^{-2}) = (2.7 \times 5) \times (10^4 \times 10^{-2})$$
$$= 13.5 \times 10^{(4 + -2)}$$
$$= 13.5 \times 10^2$$
$$= 1.35 \times 10^1 \times 10^2$$
$$= 1.35 \times 10^3 \text{ (in standard form)}$$

or $\qquad (2.7 \times 10^4) \times (5 \times 10^{-2}) = (2.7 \times 10\,000) \times (5 \div 100)$
$$= 13.5 \times 100$$
$$= 1.35 \times 1000$$
$$= 1.35 \times 10^3$$

Using a calculator:

$$(2.7 \times 10^4) \times (5 \times 10^{-2}) = 2.7 \boxed{\text{EXP}} \ 4 \ \boxed{\times} \ 5 \ \boxed{\text{EXP}} \ 2 \ \boxed{+/-}$$

(On some calculators the exponential button is $\boxed{\text{EE}}$.)

Calculator display is: $\qquad\qquad\qquad$ 1350

$$= 1.35 \times 10^3 \text{ (in standard form)}$$

If it is possible to use your calculator in scientific mode, the same sequence of operations will produce the answer in standard form displaying:

$\qquad$ 1.35 $\quad$ 03 or possibly 1.35^{03}

which is then written as 1.35×10^3.

(Some calculators round calculations to 2 s.f. displaying the answer as 1.4^{03}, which is then written as 1.4×10^3.)

Example 5

Divide (2.7×10^4) by (5×10^{-2}).

$$(2.7 \times 10^4) \div (5 \times 10^{-2}) = (2.7 \div 5) \times (10^4 \div 10^{-2})$$
$$= 0.54 \times 10^{(4 - -2)}$$
$$= 0.54 \times 10^6$$
$$= 5.4 \times 10^5 \text{ (in standard form)}$$

Using your calculator gives

$$(2.7 \times 10^4) \div (5 \times 10^{-2}) = 2.7 \ \boxed{\text{EXP}} \ 4 \div 5 \ \boxed{\text{EXP}} \ 2 \ \boxed{+/-}$$
$$= 540\,000$$
$$= 5.4 \times 10^5 \text{ (in standard form)}$$

Working in the scientific mode will give the correct answer 5.4^{05}

which is conventionally written as 5.4×10^5.

Numbers in standard form can be added and subtracted, provided the power of 10 is the same in each number.

If the powers of 10 are *not* the same, convert each number to a normal number before adding or subtracting, *or* convert to a common power of 10.

Example 6

Evaluate $(4.3 \times 10^4) - (8.7 \times 10^3)$

Method 1

$$(4.3 \times 10^4) - (8.7 \times 10^3) = 43\,000 - 8700$$
$$= 34\,300$$
$$= 3.43 \times 10^4$$

Method 2

$$(4.3 \times 10^4) - (8.7 \times 10^3) = (4.3 \times 10^4) - (0.87 \times 10^4)$$
$$= 3.43 \times 10^4$$

Exercise 2C

1   Write the following numbers in standard form:

   a 790 000   e 100 000   i $0.000\,34 \times 100$

   b 0.0046   f 0.000 282   j $0.0027 \div 1000$

   c 31 300   g 15.7×1000 k $5000 \div 10\,000$

   d 0.000 094 1   h $4700 \times 10\,000$ l $0.000\,028 \div 200$

2   Write the following numbers in normal form:

   a 7.53×10^3 d 8.37×10^{-1} g 3.192×10^6

   b 2.4×10^2 e 4.51×10^{-2} h 9.74×10^{-4}

   c 1.9×10^{-3} f 4.042×10^4 i 6.8×10^{-6}

3   Evaluate the following, giving your answers in standard form:

   a $(3 \times 10^2) \times (2 \times 10^2)$ f $(3.6 \times 10^4) \div (9.0 \times 10^5)$

   b $(1.4 \times 10^{-3}) \times (3.7 \times 10^3)$ g $(2.7 \times 10^{-2}) \times (1.6 \times 10^{-3})$

   c $(8 \times 10^3) \div (2 \times 10^2)$ h $(3.3 \times 10^4) - (4.6 \times 10^3)$

   d $(5.3 \times 10^3) + (8.2 \times 10^3)$ i $(4.5 \times 10^{-3}) - (5.0 \times 10^{-3})$

   e $(4.6 \times 10^3) \times (5.9 \times 10^{-1})$ j $(1.32 \times 10^{-1}) \div (2.2 \times 10^{-3})$

4   The weight of 100 m of fine thread is 2.3 g.

   a Find the weight of 1 m of the thread and state this weight in kg in standard form.

   b Find the length of thread which has weight 2×10^{-4} kg.

5   A mill produces 350 m of a particular material.

   a State this length (in cm) in standard form.

   One designer buys 8.34×10^3 cm.

   b How much of the material does the mill have left?

6   The revenue of the budget in West Samoa in 1990 was W$$1.21 \times 10^8$.
   The population was 1.6×10^5. What was the revenue raised per person?

7 The capacity of a computer is 40 megabytes, where 1 megabyte is 10^3 kilobytes. 1 kilobyte is 1.024×10^3 bytes. Express the capacity of the computer in bytes in standard form.

8 Before the re-unification of Germany in 1989,

the land area of West Germany was 9.6×10^4 square miles
and the land area of East Germany was 4.18×10^4 square miles.

What is the total land area of Germany after re-unification?

9 A supermarket chain buys 1.2×10^5 bottles of wine from Italy.
It pays 3.9×10^4 lire per bottle.
What is the total cost of the wine?

10 The value of a company's annual sales was $£7.2 \times 10^9$.
The company's annual profit was $£9.94 \times 10^8$.
What was the annual expenditure of the company?

11 In 1991, the population of India was 8.46×10^8
                               of Pakistan, 8.97×10^7
                    and of Bangladesh, 1.10×10^8.

What was the total population of these three countries which form the Indian sub-continent?

12 In 1992, the population of the United Kingdom was 58 million.

a Write down 58 million in standard form.

The National Debt of the United Kingdom was 1.86×10^{11} pounds.

b Calculate the National Debt per person giving your answer in
   standard form.   (SEG W95)

13 The mass of one hydrogen atom is

$$1.66 \times 10^{-27} \text{ kg}$$

In one litre of air there are 2.51×10^{22} atoms of hydrogen.

a What is the mass of the hydrogen in one litre of air?

Give your answer in standard form.

b Express this mass without using standard form.   (SEG S95)

14 In 1987 there were two hundred and forty million people in the U.S.A.

a Write this number in standard form.

The area of the U.S.A. is 3.7×10^6 square miles.

b How many people were there, on average, to each square mile?
                                                                          (SEG W93)

15 The distance from the earth to the moon is 2.38×10^5 miles.
Light travels at a speed of 3.0×10^8 metres per second.
Using 1 mile as 1.609×10^3 metres, how long does a signal sent by a
controller at the Kennedy Space Centre take to be received by an
astronaut on the moon?   (SEG W94)

2.4 The keys of a calculator

To use your calculator most effectively, you must become familiar with the keys and their functions.

The booklet that accompanies your calculator will tell you the order in which the keys are used for calculations.

For example, to find $\sqrt[4]{5}$ the keys required on most calculators are:

Other useful keys are:

a the memory keys $\boxed{\text{Min}}$ $\boxed{\text{MR}}$ $\boxed{\text{M+}}$ $\boxed{\text{M-}}$
   Can you use them correctly?

b brackets, e.g. $\dfrac{36 \times 14}{21 \times 4}$ is $36 \times 14 \div (21 \times 4)$

   or $(36 \times 14) \div (21 \times 4)$
   or $36 \times 14 \div 21 \div 4$

   not $36 \times 14 \div 21 \times 4$

c The $\boxed{\text{AC}}$ key clears the calculation (except the memories) before beginning a new calculation.

d The $\boxed{\text{C}}$ or $\boxed{\text{CE}}$ key is used to correct any entry error (i.e. if the wrong key has been pressed).

The key clears the last entry made (either a figure or an operation), provided it is pressed immediately after the error has been made.

The correct entry can then be made and the sequence continued.

e.g. $6 + 3$ $\boxed{\text{C}}$ $2 =$ will produce the answer to $6 + 2$.

Exercise 2D

Use your calculator to find:

1 $386.9 \div 32.87$ (to 1 d.p.)

2 $2756 \div 0.037\,41$ (to 3 s.f.)

3 $0.000\,356 \times 385.7$ (to 3 d.p.)

4 $\dfrac{1}{0.0345}$ (to 3 s.f.)

5 $(4.63)^2$ (to 3 s.f.)

6 $\dfrac{2.75}{1.84 + 2.91}$ (to 3 s.f.)

7 $\dfrac{7.2 + 388}{1.4 + 2.71}$ (to 3 s.f.)

8 $7.2 + \dfrac{3.88}{1.4 + 2.71}$ (to 3 s.f.)

9 $2.9 + \dfrac{3.4}{7.2} + 4.1$ (to 4 s.f.)

10 $2.4 \times (3.2)^2$ (to 3 s.f.)

11 $\dfrac{3.8 + 29.1}{7.1 + 2.3} - 4.1$

*12 Evaluate each answer in turn, then use it, without approximation, in the next calculation:

a $6 - \dfrac{1}{4}$ **b** $6 - \dfrac{1}{a}$ **c** $6 - \dfrac{1}{b}$ **d** $6 - \dfrac{1}{c}$

Write down the answer to **d** only, giving your answer correct to 4 significant figures.

*13 Evaluate each answer in turn, then use it, without approximation, in the next calculation:

a $\dfrac{1}{2}\left(2 + \dfrac{5}{2}\right)$ b $\dfrac{1}{2}\left(\mathbf{a} + \dfrac{5}{\mathbf{a}}\right)$ c $\dfrac{1}{2}\left(\mathbf{b} + \dfrac{5}{\mathbf{b}}\right)$

Write down the answer to **c** only, giving your answer correct to 4 significant figures.

*14 Repeat question 13 replacing the 2 and the 5 in the brackets with:

(i) 1 and 3 (ii) 3 and 10

*15 Evaluate $\sqrt{5}$, $\sqrt{3}$, $\sqrt{10}$, correct to 4 significant figures. Compare these answers with the answers to questions 13 and 14 and comment.

3 Fractions

3.1 Types of fraction

A fraction is a number which can be written as a ratio, with an integer divided by an integer, e.g. $\dfrac{7}{9}$ or $-\dfrac{2}{3}$.

If a shape is divided into a number of equal parts and some of those parts are then shaded, the shaded area can be written as a fraction of the whole.

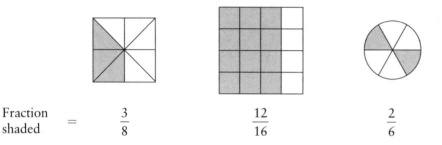

$$
\begin{array}{ccccc}
\text{Fraction} \\
\text{shaded}
\end{array}
= \quad \dfrac{3}{8} \qquad\qquad \dfrac{12}{16} \qquad\qquad \dfrac{2}{6}
$$

The **denominator** (lower integer) denotes the number of parts into which the shape was divided.

The **numerator** (upper integer) denotes the number of parts which have been shaded.

Fractions which have a smaller numerator than denominator are called **proper** fractions, e.g. $\dfrac{1}{2}, \dfrac{5}{12}, \dfrac{28}{60}$

If the numerator is larger than the denominator, the fraction is an **improper** fraction, e.g. $\dfrac{12}{5}, -\dfrac{7}{2}, \dfrac{72}{18}$

A **mixed number** is a number composed of an integer and a proper fraction, e.g. $3\dfrac{1}{2}, -24\dfrac{3}{4}, 4\dfrac{5}{6}$

3.2 Equivalent fractions

We could consider the larger square above to be divided into four columns instead of sixteen squares.

The shaded area is then $\dfrac{3}{4}$ of the whole. This means that $\dfrac{12}{16} = \dfrac{3}{4}$.

$\dfrac{12}{16}$ and $\dfrac{3}{4}$ are called **equivalent fractions** because they have the same value.

Example 1

Complete $\dfrac{6}{7} = \dfrac{?}{28}$ to give equivalent fractions.

The 7 in the denominator must be increased four times to give 28.

The numerator must be treated similarly and be increased four times:

$$\frac{6}{7} = \frac{6 \times 4}{7 \times 4} = \frac{24}{28}$$

Example 2

Reduce $\dfrac{32}{72}$ to its lowest terms.

$$\frac{32}{72} = \frac{32 \div 8}{72 \div 8} = \frac{4}{9}$$

(4 and 9 have no common factor therefore the fraction is in its lowest terms.)

Example 3

Convert $13\dfrac{2}{5}$ to an improper fraction.

$$13\frac{2}{5} = 13 + \frac{2}{5}$$
$$= \left(13 \times \frac{5}{5}\right) + \frac{2}{5}$$
$$= \frac{65}{5} + \frac{2}{5}$$
$$= \frac{65 + 2}{5}$$
$$= \frac{67}{5}$$

Example 4

Convert $\dfrac{141}{22}$ to a mixed number.

$$22\overline{)141}$$
$$ 6 \text{ remainder } 9$$

$$\frac{141}{22} = 6\frac{9}{22}$$

Exercise 3A

1 Complete each of the following to give equivalent fractions:

  a $\dfrac{4}{5} = \dfrac{?}{10}$ e $\dfrac{21}{30} = \dfrac{?}{10}$

  b $\dfrac{2}{3} = \dfrac{?}{12}$ f $\dfrac{16}{18} = \dfrac{8}{?}$

  c $\dfrac{3}{4} = \dfrac{?}{24}$ g $\dfrac{65}{39} = \dfrac{?}{3}$

  d $\dfrac{1}{4} = \dfrac{?}{36}$ h $\dfrac{20}{64} = \dfrac{5}{?}$

2 Reduce the following fractions to their lowest terms:

  a $\dfrac{10}{15}$ b $\dfrac{18}{24}$ c $\dfrac{22}{99}$ d $\dfrac{21}{63}$

  e $\dfrac{75}{100}$ f $\dfrac{40}{60}$ g $\dfrac{30}{65}$ h $\dfrac{64}{96}$

3 Convert the following mixed numbers to improper fractions:

  a $2\dfrac{2}{9}$ c $4\dfrac{11}{12}$ e $8\dfrac{3}{4}$ g $7\dfrac{3}{20}$

  b $5\dfrac{1}{6}$ d $11\dfrac{1}{10}$ f $12\dfrac{2}{3}$ h $20\dfrac{3}{7}$

4 Convert the following improper fractions to mixed numbers, in their lowest terms:

  a $\dfrac{9}{7}$ c $\dfrac{30}{9}$ e $\dfrac{153}{11}$ g $\dfrac{132}{15}$

  b $\dfrac{36}{5}$ d $\dfrac{61}{8}$ f $\dfrac{127}{12}$ h $\dfrac{94}{4}$

3.3 Operations involving fractions

Addition and subtraction

Only fractions *of the same type* can be added or subtracted, i.e. they must have the same denominator.

The method for addition or subtraction is:

  (i) Find the LCM of the denominators.
 (ii) Change each fraction to an equivalent fraction with the new denominator.
(iii) Add and/or subtract the fractions.
 (iv) If the answer is an improper fraction, convert to a mixed number.
  (v) Give the answer in its lowest terms.

Example 1

Evaluate $\dfrac{5}{8} - \dfrac{3}{4} + \dfrac{4}{5}$

The LCM of the denominators is 40.

The sum in equivalent fractions is $\dfrac{25}{40} - \dfrac{30}{40} + \dfrac{32}{40} = \dfrac{25 - 30 + 32}{40}$

$$= \frac{27}{40}$$

which is a fraction in its lowest terms.

Example 2

Evaluate $1\dfrac{2}{3} + \dfrac{4}{5} - 2\dfrac{1}{15}$

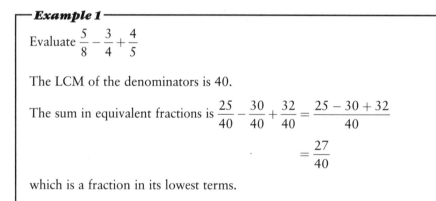

$$1\frac{2}{3} + \frac{4}{5} - 2\frac{1}{15} = 1 + \frac{2}{3} + \frac{4}{5} - \left(2 + \frac{1}{15}\right)$$

$$= 1 + \frac{2}{3} + \frac{4}{5} - 2 - \frac{1}{15}$$

The LCM of the denominators is 15.

The sum in equivalent fractions is $\dfrac{15}{15} + \dfrac{10}{15} + \dfrac{12}{15} - \dfrac{30}{15} - \dfrac{1}{15}$

$$= \frac{15 + 10 + 12 - 30 - 1}{15}$$

$$= \frac{6}{15} = \frac{2}{5} \text{ in its lowest terms}$$

Evaluate the following:

1 $\dfrac{2}{5} + \dfrac{3}{4}$

5 $4\dfrac{1}{6} - 3\dfrac{2}{3}$

9 $4\dfrac{5}{9} - 1\dfrac{1}{3} - 1\dfrac{2}{3}$

2 $\dfrac{7}{8} - \dfrac{5}{6}$

6 $3\dfrac{4}{5} + 1\dfrac{1}{4}$

10 $1\dfrac{1}{2} - 3\dfrac{1}{5} + 1\dfrac{3}{4}$

3 $\dfrac{5}{12} + \dfrac{1}{4}$

7 $3\dfrac{1}{2} - 4\dfrac{1}{4} + 2\dfrac{3}{4}$

11 $3\dfrac{5}{8} - 1\dfrac{1}{4} - 1\dfrac{1}{2}$

4 $1\dfrac{1}{2} - \dfrac{3}{4}$

8 $2\dfrac{4}{9} - 1\dfrac{2}{3} + \dfrac{5}{6}$

12 $10\dfrac{1}{12} - 8\dfrac{5}{6}$

Multiplication and division

The rules for multiplication and division of fractions are very different from those for addition and subtraction.

The fractions do not have to have the same denominator, but they must not be mixed numbers.

Answers should be given in their lowest terms and as mixed numbers, if necessary.

To divide by a fraction, invert it and then multiply by the inverted fraction.

┌─ *Example 1* ───

Multiply $\dfrac{18}{25}$ by $\dfrac{5}{6}$

$$\dfrac{18}{25} \times \dfrac{5}{6} = \dfrac{18 \times 5}{25 \times 6} = \dfrac{90}{150} = \dfrac{3}{5} \qquad or \qquad \dfrac{18}{25} \times \dfrac{5}{6} = \dfrac{\overset{3}{\cancel{18}} \times \overset{1}{\cancel{5}}}{\underset{5}{\cancel{25}} \times \underset{1}{\cancel{6}}} = \dfrac{3 \times 1}{5 \times 1} = \dfrac{3}{5}$$

┌─ *Example 2* ───

Evaluate $1\dfrac{3}{4} \times 2\dfrac{2}{7}$

The mixed numbers are converted to improper fractions:

$$\dfrac{7}{4} \times \dfrac{16}{7} = \dfrac{7 \times 16}{4 \times 7} = \dfrac{112}{28} = 4 \qquad or \qquad \dfrac{7}{4} \times \dfrac{16}{7} = \dfrac{\overset{1}{\cancel{7}} \times \overset{4}{\cancel{16}}}{\underset{1}{\cancel{4}} \times \underset{1}{\cancel{7}}} = \dfrac{1 \times 4}{1 \times 1} = 4$$

Example 3

Evaluate $\dfrac{5}{9} \div \dfrac{2}{3}$

$$\frac{5}{9} \div \frac{2}{3} = \frac{5}{9} \times \frac{3}{2} = \frac{15}{18} = \frac{5}{6}$$

or

$$\frac{5}{9} \div \frac{2}{3} = \frac{5}{9} \times \frac{3}{2} = \frac{5 \times \overset{1}{\cancel{3}}}{\underset{3}{\cancel{9}} \times 2} = \frac{5 \times 1}{3 \times 2} = \frac{5}{6}$$

Example 4

Divide $3\dfrac{1}{6}$ by $\dfrac{2}{3}$

$$3\frac{1}{6} \div \frac{2}{3} = \frac{19}{6} \div \frac{2}{3} = \frac{19}{6} \times \frac{3}{2} = \frac{57}{12} = 4\frac{9}{12} = 4\frac{3}{4}$$

or

$$3\frac{1}{6} \div \frac{2}{3} = \frac{19}{6} \div \frac{2}{3} = \frac{19}{\underset{2}{\cancel{6}}} \times \frac{\overset{1}{\cancel{3}}}{2} = \frac{19 \times 1}{2 \times 2} = 4\frac{3}{4}$$

Exercise 3C

Evaluate the following:

1   $\dfrac{7}{8} \times \dfrac{4}{5}$

2   $1\dfrac{2}{3} \times \dfrac{4}{5}$

3   $6 \div \dfrac{1}{2}$

4   $\dfrac{5}{6} \div \dfrac{3}{4}$

5   $2\dfrac{1}{5} \times 3\dfrac{1}{4}$

6   $5\dfrac{1}{7} \div 3$

7   $2\dfrac{1}{7} \times 1\dfrac{3}{5} \times 2\dfrac{1}{3}$

8   $1\dfrac{5}{9} \div 1\dfrac{1}{3} \times \dfrac{3}{4}$

9   $2\dfrac{6}{11} \div \dfrac{7}{11} \div 3\dfrac{2}{5}$

10   $4\dfrac{3}{8} \div \left(1\dfrac{5}{8} - \dfrac{3}{4}\right)$

11   $6\dfrac{2}{3} \times \left(\dfrac{1}{4} + \dfrac{1}{5}\right)$

3.4   The conversion of fractions to decimal fractions

The fraction $\dfrac{7}{8}$ may be stated as $7 \div 8$.

Using a calculator, $7 \div 8 = 0.875$, and this is the **decimal fraction** which is equivalent to $\dfrac{7}{8}$.

> **Example**
>
> Convert $3\frac{5}{6}$ to a decimal fraction.
>
> The integer part of the mixed number remains the same. Only the fractional part needs to be converted.
>
> On the calculator $5 \div 6 = 0.833\,333\,3$ which is a recurring decimal.
>
> $$3\frac{5}{6} = 3.8\dot{3} \text{ or } 3.83 \text{ (to 3 s.f.)}$$
>
> All fractions convert to either a terminating or a recurring decimal. The dot above the 3 indicates that the 3 is a recurring decimal.

Exercise 3D

1   Write down the shaded area as   (i) a fraction,   (ii) a decimal fraction of the whole area.

a 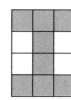   b 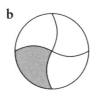   c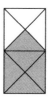

2   Convert the following fractions to decimals:

a $\dfrac{1}{10}$ c $\dfrac{3}{4}$ e $4\dfrac{21}{25}$ g $7\dfrac{4}{9}$ i $\dfrac{9}{11}$

b $\dfrac{1}{2}$ d $1\dfrac{9}{20}$ f $2\dfrac{5}{6}$ h $2\dfrac{8}{9}$ j $3\dfrac{1}{7}$

4 Ratio and Proportion

4.1 Ratio

Carmen's parents give her a weekly allowance of £3.60.
Her younger brother, Leroy, is given an allowance of £1.20. Carmen receives three times as much allowance as Leroy. Here is another way of saying the same thing:

The ratio of Carmen's allowance to Leroy's allowance is **3 : 1**

Ratios can be written with a colon between the amounts, like this:

$$\text{First quantity : Second quantity}$$

or as a fraction, like this:

$$\frac{\text{First quantity}}{\text{Second quantity}}$$

Example 1

When Leroy is older, his parents decide that the ratio between his allowance and his sister's should now be 2 : 3.
If Carmen receives £3.90 per week, how much should they give Leroy?

Leroy's allowance : Carmen's allowance $= 2 : 3 = \dfrac{2}{3}$

Leroy should receive $\dfrac{2}{3}$ of Carmen's allowance.

Leroy's allowance $= \dfrac{2}{3} \times £3.90$

$\qquad\qquad\qquad = £2.60$

Example 2

A large jar of coffee costs £2.38 and a small jar costs 84p.
Express these prices as a ratio in its lowest terms.

Converting both prices to pence gives the ratio:

$$\text{Large jar : small jar} = 238 : 84$$

The ratio can be reduced if 238 and 84 can both be divided by the same number (called a **common factor**).

The largest factor which is common to 238 and 84 may not be immediately obvious, in which case the reduction to lowest terms can be carried out in stages.

2 is a common factor of 238 and 84. Dividing by 2 reduces the ratio to 119 : 42.

Possible factors of 42 are 2, 3, 6 and 7.
Only 7 is also a factor of 119.

Dividing by 7 gives the ratio in its lowest terms

$$= 17 : 6$$

Exercise 4A

In questions 1–3, write all ratios in their lowest terms.

1   Two brothers have £20 and £24 in their respective savings accounts.
    Express these amounts as a ratio.

2   Miss Morgan has £320 in her current account, £400 in her deposit account, and £800 in her savings account.
    Express these amounts as a ratio.

3   A pound of grapes cost £1.60 and a pound of pears 72p. Write these prices as a ratio.

4   Write each pair of quantities as a ratio in its lowest terms:

    a   60 m, 40 m   f   15 cm, 10 cm

    b   £2, 20p   g   750 g, 2 kg

    c   0.6 cm, 0.05 cm   h   39 litres, 26 litres

    d   0.4 m, 1.6 m   i   32 mph, 48 mph

    e   1 ft, 9 in

5   Complete the following ratios:

    a $3 : 4 = 6 : ?$ d $240 : 400 = ? : 1$

    b $18 : 9 = ? : 1$ e $20 : 1 = 64 : ?$

    c $? : 1 = 12 : 10$ f $1 : ? = 5 : 13$

6   A line is trisected (i.e. divided in the ratio 1 : 2). If the smaller length is 14 centimetres, how long is the line?

7   The circumference of a circle and its diameter have lengths which are approximately in the ratio 3 : 1.

    a   What is the approximate length of the circumference of a circle with a diameter of 4.5 centimetres?

    b   What is the approximate length of the diameter of a circle with a circumference of 24 inches?

8   E3 and E10 size soap packets contain powder with weight in the ratio 3 : 10.
    The E3 size contains 1.05 kilograms of powder.

    a   What weight of powder does the E10 size contain?

    b   What weight of powder does the E15 size contain?

9   Spring bulbs are planted in a border in the ratio of 3 yellow tulips to 2 pink tulips to 5 grape hyacinths.
    If 615 yellow tulip bulbs are planted, how many pink tulip bulbs and how many grape hyacinth bulbs are planted?

10   On their birthdays David and Emily are given money in the ratio of their ages.
    When David is 15 years old, Emily is 12.

    a   Express the children's ages as a ratio.

    b   If David is given £12.50, how much does Emily receive?

    c   When the children are three years older, Emily is given £12.50. How much will David receive?

4.2 Division in a given ratio

FAIR SHARES FOR ALL!!

This does not necessarily mean equal shares for all.

For example, if three partners invest different amounts of money in a business, they might expect the profits to be shared in proportion to their investment.

Example 1

Divide £672 between Emily, Faye and Geoff in the ratio $7 : 5 : 9$ respectively. How much does each person receive?

Method
 (i) Find the total number of shares.
 (ii) Find the amount of one share.
 (iii) Find the amount each receives.

Calculation
 (i) Total number of shares $= 7 + 5 + 9 = 21$

 (ii)   Amount of one share $= \dfrac{£672}{21} = £32$

 (iii)   Emily receives $£32 \times 7 = £224$
                   Faye receives $£32 \times 5 = £160$
                  Geoff receives $£32 \times 9 = £288$

(*Check:* $224 + 160 + 288 = 672$.)

Example 2

Three partners, A, B and C, invest money in a small business.
The amounts they invest are £10 000, £12 000 and £6000, respectively.

At the end of the first year of trading the profits from the business are £14 350.

They each receive profits in proportion to their investment.
How much does each partner receive?

The investments are in the ratio $10\,000 : 12\,000 : 6000$
$$= 5 : \quad 6 \quad : 3$$

The total number of shares	$= 5 + 6 + 3$	$= 14$
One share of the profits	$= £14\,350 \div 14$	$= £1025$
A receives 5 shares	$= £1025 \times 5$	$= £5125$
B receives 6 shares	$= £1025 \times 6$	$= £6150$
C receives 3 shares	$= £1025 \times 3$	$= £3075$

(*Check:* $£5125 + £6150 + £3075 = £14\,350$.)

Exercise 4B

1 Divide £650 in the ratio 2 : 3.

2 Divide £12 000 in the ratio 1 : 3 : 4.

3 Divide £104 in the ratio 6 : 4 : 3.

4 Mrs Chandra shared £4000 among her three children in the ratio 7 : 5 : 4.
   How much did each receive?

5 The sum of £1000 is invested in unit trusts for income and for capital growth in the ratio 3 : 5.
   How much is invested in each type of trust?

6 Three sisters win a prize in a Mathematics competition for the best Statistics project. They decide to share the prize of £200 according to the amount of work each contributed.
   They calculate that this was in the ratio 3 : 3 : 2.

   a How much should each of the sisters who contributed most receive?

   b How much should the sister who contributed least receive?

7 a Whenever Grandma comes to visit she gives her grandchildren, Emily and David, money in the ratio of their ages.
      When David is 10 years old, Emily is 6 years old.
      (i) What is the ratio of their ages, in its simplest form?
      (ii) If Grandma gives the children £12 to share, how much does each child receive?

   b The next time Grandma visits the children have both had a birthday. If they again share £12 between them, in the ratio of their ages, how much (to the nearest 1p) do they each receive?

8 Four office workers run a pools syndicate and each week pay £4.95, £6.60, £3.30 and £4.95 respectively for their entry. When they win £104 616 they divide the winnings in the ratio of their weekly contribution.
   How much does each receive?

9 The Shang Dynasty in China was making bronze artifacts more than three thousand years ago.

   The bronze they used was an alloy of copper and zinc in the ratio (by weight) of 17 : 3.
   What weights of copper and zinc were used to make a bronze bowl weighing 1.6 kg?

10 Sharon, Emily and Ray have each worked as babysitters in the last term. They share their pay of £183 in the ratio of the hours which they have each worked.
   Sharon has worked 28 hours, Emily 35 hours and Ray 21 hours.
   How much does each person receive?

   (SEG S95)

4.3 Direct proportion

If two quantities increase, or decrease, at the same rate, they are said to be in **direct proportion**.

For example, if you double your speed of walking you will travel twice as far in the same period of time.

Example

Mrs Wall usually buys 12 pints of milk each week and pays the milkman £3.72. In a week when she has visitors, she buys 3 extra pints. How much is her milk bill for that week?

$$12 \text{ pints of milk cost } £3.72$$

$$1 \text{ pint of milk costs } \frac{£3.72}{12} (= 31p)$$

$$(12 + 3) \text{ pints of milk cost } \frac{£3.72}{12} \times 15$$

$$= £4.65$$

Exercise 4C

1   Find the cost of 6 lb of apples if 4 lb cost £2.12.

2   The baker's shop sells cheese biscuits for £1.52 per quarter pound.
    How much would 7 oz cost? ($\frac{1}{4}$ lb = 4 oz).

3   Pic'n'Mix sweets cost 56p per quarter pound. How much would you be charged for 9 oz?

4   Anemone bulbs cost £1.15 for 25.
    How much should be charged for 40 bulbs?

5   A shop stocks batteries in packs of five with a retail price of £2.99 per pack.
    To try and sell the batteries more quickly, the shop offers them for sale in packs of three. How much should the shop charge for the new packs?

6   It takes a hiker $3\frac{1}{2}$ hours to walk $10\frac{1}{2}$ miles. How long will it take her to walk 12 miles, at the same speed?

7   Fruit cakes costing £1.68 per cake are each cut into ten slices to be sold at a charity tea party. The organisers decide to make a profit of 52p on each cake.
    How much will they charge for 4 slices of cake?

8   a   A tin of custard is 11.5 cm high and contains 440 g of custard. The manufacturer increases the height of the tin to 13.5 cm. How many grams does a new tin hold?

    b   If the original tin cost 29p, how much should be charged for the new size?

    c   The manufacturer claims that the new tins hold 20% more. Is he correct?

4.4   Inverse proportion

Michael Bates is slowly and carefully laying a concrete path when he suddenly notices a dark cloud approaching. Michael immediately doubles his rate of working so that he can finish the job in half the time (before it rains).

If an increase in one quantity causes a decrease, at the same rate, in another, the two quantities are **inversely proportional**.

Example

Alice, working at a rate of 20 questions per minute, completes a mental arithmetic test in 10 minutes.
Carol works at a rate of 25 questions per minute.
How long will it take Carol to complete the test?

At 20 questions per minute it takes 10 minutes.

At 1 question per minute it takes 10×20 minutes.

At 25 questions per minute it takes $\dfrac{10 \times 20}{25}$ minutes
$= 8$ minutes.

It takes Carol 8 minutes to finish the test.

---

1 Eight people share a prize and receive £125 each. How much would each have received if there had been only five prize winners?

2 It takes two combine harvesters $4\frac{1}{2}$ hours to clear a field of wheat.
How long would it take three combine harvesters to clear the same field?

3 If a job takes eight people six days to complete,

   a how long would it take five people to complete?

   b how many people would be needed if the job had to be completed in four days?

4 When eight checkout points are staffed at a supermarket, it takes an average of 54 minutes to deal with 100 customers.
If all 12 checkout points are staffed, how long does it take to deal with 100 customers?

5 It usually takes five farm workers six days to harvest the first strawberry crop. Because of wet weather, the crop needs to be harvested in two days. How many extra pickers should the farmer employ?

6 Five cousins expect to inherit £8500 each from their grandfather's estate. Unfortunately, one cousin suffers a fatal accident before he can inherit.
How much do the four surviving cousins each receive?

4.5 Measures of rate

Rate

In the previous two exercises (4C and 4D), an intermediate step in the calculation was to find the **cost of one unit**: e.g. one pound, the time taken to make one article, the amount of work done in one day.

If the information given is already in this form, it is called a **rate**. For example:

A speed of 50 miles per hour is a rate of travel
French francs 8.93 per £1   is a rate of exchange
£423 per week   is a rate of pay
Time and a half   is an overtime rate
5 kg per m^3 is a density

Given the basic rate, it is a simple calculation to find any other amount required.

Example 1

A firm pays an overtime rate of time and a half for work done on a Saturday.
The firm's basic rate of pay is £3.42 per hour.
How much is paid for $6\frac{1}{2}$ hours of work on a Saturday?

Basic rate of pay $= £3.42$ per hour
Overtime rate of pay $= 1.5 \times £3.42$ per hour
                 $= £5.13$ per hour
Saturday pay $= £5.13 \times 6.5$
                 $= £33.35$

Example 2

An electric fire uses 18 units of electricity over a period of $7\frac{1}{2}$ hours.
What is the hourly rate of consumption of electricity?

$$\text{Hourly rate of consumption} = \frac{\text{Number of units used}}{\text{Time}}$$

$$= \frac{18 \text{ units}}{7.5 \text{ hours}}$$

$$= 2.4 \text{ units per hour}$$

Example 3

The density of glass is 2.74 g/cm^3
Find the weight of 740 cm^3 of glass.

Weight is $2.74 \times 740 \text{ g}$
$\quad\quad = 2027.6 \text{ g}$
$\quad\quad = 2.03 \text{ kg}$

Exercise 4E

1   Carrots cost 18p per pound.
How much would $4\frac{1}{2}$ lb of carrots cost?

2   A man earns a basic wage of £10.70 per hour.
What is his basic weekly pay for a 35-hour week?

3   A car uses petrol at the rate of 36 miles per gallon.

   a   How many miles can the car travel on $2\frac{1}{2}$ gallons of petrol?

   b   How many gallons of petrol (to the nearest gallon) would be needed for a journey of 240 miles?

   c   One gallon is approximately 4.55 litres. What is the car's rate of consumption in miles per litre?

4   At the current rate of exchange there are 8.93 French francs to the pound sterling. How much would 1800 francs be worth in pounds?

5   The speed limit on a motorway is 70 mph. What is the shortest possible journey time from Dover to Glasgow, a distance of 470 miles?

6   Lawn sand should be applied at a rate of 136 g per square metre.
How much lawn sand is required for a lawn 9.5 m × 7.5 m? (Give your answer to the nearest 0.1 kg.)

7   a   The toll charged for a car travelling on a motorway was £33.60 for a journey of 420 kilometres.
What was the rate per kilometre?

   b   Cars with trailers are charged double. How much would it cost for a car and caravan to travel 264 kilometres?

8   A sheet of perspex of thickness 3 cm measures 1 m by 70 cm, and weighs 24.6 kg.
Find the density of perspex in grams per cubic centimetre.

4.6 Speed or velocity

Speed is the rate of travel.
Common measures of speed are miles per hour (mph); kilometres per hour (km/h); metres per second (m/s).

To calculate speed you need to know the distance travelled and the time taken:

$$\text{Speed} = \frac{\text{Distance travelled}}{\text{Time taken}}$$

Unless the speed is constant over the whole distance, this will be an **average speed** for the journey.

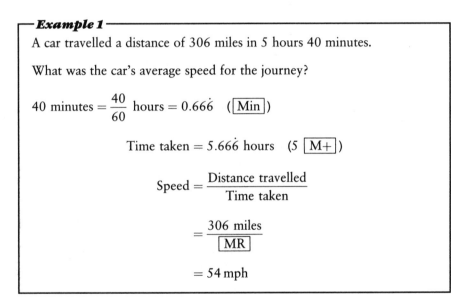

Example 1

A car travelled a distance of 306 miles in 5 hours 40 minutes.

What was the car's average speed for the journey?

$$40 \text{ minutes} = \frac{40}{60} \text{ hours} = 0.66\dot{6} \quad (\boxed{\text{Min}})$$

$$\text{Time taken} = 5.66\dot{6} \text{ hours} \quad (5 \; \boxed{\text{M+}})$$

$$\text{Speed} = \frac{\text{Distance travelled}}{\text{Time taken}}$$

$$= \frac{306 \text{ miles}}{\boxed{\text{MR}}}$$

$$= 54 \text{ mph}$$

Example 2

A car travelling at a speed of 65 kilometres per hour takes $2\frac{1}{2}$ hours for a journey.

What distance has the car covered in this time?

In 1 hour the car travels 65 km

In $2\frac{1}{2}$ hours the car travels $65 \times 2\frac{1}{2}$ km

$$= 162.5 \text{ km}$$

or $\qquad$ $\text{Distance} = \text{Speed} \times \text{Time} = 65 \times 2\frac{1}{2}$ km

$$= 162.5 \text{ km}$$

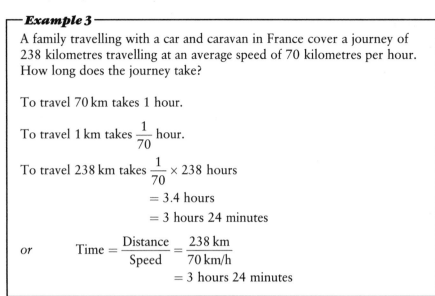

Example 3

A family travelling with a car and caravan in France cover a journey of 238 kilometres travelling at an average speed of 70 kilometres per hour. How long does the journey take?

To travel 70 km takes 1 hour.

To travel 1 km takes $\dfrac{1}{70}$ hour.

To travel 238 km takes $\dfrac{1}{70} \times 238$ hours

$= 3.4$ hours

$= 3$ hours 24 minutes

or $\qquad \text{Time} = \dfrac{\text{Distance}}{\text{Speed}} = \dfrac{238 \text{ km}}{70 \text{ km/h}}$

$= 3$ hours 24 minutes

Exercise 4F

1 Find the average speed, in the most appropriate units, for each of the following:

 a a journey of 162 miles which took 3 hours

 b a 1500 m race which was won in a time of 2 min 54.4 s

 c a journey of 286 km which took $3\frac{1}{4}$ hours.

2 Find the distance covered:

 a by a car travelling at 85 km/h for 2.4 hours

 b by an aircraft travelling at 600 mph for 4 hours 36 minutes

 c by a satellite travelling at 7850 m/s for 12.5 s.

3 Find the time taken

 a to travel 186 km at an average speed of 62 km/h

 b to travel 306 miles at an average speed of 40 mph

 c to travel 91.2 nautical miles at an average speed of 24 knots.
 (1 knot = 1 nautical mile per hour.)

4 Andy goes running every evening for 45 minutes. If he runs at a speed of 10 mph, how far does he run?

5 A cheetah is the fastest animal over short distances. It can travel at a speed of 96 km/h. At this speed, how much ground can a cheetah cover in 30 seconds?

6 a A race for small yachts has three 'legs' of distances 8 km, 6 km and 10 km. The average speed for the winning yacht was 6.2 kph. How long did it take the winning yacht to complete the course?

 b The second yacht finished 8 minutes after the winner.
 What was its average speed?

7 A car travels 130 miles in 2 hours 40 minutes. What is its average speed? (SEG W94)

8 A ferry operates from Plymouth to Roscoff, a distance of 115 miles.

 a How long would the journey take in hours and minutes, if the average speed of the ferry is 19.7 miles per hour?

 One day, the ferry left Plymouth late. In order to arrive in Roscoff on time, the crossing must take only $5\frac{1}{4}$ hours.

 b At what average speed should the ferry travel? (SEG S95)

9 The distance from London to Newcastle is 285 miles. Derek takes $4\frac{1}{2}$ hours to drive this distance.

 a Calculate Derek's average speed.

 A train takes 3 hours 20 minutes to travel this distance.

 b Calculate the train's average speed.
 (SEG W95)

Average speed

If a journey is divided into sections of different lengths travelled at different average speeds, then to calculate the average speed for the **whole** journey you need to know the **total** distance travelled and the **total** time taken.

$$\text{Average speed} = \frac{\text{Total distance travelled}}{\text{Total time taken}}$$

Example

A car travelled from Newbury to Aylesbury, a distance of 50 miles. Twenty miles of the journey was on dual carriageway and the car averaged a speed of 50 mph. The remainder of the journey was on single carriage roads and the average speed dropped to 36 mph.

What was the average speed for the whole journey?

$$\text{Total distance travelled} = 50 \text{ miles}$$

To find the total time taken we need to calculate the time taken for each part of the journey.

$$\text{Time taken} = \frac{\text{Distance travelled}}{\text{Average speed}}$$

On dual carriageway:

$$\text{Take taken} = \frac{20 \text{ miles}}{50 \text{ mph}}$$

$$= 0.4 \text{ hours}$$

$$= 24 \text{ minutes}$$

On single carriageway:

$$\text{Time taken} = \frac{30 \text{ miles}}{36 \text{ mph}}$$

$$= \frac{30}{36} \times 60 \text{ minutes}$$

$$= 50 \text{ minutes}$$

$$\text{Total time taken} = (50 + 24) = 74 \text{ minutes} = 1.23\dot{3} \text{ hours} \; (\boxed{\text{Min}})$$

$$\text{Average speed} = \frac{\text{Total distance travelled}}{\text{Total time taken}}$$

$$= \frac{50 \text{ miles}}{\boxed{\text{MR}}}$$

$$= 40.5 \text{ mph}$$

Exercise 4G

1 Two friends, on holiday in Jersey, cycled from St Helier to St Aubin (a distance of 5 km) and back again. The ride to St Aubin took them 11 minutes and the return ride took 14 minutes.

What was their average speed for the whole journey?

2 a A car travelling along the M1 takes 40 minutes to travel between junction 18 and junction 21, a distance of 30 miles. What is the car's average speed?

 b Between junction 21 and junction 22, a distance of 13 miles, the car travels at a speed of 65 mph. How long does it take?

 c What is the car's average speed between junction 18 and junction 22?

3 Two friends are youth hostelling in the Northumbrian hills. At 9 o'clock one morning they set off from Rothbury to walk to Redesdale. They walk at a steady rate of $3\frac{1}{3}$ mph until noon, when they stop for lunch for 1 hour. They then walk the remaining 14 miles and arrive at their destination at 5 pm.

 a How far did they walk before lunch?

 b At what average speed did they walk after lunch?

 c What was their average speed for the whole journey?

4 A family, on holiday in Italy, travelled from Florence to Bologna at an average speed of 68 km/h and from Bologna to Verona at an average speed of 81 km/h.
The journey from Florence to Bologna took them 1 hour 15 minutes and the total distance travelled was 220 km.

 a How far is it from Florence to Bologna?

 b How far is it from Bologna to Verona?

 c How long did the journey from Bologna to Verona take?

 d What was the average speed for the 220 km journey?

5 Measurement

5.1 Metric and imperial units

In 1971 the British monetary system was decimalised. We stopped using pounds, shillings and pence (£ s d) and started using pounds and 'new' pence (£ p). Since then, many more units have changed from imperial to metric units:

We can now buy petrol in litres not gallons.
We buy material in metres instead of yards.
The standard length of a ruler is 30 centimetres not 12 inches.
The weather forecast gives temperatures in degrees Celsius rather than in degrees Fahrenheit.

We do, however, still use imperial measures:

Distances are given in miles and speed restrictions in miles per hour.
Although most grocery items may still be labelled with weights in both grams and ounces, fruit, vegetables and sweets may still be bought in pounds and ounces.
Milk and beer are still sold in pints.
Carpets are often sold in feet and yards.

The advantage of the metric system over the imperial is the ease with which one unit can be converted to another.

Metric units

The metric system is a decimal system and units are converted by multiplying or dividing by powers of 10 (i.e. 10, 100, 1000).

Each prefix to a standard measure (e.g. metre, gram, litre) indicates the relative size:

$$\begin{array}{llll} \text{kilo (k)} & \text{means } 1000\times & \text{deci (d)} & \text{means } \dfrac{1}{10}\times \\[2mm] \text{hecto (h)} & \text{means } 100\times & & \\[2mm] \text{deca (da)} & \text{means } 10\times & \text{centi (c)} & \text{means } \dfrac{1}{100}\times \\[2mm] & & \text{milli (m)} & \text{means } \dfrac{1}{1000}\times \end{array}$$

The most common metric units in everyday use are:

Length Metre (m)	Weight Gram (g)	Capacity Litre (l)
1 kilometre = 1000 metres	1 kg = 1000 g	1 litre = 100 cl
1 centimetre = 10 millimetres	1 g = 1000 mg	1 litre = 1000 ml = 1000 cm³
1 metre = 100 centimetres	1 tonne = 1000 kg	1 cl = 10 ml
1 metre = 1000 millimetres		

Example

a Convert 2.75 kg to g.

b Convert 6.3 ml to cl.

c Convert 56 cm to km.

a 1 kg = 1000 g, so

$$2.75 \text{ kg} = 1000 \times 2.75 \text{ g} = 2750 \text{ g}$$

b $1 \text{ ml} = \dfrac{1}{10} \text{ cl},$ so

$$6.3 \text{ ml} = \dfrac{1}{10} \times 6.3 \text{ cl} = 0.63 \text{ cl}$$

c $1 \text{ cm} = \dfrac{1}{100} \text{ m} = \dfrac{1}{1000} \times \dfrac{1}{100} \text{ km},$ so

$$56 \text{ cm} = \dfrac{1}{1000} \times \dfrac{1}{100} \times 56$$

$$= 0.000\,56 \text{ km}$$

Exercise 5A

1 Change:

a 1232 m to km

b 0.032 m to mm

c 626 g to kg

d 0.731 litres to cl

e 1.62 cl to ml

f 12.7 km to m

g 59.1 ml to litres

h 3.4 t to kg

i 1.39×10^{-5} kg to mg

j 8.5×10^{7} mg to kg

2 The weight of sliced meat is marked on the packet as 0.23 kg.
   What is the weight in grams?

3 A load of 2 tonnes of potatoes is to be split into 10 kg bags.
   How many bags can be filled from this load?

4 The weight of a packet of crisps is 25 g.
   How many milligrams is this?

5 Two pieces of wood of lengths 2.56 m and 37 cm are joined together.
   What is the total length of wood

   a in metres,

   b in centimetres?

6 A quantity of sugar is weighed out on a scale and the reading is 250 g. A spoonful of sugar is removed and the new reading is 221.6 g.

   a How many grams of sugar are removed?

   b What is the weight in milligrams?

   c Write your answer to **b** in standard form.

7 A bottle of lemonade is emptied into eight cups, each of which holds 150 ml.
   How many litres of lemonade were in the bottle?

8 A bottle of weedkiller holds 1 litre of concentrate. To make up a solution, 5 capfuls are added to 2 litres of water. A capful is 15 ml.
   How many litres of solution can be made from the bottle?

Imperial units

The most common imperial units in everyday use are:

Length		Weight	Capacity
1 mile = 1760 yards	1 ton	= 20 hundredweight (cwt)	1 gallon = 8 pints
1 yard = 3 feet (ft or ')	1 cwt	= 112 pounds (lb)	1 pint = 20 fluid
1 foot = 12 inches (in or ")	1 lb	= 16 ounces (oz)	ounces
	1 stone	= 14 pounds	(fl oz)

Example

a Convert 76″ to feet and inches.

b Convert 6 lb 5 oz to ounces.

c Convert 18 pints to gallons.

a 12 in = 1 ft,   so

$$76'' = \frac{76}{12} \text{ ft} = 6\frac{4}{12} \text{ ft} = 6' \, 4''$$

b 1 lb = 16 oz,   so

$$6 \text{ lb } 5 \text{ oz} = (6 \times 16) + 5 \text{ oz} = 101 \text{ oz}$$

c 8 pt = 1 gal,   so

$$18 \text{ pt} = \frac{18}{8} \text{ gal} = 2.25 \text{ gal}$$

Exercise 5B

1 Change:

    a 39 in to feet

    b 40 oz to lb and oz

    c $16\frac{1}{2}$ ft to yards

    d 2 ft 3 in to inches

    e 3 lb 7 oz to oz

    f 20 pints to gallons

    g $3\frac{1}{2}$ gallons to pints

    h 0.7 pint to fluid ounces.

2 A milk churn holds 5 gallons of milk. How many pint bottles of milk can be filled from the churn?

3 A recipe requires 15 fluid ounces of water, but the measuring jug is calibrated in pints. How many pints of water should be measured?

4 Coal is delivered in hundredweight bags. A load of twenty-four bags is delivered to a house. What is the weight of the load in tons?

5 A bowl of honey is weighed. The reading on the scale is 1 lb 5 oz. The bowl is known to weigh $12\frac{1}{4}$ oz. How much does the honey weigh?

6 Nina measures her bedroom and finds that the length is 12′ 9″ and the breadth is 10′ 6″. What is the perimeter of the room
    a in feet   **b** in yards?

7 A grocer divides a block of cheese weighing 4 lb $9\frac{1}{2}$ oz into twelve equal portions. How many ounces does each portion weigh?

8 A bottle of concentrated orange juice holds 8 fluid ounces. To make up the juice for drinking, three times this amount of water is added. If 1 gallon of diluted juice is required, how many bottles of concentrated juice should be used?

5.2 Conversion between metric and imperial units

It is often necessary, for comparison, to convert from imperial to metric units or vice versa.

If a rough comparison is all that is required, an approximate conversion factor can be used. For large quantities, or where a correct comparison is required, a conversion factor of the appropriate degree of accuracy should be used.

Approximate conversion	More accurate conversion
5 miles ≈ 8 km	1 mile = 1.609 km
1 metre ≈ 39 inches	1 yard = 0.914 m
1 foot ≈ 30.5 cm	1 foot = 30.48 cm
1 litre ≈ 1.75 pints	1 inch = 2.54 cm
1 litre ≈ 0.2 gallons	1 kg = 2.205 lb
1 kg ≈ 2.2 lb	1 litre = 1.76 pt
	1 gallon = 4.55 litres

Please note that in the SEG Modular Mathematics syllabus, this topic is examined as part of the Terminal module.

Example 1

The contents of a packet of sugar weigh 250 g.
What weight, in ounces, should be printed on the packet?

Merchandise must carry an accurate description, and therefore the weight given must be correct to a reasonable degree of accuracy.
The weight stated on a packet, tin, etc., is usually the minimum weight of the contents.

The conversion used is
$$1\,\text{kg} = 2.205\,\text{lb}$$
$$\therefore \quad 1000\,\text{g} = 2.205 \times 16\,\text{oz} = 35.28\,\text{oz}$$
$$1\,\text{g} = 0.035\,28\,\text{oz}$$
$$250\,\text{g} = 0.035\,28 \times 250\,\text{oz}$$
$$= 8.82\,\text{oz}$$

The minimum weight of the sugar is 8.8 oz.

Example 2

A family, on a self-catering holiday in Europe, wish to buy the equivalent in weight of 3 lb of potatoes and $\frac{1}{2}$ lb of butter.
How much of each item should they buy?

Exact weights are not required.

The conversion used is $2\,\text{lb} \approx 1\,\text{kg}$
or $1\,\text{lb} \approx 0.5\,\text{kg}$
so $3\,\text{lb} \approx 1.5\,\text{kg}$
and $\frac{1}{2}\,\text{lb} \approx 0.25\,\text{kg} = 250\,\text{g}$

They should buy $1\frac{1}{2}$ kg of potatoes and 250 g of butter.

Exercise 5C

1   Vegetables are sold in tins of various sizes. What weight, in ounces to the nearest 0.1 oz, should be printed on a tin if the contents weigh:

   a 425 g   **b** 440 g   **c** 415 g?

2   A shop prepacks its groceries and labels the packets with both metric and imperial weights. For each of the following items, calculate the missing weight:

   a 1 lb of mince

   b 150 g of cooked ham

   c 5 lb of potatoes

   d 0.74 lb of cheese

   e 14.2 g of dried herbs.

3   A set of kitchen scales is calibrated in both imperial and metric weights.
Calculate the number of pounds and ounces (to the nearest ounce) which can be read alongside the metric weights of:

   a 1 kg   **b** 0.4 kg   **c** 1.5 kg   **d** 2.1 kg

4   **a** What is the equivalent of 1 litre of petrol in gallons?

   b A driver buys 20 litres of petrol. Approximately, how many gallons of petrol does he buy (correct to 1 d.p.)?

5   The specifications for a particular make of car give the petrol consumption as 40 miles per gallon. How many miles per litre is this?

6   A couple on touring holiday in France cover a certain distance each day.
Calculate the approximate number of miles which are equivalent to:

   a 120 km   **c** 429 km   **e** 370 km

   b 324 km   **d** 56 km

7   Gary is shopping for shelving for his collection of CDs. He finds some which is 6 inches wide. His CDs are 14.1 cm wide. Is the shelving wide enough?

8   Omar has bought six pints of milk. Dominique buys three litres of milk. Who has bought the most?
You *MUST* show your working.   (SEG W95)

6 Percentages

'*Inflation now stands at 7.3%.*' '*Ford have given their workforce a 21.3% rise.*' '*Unemployment in Winchester is less than 2%.*'

Percentages are a part of our everyday lives. They are often quoted in the media, particularly in connection with money matters.

Percentages often help us make comparisons between numbers, but we must know exactly what a 'percentage' is.

6.1 Percentages

A percentage is a fraction with a particular number divided by 100:

$$20\% \text{ means } \frac{20}{100}.$$

A decrease of 20% would be a decrease of $\frac{20}{100}$, which is the same thing as a decrease of $\frac{1}{5}$, a fifth.

$$\frac{1}{4} = \frac{1}{4} \times \frac{25}{25} = \frac{25}{100} \text{ which is } 25\%$$

To convert a fraction to a percentage multiply by 100.
To convert a percentage to a fraction, divide by 100.

Example 1

Convert **a** $\frac{3}{5}$, **b** $\frac{9}{11}$ to percentages.

a $\frac{3}{5}$ as a percentage $= \frac{3}{5} \times 100 = 60\%$

b $\frac{9}{11}$ as a percentage $= \frac{9}{11} \times 100 = 81.82\%$ (to 2 d.p.)

Example 2

Convert 65% to a fraction in its lowest terms.

$$65\% = \frac{65}{100} = \frac{13}{20}$$

Exercise 6A

1 Convert the following fractions to percentages:

a $\dfrac{1}{5}$ c $\dfrac{7}{10}$ e $\dfrac{2}{3}$ g $1\dfrac{3}{4}$

b $\dfrac{1}{8}$ d $\dfrac{13}{20}$ f $\dfrac{9}{25}$ h $2\dfrac{1}{2}$

2 Convert the following percentages to fractions:

a 60%   c 10%   e 15%   g $37\dfrac{1}{2}\%$

b 25%   d 85%   f 130%   h $33\dfrac{1}{3}\%$

3 Copy and complete the following table to give each quantity in its fractional, decimal and percentage form.

	Fraction	Decimal	Percentage
a	$\frac{3}{4}$		
b		0.5	
c	$\frac{1}{8}$		
d			$33\frac{1}{3}$
e		0.375	
f	$\frac{7}{10}$		
g			35
h		$0.\dot{6}$	
i	$\frac{3}{5}$		
j			62.5

6.2   Finding a percentage of an amount

Example

Julian reads in the newspaper that the average pocket money for 12-year-olds has increased nationally by 14% in the last year. Julian's 12-year-old daughter Gilly has been given £1.60 per week for the last two years. How much more per week should he provide for a 14% increase?

$$14\% = \frac{14}{100}$$

$$14\% \text{ of } £1.60 = \frac{14}{100} \times £1.60$$

$$= 0.14 \times £1.60$$

Increase in pocket money = 22.4p = 22p (to the nearest 1p)

Note. See Section 2.1 for a further explanation of approximations.

Exercise 6B

1 Calculate the following percentages to the nearest 1p.

a 10% of £13.75

b 50% of £637.24

c 7% of £316

d 15% of £92.72

e 123% of £4.20

f $12\frac{1}{2}$% of £69.80

g 10.80% of £900

h 34.4% of £128.50

6.3 Increasing an amount by a given percentage

Example

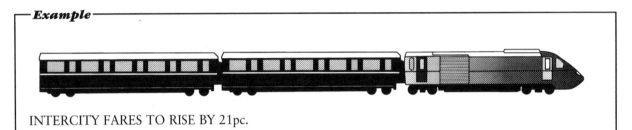

INTERCITY FARES TO RISE BY 21pc.

What does this mean in cash terms to the 20 000 long-distance commuters who travel to London every day?

If an InterCity season ticket costs £2632 now, how much will it cost after the rise?

Method 1
To find the new cost of a ticket we can find 21% of £2632 and then add this to the original:

$$21\% \text{ of } £2632 = \frac{21}{100} \times £2632$$

$$\text{Increase in fare} = £552.72$$

$$\text{New cost of fare} = £2632 + £552.72$$

$$= £3184.72$$

Method 2
Consider the original amount of £2632 as 100%. Increasing it by 21% is the same as finding 121% of the original cost.

$$121\% = \frac{121}{100} = 1.21$$

Therefore the quickest way of increasing the original fare by 21% is to multiply it by 1.21.

$$121\% \text{ of } £2632 = 1.21 \times £2632$$

$$\text{New cost of fare} = £3184.72$$

Exercise 6C

Give all answers to the nearest 1p.

1 Increase the following rail fares by 10%.

a £9.20

b £3.70

c £5.00

d £9.81

e £9.13

2 Increase the given amount by the required percentage.

a £72.12 by 50%

b 95p by 10%

c £360 by 120%

d £220 by $6\frac{1}{4}$%

e £124.80 by 25%

f £19.99 by $8\frac{1}{2}$%

6.4 Decreasing an amount by a given percentage

Example

The marked price of a sweater is £24.90.
What is its sale price?

SALE
20% OFF ALL
MARKED
PRICES

Method 1

The reduction = 20% of £24.90

$$= \frac{20}{100} \times £24.90$$

$$= £4.98$$

The sale price = £24.90 − £4.98

$$= £19.92$$

Method 2

£24.90 is the equivalent of 100% and so decreasing the price by 20% is equivalent to finding 80% of the original price.

$$\frac{80}{100} \times £24.90 = 0.80 \times £24.90$$

Sale price = £19.92

Exercise 6D

Give all answers to the nearest 1p.

1 Reduce the following marked prices by 20% to find the sale prices:

  a £30.00   **d** 45p

  b £10.50   **e** £12.99

  c £17.60

2 Decrease the amount by the required percentage:

  a £54.10 by 8%   **d** £27.15 by $12\frac{1}{2}$%

  b 84p by 30%   **e** £99.05 by 40%

  c £128 by 60%   **f** £1.62 by 33%

6.5 Expressing one quantity as a percentage of another

In a survey of insurance companies it was found that the most common type of car accident was one car running into the back of another.

Out of 35 000 claims, 6280 were for this type of accident.

Information of this type is usually quoted as a percentage.

Example 1

Find 6280 as a percentage of 35 000:

First express 6280 as a fraction of 35 000:

$$\frac{6280}{35\,000}$$

Then multiply this fraction by 100 to express it as a percentage:

6280 as a percentage of 35 000

$$= \frac{6280}{35\,000} \times 100\%$$

$$= 17.9\%$$

This means that almost 18% of car accidents are caused by cars running into the backs of other vehicles.

Note. See Section 2.1 for a further explanation of significant figures.

Example 2

A shop buys wallpaper from a wholesaler at £6 per roll and sells it to customers at £8.20 per roll.
What is the percentage increase in price?

The **increase** in price is £8.20 − £6.00 = £2.20

This is $\dfrac{£2.20}{£6.00}$ as a fraction of the **original** price.

$$\text{Percentage increase} = \frac{£2.20}{£6.00} \times 100\%$$

$$= 36.7\%$$

Example 3

By what percentage has the marked price of £4.70 been decreased to give a sale price of £3.80?

The **decrease** in price is expressed as a fraction of the **original** price and then multiplied by 100.

Decrease in price = 90p

$$\text{Percentage decrease} = \frac{90\text{p}}{£4.70} \times 100\%$$

$$= \frac{£0.90}{£4.70} \times 100\% \text{ (both quantities must be in the same units)}$$

$$= 19.1\%$$

Exercise 6E

Give all answers correct to 3 significant figures.

1   Express the first quantity as a percentage of the second:

   a 20, 25   **d** 54, 108

   b 3, 87   **e** 60p, £1.10

   c 140, 80   **f** £16.25, £12.50

2   Find the percentage by which the first amount is increased or decreased to give the second amount:

   a £65, £80   **d** £499, £399

   b £250, £300   **e** £1.23, 67p

   c 20p, 95p   **f** £24.50, £138.20

Exercise 6F

1 Restaurants often add a service charge of 10% to your bill. If a meal for two costs £28.60, how much service charge will be added?

2 In a survey of 348 dog owners, it was found that 103 bought Pedigree Chum dog food. What percent of the market prefer Pedigree Chum?

3 Your employer offers to raise your weekly wage by 10% or £10. Which should you choose and why?

4 In the survey of 35 000 car accidents, 17.7% happened on a Friday.
How many of the accidents occurred on a Friday?

5 In the same survey, 3213 accidents happened in November. What percentage of the total number of car accidents occurred in November?

6 A survey on teenage smoking found that 70% of girls of secondary school age tried smoking and that 36% of those who tried it became addicted to cigarettes.

In a secondary school with 580 female students:

a how many girls would you expect to find had tried smoking?

b how many girls would you expect to find had become addicted to smoking?

7 a In a village with 250 on the electoral roll, 80% voted in the local elections.
How many voted?

b Of those who voted, $37\frac{1}{2}$% voted Labour. How many votes did Labour receive?

c What fraction of those eligible to vote, voted Labour?

8 A holiday company offers a 5% discount on all holidays booked before 31 December the previous year. How much will a family pay for a holiday whose advertised price is £996?

9 A wrist watch has a MRRP (maker's recommended retail price) of £32.99, but a jeweller's shop advertises it for £28.99.
By what percentage has the shop reduced the price? (Give your answer to the nearest whole number.)

10 Whilst on holiday in America, Jenny and Robert find jeans on sale at $19.99 a pair. The exchange rate is $1.47 to the pound.

a Calculate the price of a pair of jeans in pounds.

Jenny and Robert buy 80 pairs of jeans to sell in England. They have to pay 10% duty on them when they arrive back in England.

b Calculate the total cost of the jeans including duty.

They sell $\frac{4}{5}$ of the jeans at £27 a pair.

c How much do they receive from this sale?

Jenny sells the remainder of the jeans at £5 a pair.

d What is the total amount of money they receive from the sale of all 80 pairs of jeans?

e Express the profit made as a percentage of the total expenditure.   (SEG W95)

7 Accuracy

7.1 Degrees of accuracy

Whenever you solve a numerical problem, you must consider the accuracy required in your answer, especially if you cannot obtain an exact answer. In Section 2.1, you saw how to correct an answer to a number of significant figures or decimal places. Unit 7 shows you how to select an appropriate degree of accuracy.

Sometimes the answer to a numerical problem is an integer, and this gives you an exact answer. Fractions are also exact, but their decimal equivalents are often inexact. For example, suppose you were able to buy sixteen pencils for £3 but wanted to buy only one. Each pencil would cost £$\frac{3}{16}$.

As a decimal this would be 18.75 pence. Obviously you cannot pay 18.75 pence for an individual pencil. The shopkeeper would work in the smallest unit of currency, which is one penny. To make sure that she did not lose money she would **round up** the price to 19p.

Example 1

Sara makes a 12 minute local phone call, at a rate of $4\frac{1}{2}$ minutes per unit of charge. British Telecom always round up the number of units.

How many units will Sara be charged for?

No. of units charged for is $\dfrac{12}{4.5} = 2.6666$.

The 2.666 is rounded up to the next integer.

$\therefore$ Sara is charged for 3 units.

Sometimes you will need to **round down** a mathematical answer as the following example shows.

Example 2

Pam changes 7254 French francs into pounds at the rate of 8.4 francs to £1. The bank, not wishing to be generous, decides to round down to the nearest penny the money it will give Pam.

How much does Pam receive?

$$7254 \text{ French francs} = \frac{£7254}{8.4} = £863.571\,43$$

$\therefore$ Pam receives £863.57

---

1   Give 38.4578 to 2 decimal places.

2   Give 452.89 to 2 significant figures.

3   Twelve pens cost £6.20. What would be charged for one pen?

4   Fifteen oranges are sold for £1. What is the cost of one orange?

5   A local phone call takes 20 minutes. At $4\frac{1}{2}$ minutes per unit, how many units are charged for?

6   A phone call lasts 4 minutes. The time allowed per unit is 25 seconds. How many units are charged for?

7   Eleven students receive a bill for £120 after an evening out. How much should each pay?

8   A booking agency normally sells a block of four tickets for £15.43. It agrees to sell them individually. How much should it charge for one ticket if it does not wish to lose money through individual sales?

*7.2   Recurring decimals

On page 31, you found that some fractions convert to a recurring decimal. For example:

$$\frac{1}{3} = 0.3\dot{3}, \text{ which is the shorthand way of writing}$$

$$0.3333333\ldots$$

Sometimes a group of digits is repeated. For example:

$$\frac{1}{11} = 0.09090909\ldots$$

The dots are placed above the repeating digits:

$$\frac{1}{11} = 0.\dot{0}\dot{9} \text{ or } 0.09\dot{0}\dot{9}$$

(The repeating digits are sometimes written twice, with the dots over the digits on the second time they are written.)

Note. Because of rounding errors, a calculator will sometimes display an answer as a recurring decimal when the correct answer should be an integer or a terminating decimal. Carry out the following operation on your calculator:

$$\boxed{1}\ \boxed{\div}\ \boxed{9}\ \boxed{=}\ \boxed{\times}\ \boxed{9}$$

What answer is given?
Many calculators will display the answer 0.9999999.
However, $\frac{9}{9}$ is clearly 1 and the answers should be given as the exact value 1.

Any fraction with a denominator which is not 2 or 5, or a power of 2 and/or 5, is a recurring decimal, even though the calculator display does not show this.

For example, $\frac{1}{40} = 0.025$,   an exact decimal, as $40 = 2^3 \times 5$,

i.e.   40 has prime factors which are 2 and 5 only.

43 cannot be expressed in terms of powers of 2 and/or 5 only.

$\therefore$ $\frac{1}{43}$ is not a terminating decimal.

$\frac{1}{43}$ is a recurring decimal.

From the calculator display,

$$\frac{1}{43} = 0.023\,255\,813$$

However, $\frac{1}{43}$ is really the recurring decimal: $0.\dot{0}2\dot{3}\,\dot{2}5\dot{5}\,\dot{8}1\dot{3}\,\dot{9}5\dot{3}\,\dot{4}8\dot{8}\,\dot{3}7\dot{2}\,\dot{0}9\dot{3}$,

with all 21 digits repeating in this order. It is possible to show this with a repeating dot over the first and last digits which recur, i.e. as $0.\dot{0}23\ldots09\dot{3}$ but, in this notation, you must be careful not to miss a dot and just have the 3 recurring.

■ *Exercise 7B* ■

In questions **1–10** write the decimal equivalents.

1 $\dfrac{7}{9}$ **3** $\dfrac{7}{11}$ **5** $\dfrac{5}{14}$ **7** $\dfrac{1}{12}$ **9** $\dfrac{1}{18}$

2 $\dfrac{3}{11}$ **4** $\dfrac{1}{7}$ **6** $\dfrac{18}{9}$ **8** $\dfrac{4}{15}$ **10** $\dfrac{3}{12}$

11 What is the maximum number of digits which could be repeated in a recurring decimal?

Conversion of recurring decimals to fractions

The fact that $\frac{1}{9} = 0.\dot{1}$ can help you to convert recurring decimals into fractions:

$$0.\dot{7} = 7 \times 0.\dot{1} = 7 \times \frac{1}{9} = \frac{7}{9}$$

Similarly,

$$0.00\dot{7} = 7 \times 0.00\dot{1} = 7 \times \frac{1}{100} \times 0.\dot{1}$$

$$= 7 \times \frac{1}{100} \times \frac{1}{9} = \frac{7}{900}$$

The conversion $\frac{1}{99} = 0.\dot{0}\dot{1}$ enables decimals including two recurring digits to

be turned into fractions: $0.\dot{3}\dot{4} = 34 \times 0.\dot{0}\dot{1} = 34 \times \frac{1}{99} = \frac{34}{99}$.

This method can be extended to numbers which include more than two recurring digits.

―Example 1―

Convert $0.34\dot{8}$ to a fraction.

$$0.34\dot{8} = 0.34 + 0.00\dot{8}$$

$$= \frac{34}{100} + \frac{1}{100} \times \frac{8}{9}$$

$$= \frac{34}{100} + \frac{8}{900}$$

$$= \frac{9 \times 34 + 8}{900}$$

$$= \frac{314}{900} \quad \text{or} \quad \frac{157}{450}$$

Note. You can check your answer by using your calculator to divide 157 by 450.

An alternative method of converting a recurring decimal into a fraction is as follows:

Example 2

Convert $0.\dot{7}8\dot{6}$ to a fraction.

$$\text{Let } x = 0.\dot{7}8\dot{6}$$

$$\therefore \quad 1000x = 786.\dot{7}8\dot{6} \quad \text{(There are 3 recurring decimals, so find } 1000x.$$
$$\text{If there are 2 recurring decimals find } 100x.$$
$$\text{For 4 recurring decimals, find } 10\,000x, \text{ etc.)}$$

Subtracting: $\qquad 1000x - x = 786.\dot{7}8\dot{6} - 0.\dot{7}8\dot{6}$

$$999x = 786$$

$$\therefore \qquad\qquad x = \frac{786}{999}$$

Often you can divide out a common factor to give the fraction in its simplest form:

$$x = \frac{786}{999}$$

$$= \frac{262}{333} \quad \text{(dividing by the common factor 3)}$$

Exercise 7C

Convert the following recurring decimals into fractions, expressed in their simplest form:

1	$0.\dot{6}$	6	$0.7\dot{4}$	11	$1.2\dot{8}\dot{6}$
2	$0.\dot{5}$	7	$0.\dot{2}\dot{4}$	12	$0.3\dot{6}7\dot{8}$
3	$0.0\dot{4}$	8	$0.3\dot{9}$	13	$0.1\dot{7}\dot{3}$
4	$0.00\dot{3}$	9	$0.58\dot{4}$		
5	$0.2\dot{3}$	10	$0.3\dot{2}\dot{7}$		

*7.3   Lower bounds and upper bounds

When you measure a length, you will always find an approximate answer. If your garden path is 5.218 174 m long, you will say its length is 5 m or 5.2 m or even possibly 5 m 22 cm. As in Section 2.1, you will give the answer to a sensible degree of accuracy.

Hence, if your friend says his garden path is 7 m long, you will assume that the length is 7 m to the nearest metre.

The actual length of the path can be anything from 6.5 m to 7.4999... m (which is virtually 7.5 m).

The **lower bound** is the smallest value which the number could be (in this case 6.5 m), and the **upper bound** is the largest value which the number could be (in this case 7.5 m).

Example 1

A weight is 17 kg correct to the nearest integer.
What are its upper and lower bounds?

The upper bound is 17.5 kg; the lower bound is 16.5 kg.

Example 2

A tree is 12.3 m high (to the nearest 0.1 of a metre).
What are the upper and lower bounds for its height?

The upper bound is 12.35 m; the lower bound is 12.25 m.

In some cases, e.g. in money, the value given goes up in steps.

For example: an item priced at £3.60 to the nearest 10p can have an actual value between £3.55 and £3.64.

This is because:

(i) an item cannot be priced at £3.641 (64.1 pence does not exist)

(ii) a price of £3.65 would round up to £3.70.

Hence £3.64 is its largest possible price.

Thus if the measurement can take *all* possible values, a measurement of 12 to the nearest integer can have an actual value between 11.5 and 12.5.

If the measurement goes up in steps, a measurement of 12.5 would be rounded to 13. Whenever 12.4999 is a possible measurement, the data is continuous.

Although age is continuous, people give their age in steps of one day. Age is recorded as going from, for example, 17 years 0 days to 17 years 1 day. Thus age can be treated as a discrete variable.

Example 3

Victoria is aged 17. What are the upper and lower bounds of her age?

The exact age of a person aged 17 years can be between 17 years 0 days and 17 years 364 days (unless it is a leap year!)

∴ The upper bound is 17 years 364 days and the lower bound is 17 years.

Give the lower and upper bounds for:

1   the length of a lorry measured as 12 m to the nearest metre

2   the length of a car measured as 4.7 m to the nearest 0.1 m

3   the weight of a lorry measured as 38 tons to the nearest ton

4   the cost of a TV given as £290 to the nearest £10

5   the age of Caroline who is 16 years old (it is not a leap year)

6   the weight of a packet of cornflakes given as 350 grams to the nearest 5 grams

Calculations involving lower or upper bounds

In cases where two or more values are combined, you must consider carefully whether or not the values have been rounded. You may need to use the upper bounds of these values to calculate the upper bound of the combined values, and vice versa, or possibly a combination of both the upper and lower bounds (see Example 4).

Example 1

A rectangular field is 25 m by 15 m (each measurement to the nearest metre).
What are the lower and upper bounds for its area?

Lower bound for its area = Minimum possible length
$\qquad\qquad\qquad\qquad\qquad$ × Minimum possible width.

The minimum possible length of the field is 24.5 m
The minimum possible width of the field is 14.5 m.
∴   The minimum possible area is $24.5 \times 14.5 \text{ m}^2 = 355.25 \text{ m}^2$
∴   The lower bound of the area is 355.25 m^2.

Upper bound for its area = Maximum possible length
$\qquad\qquad\qquad\qquad\qquad$ × Maximum possible width.

The upper bounds for the dimensions are 25.5 and 15.5 m.
∴   The upper bound for the area is $25.5 \times 15.5 \text{ m}^2 = 395.25 \text{ m}^2$

Example 2

Joanna buys a TV, a video and a pack of video tapes. The video costs £310 and the tapes cost £20. Joanna spends £580 in total. All the prices are given to the nearest £10.
Find the maximum price Joanna could have paid for the TV.

$\qquad$ Maximum TV price = Maximum price Joanna could have paid in total
$\qquad\qquad\qquad\qquad\quad$ − Minimum price she could have paid for the video
$\qquad\qquad\qquad\qquad\quad$ − Minimum price she could have paid for the tapes.
$\qquad\qquad\qquad\qquad$ = £584.99 − £305 − £15
$\qquad\qquad\qquad\qquad$ = £264.99

Example 3

The cost of a table is £300 and the cost of one chair is £20, both prices being given to the nearest ten pounds.
Find the least and greatest possible total costs of the table and four chairs.

The least cost is the addition of the minimum possible cost of the table and each of the chairs.
The least possible cost of the table is £295
The least possible cost of each chair is £15.
∴ The least possible cost of the furniture is: £295 + (4 × £15) = £355

Similarly, for the greatest possible cost:
The maximum possible cost of the table is £304.99.
(£305 would round to £310 as the cost is a discrete variable –
see page 58.)
The maximum possible cost of each chair is £24.99.
∴ The greatest possible cost of the furniture is
£304.99 + (4 × £24.99) = £404.95

Example 4

The area of a field is 280 ft^2 and its length is 25 ft, each number being given to the nearest integer.
What are the maximum and minimum widths of the field?

The true value of the area lies between 280.5 ft^2 and 279.5 ft^2.
The true value of the length is between 25.5 ft and 24.5 ft.
The greatest possible value of the width is obtained by dividing the *largest* possible area by the *smallest* possible length.
∴ The upper bound of the width $= \dfrac{280.5}{24.5}$ ft $= 11.45$ ft

The minimum possible value of the width is obtained by dividing the *smallest* possible area by the *largest* possible length.
∴ The lower bound of the width $= \dfrac{279.5}{25.5}$ ft $= 10.96$ ft

---

Exercise 7E

In questions 1–8 find the upper and lower bounds of:

1   the perimeter of a field of length 17 m and width 11 m (measured to the nearest metre)

2   the area of a patio of length 21 ft and width 12 ft (measured to the nearest foot)

3   the perimeter of a table of length 3.43 m and width 2.70 m (measured to the nearest centimetre)

4   the area of a rectangular dance floor of length 25.2 m and width 17.1 m (measured to the nearest 0.1 m)

5   Find the total cost of a rhododendron and an azalea costing £35 and £17 respectively, each cost being given to the nearest pound.

6   Find the length of a lawn of area 342 m^2. Its width is 12 m. Each measurement is given to the nearest integer.

7   Find the height of a box of volume 240 in^3. The length and width are 21 inches and 11 inches. All these measurements are given to the nearest integer.

8   Find the volume of sand on an artificial beach of length 340 m, to the nearest integer. The cross-sectional area is 200 m^2, given to the nearest 10 m^2.

9 A vase made in a pottery class weighed 8 ounces, to one significant figure. To cover overheads, the college charges students 46p per ounce for the finished item. Find the maximum and minimum prices which could be charged for the vase. (SEG W95)

10 Fred's car has a petrol tank which can hold 12 gallons and the car does 42 mpg, both figures being given to 2 significant figures.

a What is the maximum distance which the car could travel on a full tank of petrol?

b Calculate the range of miles in which the car could run out of petrol. (SEG W95)

11 Charlotte makes model animals. Each model badger weighs 20 grams and each model deer weighs 40 grams, both weights being given to the nearest ten grams.

a What is
(i) the greatest possible weight of a badger?
(ii) the smallest possible weight of a badger?

Jenny buys three badgers and six deer. The cost of posting them to her daughter in America is 41p for the first 10 grams, plus 17p for every additional 10 grams or part thereof. Jenny calculates the cost as £5.34.

b What is the maximum error which Jenny could have made? (SEG W95)

12 The large and family size packets of Krunchie Kornflakes are shown below.

a The 500 g contents of the large packet have been given to the nearest 5 g. What is the smallest number of grams it can contain?

b The 750 g contents of the family packet have been given to the nearest 10 g.
(i) What is the largest number of grams it can contain?
(ii) A shop sells the family size for £2.70. This price is given to the nearest 10p.

Find a the maximum and b the minimum cost per 100 g of Krunchie Kornflakes. (SEG S96)

13 Each lorry in a queue of twelve measures 38 feet to the nearest 6 inches. Can you assume that all twelve lorries will fit into one lane of length 460 feet on a ferry? Explain clearly how you obtained your answer. (SEG S94)

14 An aeroplane flies the 6400 miles from London Heathrow to San Francisco in 13 hours, both the distance and the time being given to two significant figures.

a What is the maximum possible average speed of the plane?

b By finding the minimum possible average speed of the plane, find the range of the possible speeds of the plane. Give your answer to two significant figures. (SEG W93)

15 In his garden, Terry has a water tank which is 8 feet by 6 feet by 5 feet. All these measurements are to the nearest foot.

a Find the maximum possible volume of water in the tank.

The tank leaks at the rate of 40 gallons per day, to the nearest integer. Assume one cubic foot is exactly 6.23 gallons.

b Find the maximum and minimum number of days in which the water tank will empty assuming that there is no rain. (SEG W94)

8 Wages and Salaries

8.1 Basic pay

All employees receive a wage or salary as payment for their labour.

A **wage** is paid weekly and is calculated on a fixed hourly rate.
A **salary** is paid monthly and is calculated on a fixed annual amount.

Many wage earners are required to work a fixed number of hours in a week, and they are paid for these hours at the basic hourly rate.

Example 1

Simon works in a hairdresser's. His basic pay is £3.75 per hour for a 40-hour week.
Calculate his weekly wage.

Weekly wage = Rate of pay × Hours worked

$= £3.75 × 40$

$= £150$

Example 2

Layla's gross weekly wage (i.e. her wage before deductions) is £166.95. She works a 35-hour week.
What is her hourly rate of pay?

Hourly rate of pay $= \dfrac{\text{Weekly wage}}{\text{Hours worked}}$

$= \dfrac{£166.95}{35}$

$= £4.77$

Exercise 8A

1 Calculate Donna's gross weekly wage if she works for 42 hours per week at a rate of pay of £4.10 per hour.

2 Daniel's yearly salary is £5756. How much is he paid per month?

3 A basic working week is 36 hours and the weekly wage is £231.48. What is the basic hourly rate?

4 An employee's gross monthly pay is £965.20. What is his annual salary?

5 The basic week in a factory is 35 hours. How much is the weekly wage for someone whose basic rate per hour is:

a £7.42   b £4.45   c £12.60?

8.2 Overtime rates

An employee can increase a basic wage by working longer than the basic week, i.e. by doing overtime.
A higher hourly rate is usually paid for these additional hours.
The most common rates are **time and a half** and **double time**.

Example

Mr Arkwright works a basic 36-hour week for which he is paid a basic rate of £5.84 per hour. In addition, he works 5 hours overtime at time and a half and 3 hours overtime at double time.

Calculate his gross weekly wage.

36 hours basic pay	$= £5.84 \times 36$	$= £210.24$
5 hours overtime at time and a half	$= (£5.84 \times 1.5) \times 5 =$	$£43.80$
3 hours overtime at double time	$= (5.84 \times 2) \times 3$	$= £35.04$

$\therefore$ Gross pay $= £289.08$

Exercise 8B

1 A garage mechanic works a basic $37\frac{1}{2}$-hour week at an hourly rate of £4.90. He is paid overtime at time and a half.
How much does he earn in a week in which he does 9 hours overtime?

2 Mr Cooper's basic wage is £6.20 per hour, and he works a basic 5-day, 40-hour week. If he works overtime during the week, he is paid at time and a half. Overtime worked at the weekend is paid at double time.

Calculate his gross wage for a week when he worked five hours overtime during the week and four hours overtime on Saturday.

3 The basic working week in a small factory is 35 hours (i.e. 7 hours per day) and the basic rate of pay is £3.98 per hour. The overtime rate is time and a half from Monday to Friday and double time on Saturdays.
The table below shows the hours worked by five employees. For each employee, calculate the gross weekly pay.

	Mon	Tue	Wed	Thu	Fri	Sat
Andrews A J	7	7	8	9	8	0
Collins F	8	8	9	9	7	5
Hammond C	9	9	9	10	10	0
Jali Y	8	10	11	11	9	4
Longman B H	9	10	10	9	7	6

4 The number of hours worked by an employee is often calculated from a clock card similar to the one shown below. Mr Meyer works a basic eight-hour day, five days a week and his basic hourly rate is £4.15.

Day	In	Out	In	Out	Clock hours	O/T hours
SAT	0730	1200	1230	1400	6	6
SUN	0800	1200				
MON	0730	1200	1300	1700		
TUES	0730	1200	1300	1730		
WED	0800	1230	1300	1730		
THUR	0730	1200	1245	1815		
FRI	0730	1200	1230	1600		

Overtime is paid at the following rates:
Time and a quarter for Monday to Friday
Time and a half for Saturday
Double time for Sunday.
Calculate Mr Meyer's gross pay for this week.

5 a The basic weekly wage of employees in a small firm is £133 for a 38-hour week. What is the basic hourly rate?

b All overtime is paid at time and a half. Calculate the number of hours of overtime worked by Ms Wiley during a week when her gross pay was £154.

6 Mustafa is paid £3.80 per hour as his basic rate of pay.
In one week he works 3 hours and 40 minutes overtime at time and a half and $5\frac{1}{2}$ hours overtime at time and a quarter.
How much does Mustafa earn in overtime that week? (SEG S95)

7 Richard is paid £4.70 per hour for his basic 37-hour week. In one week Richard also works overtime at time and a half. His total pay is £195.05.
How many hours overtime does Richard work? (SEG W95)

8.3 Commission

People who are employed as salespersons or representatives, and some shop assistants, are paid a basic wage plus a percentage of the value of the goods they have sold.

Their basic wage is often small, or non-existent, and the **commission** on their sales forms the largest part or all of their gross pay.

Example

A salesman earns a basic salary of £690 per month plus a commission of 5% on all sales over £5000.
Find his gross income for a month in which he sold goods to the value of £9400.

He earns commission on (£9400 − £5000) worth of sales.

$$\text{Commission} = 5\% \text{ of } (£9400 - £5000)$$
$$= 0.05 \times £4400$$
$$= £220$$

$$\text{Gross salary} = \text{Basic salary} + \text{Commission}$$
$$= £690 + £220$$
$$= £910$$

Exercise 8C

1   An estate agent charges a commission of $1\frac{1}{2}\%$ of the value of each house he sells.

How much commission is earned by selling a house for £104 000?

2   Calculate the commission earned by a shop assistant who sold goods to the value of £824 if her rate of commission is 3%.

3   A car salesman is paid $2\frac{1}{2}\%$ commission on his weekly sales over £6000. In one particular week he sold two cars for £7520 and £10 640.

What was his commission for that week?

4   An insurance representative is paid a commission of 8% on all insurance sold up to a value of £4000 per week. If the value of insurance sold exceeds £4000, he is paid a commission of 18% on the excess.

Calculate his total gross pay for the 4 weeks in which his sales were £3900, £4500, £5100 and £2700.

5   Two firms place adverts for an insurance representative. Firm A offers an annual salary of £5000 plus a company car (worth £2500 per year) and 4% commission on sales over £200 000 per annum.

Firm B offers an annual salary of £7000 and 3.5% commission on sales over £150 000 per annum.

a   Which is the better job if sales of £350 000 per year can be expected?

b   If, in a good year, sales rose to £500 000 per year, which job would pay the higher amount and by how much?

8.4 Piecework

Some employees, particularly in the manufacturing and building industries, are paid a fixed amount for each article or piece of work they complete. This is known as **piecework**.

They may also receive a small basic wage and, in addition, some are paid a **bonus** if production exceeds a stipulated amount.

Example

Workers in a pottery firm who hand-paint the plates are paid a basic weekly wage of £120 and a piecework rate of 80p for every plate over 30 which they paint in a day.

Calculate the weekly wage of an employee whose daily output was as follows:

Day 1   35 plates   Day 2   38 plates   Day 3   40 plates
Day 4   45 plates   Day 5   39 plates

No. of plates over 30 painted = 5 + 8 + 10 + 15 + 9
$$= 47$$

Piecework bonus = 80p × 47   Weekly gross pay = £120 + £37.60
$$= £37.60 \qquad\qquad\qquad\qquad = £157.60$$

Exercise 8D

1   A firm employs casual labour to deliver advertising leaflets door to door. The rate of pay is £4.80 for every 100 leaflets delivered. How much is earned by someone who delivers 1230 leaflets?

2   Mrs Mason works a basic 40-hour week at £2.74 per hour plus a bonus of 66p for every skirt she makes over 15 per day. In five successive days she makes 18, 29, 16, 17, and 19 skirts. What is her gross wage?

3   A bricklayer receives a bonus of 54p for every 10 bricks laid in excess of 350 per day. What bonus does he receive if he lays 3070 in a six-day week?

4   A perfumery pays its packers a basic wage of £35 per week plus 2p for each bottle of perfume packaged up to a limit of 1200 per day and 4p per day for each bottle packaged over 1200.

Calculate the gross pay for a packer whose output on five successive days of one week was 1210, 912, 1311, 1232 and 1043.

5   **a** A firm producing knitting patterns employs outworkers to knit sample sweaters and cardigans. The firm pays a basic wage of £20 per week plus £22 per garment. Mrs English knits 20 garments per month and Mrs Beckett knits 16 garments per month. Calculate how much they will each earn in a month. Assume 1 month = 4 weeks.

b In order to increase productivity among its knitters, the firm decides to abolish the basic weekly wage, but increase the piecework rate to £26.50 per garment. How much will the two ladies earn per month under the new system?

c Comment on the consequences of the new system.

8.5 Deductions from pay

Employees do not usually receive all the money they have earned.

Certain amounts of money are **deducted** from the gross pay and the pay the employee receives is the **net pay** or **take-home pay**.

The main deductions are:

- **Income Tax**
- **National Insurance**
- **Pension** (also called **Superannuation**)

Income tax (lowest rate)

Income tax is used to finance government expenditure.

It is a tax based on the amount a person earns in a tax year that begins on 6 April and ends on 5 April the following year.

Most people have income tax deducted from their pay, before they receive it, by their employer, who then pays the tax to the Government. This method of paying income tax is called **PAYE** (Pay As You Earn).

Certain amounts of each person's income are not taxed. These amounts are called **tax allowances**.

The Tax Office sends the employee and the employer a PAYE Code which tells them the value of the allowances. For example, a tax code of 0342L would be given to a single person who has allowances of £342 × 10 = £3420.

$$\text{Gross income} - \text{Tax allowances} = \text{Taxable income}$$

The **lowest** rate of income tax in 1999–2000 was 10p in the pound on taxable income up to £1500.

The Personal tax allowance (in 1999–2000) was £4335.

In addition you may have a tax allowance by reason of expenses connected with your work (eg union contributions) or caring for a dependent relative.

Example

Mr Brown earns £5500 per year, which is his only source of income.

Find **a** his tax allowance, **b** his taxable income,
c the tax paid per annum.

a Personal tax allowance = £4335

b Taxable income = Gross income – Tax allowance
$$= £5500 - £4335$$
$$= £1165$$

c Tax paid $= \dfrac{1.0}{100} \times £1165$

$$= £116.50$$

Assume that the earnings stated in the following questions are the only source of income.

1 Find
  (i) the total tax allowance,
  (ii) the taxable income,
  (iii) the yearly tax paid, for:
  a someone earning £4500 per annum
  b someone earning £4950 per annum
  c someone earning £67 a week.

2 Find the yearly income tax payable by Mrs Bailey, who earns £5705 per annum and has a total tax allowance of £4335.

3 Miss Davis earns £6462 per year. She is entitled to the personal allowance plus an allowance of £960 per year for expenses necessary in her work. Calculate her monthly tax bill.

Income tax (lowest rate and basic rate)

Most people who pay income tax earn more than £5835, and hence have a taxable income in excess of £1500.

Taxable income between £1500 and £28 000 is taxed at 23p in the pound, i.e. 23%. Since most people pay the majority of their income tax at this rate, the 23p rate is known as the basic rate.

Example

Mr Jackson is an engineer and earns £21 250 per year. His personal tax allowance is £4557. How much income tax does he pay per year?

Annual taxable income = Annual gross pay − Total allowance
$$= £21\,250 - £4557 = £16\,693$$

Income taxable at basic rate = Taxable income − £1500 (maximum at lowest rate)
$$= £15\,193$$

Income tax at lowest rate $= 10p \times 1500 = £150$
Income tax at basic rate $= 23p \times 15\,193 = £3494.39$
Total annual income tax $= £3644.39$

---

1 Find the yearly tax paid by:
  a Andrea, a single woman, earning £12 050 a year.
  b Philip, a single man, earning £182 per week.
  c Jennie and Carl, a married couple, earning £21 470 between them. Both earn more than £10 000.

2 Find the monthly income tax paid by Caroline who has a tax allowance of £3920 and earns £1450 per month.

3 Asif earns £320 per week as his basic wage, and he has a second income of £120 per month. His tax allowance is £3765. How much tax does he pay each month?

4 a Jane and Denis live together. Denis earns £13 185 per annum and Jane earns £10 585 per annum.

   What was their total tax bill for the year 1999–2000, if they claimed only the standard allowances?

  b How much tax would they have paid if they had been married?

5 If, in the next budget, the Chancellor changes the basic rate of tax from 23% to 22%, but wages remain the same, how much less tax will be paid by the employees in question **1** above?

*Income tax (higher rate)

Anyone whose taxable income is greater than the amount set by the
Chancellor of the Exchequer for basic rate tax payers, has to pay a higher rate
of tax on the **excess income**.

For the 1999–2000 tax year the rates payable are:

- 10% lowest rate on a taxable income up to £1500
- 23% basic rate on a taxable income from £1500 up to £28 000
- 40% higher rate on a taxable income over £28 000

Example

Mrs Carpenter is a director of a small company and earns £37 175 per year.
How much income tax does she pay per month?

Her personal allowance   = £4335

Annual taxable income   = Annual gross pay – Total allowance
                              = £37 175   – £4335
                              = £32 840

Income tax at higher rate = Taxable income – £28 000
                                  (maximum at lowest or basic rate)
                                  = £32 840 – £28 000
                                  = £4840

Income tax at lowest rate = 10% of £1500   =   £150
Income tax at basic rate   = 23% of £26 500 = £6095
Income tax at higher rate = 40% of £4840   = £1936
                                              £8181

Monthly income tax   = £8181 ÷ 12 = £681.75

Exercise 8G

1   Calculate (i) the yearly, (ii) the monthly income
tax paid in each of the following cases:

   a   Anthea Brown earns a salary of £37 000 each
year.

   b   John Knight's annual salary is £38 300. He is
married, but his wife is not in paid
employment.

   c   Mr Woolfe's monthly salary is £3010, and
his yearly tax allowance is £5060.

2   **a**   David has a full-time job and earns £36 000
per annum. He also has a taxable income of
£2000 from shares. His wife Erica has a
part-time job which pays £2500 per annum.
How much income tax do they pay per
year?

   b   The income from the shares could be
transferred to Erica.

How much tax will they save if the £2000 is
added to Erica's income instead of David's?

3   Paul and his wife Rosemary have a total earned
income of £76 600 per annum. Paul's salary is
£39 500 per annum.
How much income tax do they pay?

4   If, in a budget, the Government decides to have
four income tax bands as follows:

10% on the first £3000 of taxable income
22% on the remaining taxable income up to
£29 000
35% on the next £5500 of taxable income
45% on the remaining taxable income,

   a   calculate the yearly income tax which would
be payable by Mr Woolfe (see question 1)

   b   would he benefit under the new system?

*8.6 National Insurance

National Insurance (NI) is paid by both the **employee** and the **employer**. It helps to pay for:

National Health Service
Statutory sick pay
Pensions
Unemployment benefits

The amount paid in National Insurance depends on your gross pay and whether or not you are 'contracted out' of the State Pension Scheme.

Employees who are 'contracted out' pay less National Insurance. On retirement, these employees can claim only the basic State Pension, but they will receive an employment pension.
Employees who are not 'contracted out' receive the basic State Pension plus an earnings-related state pension.

The structure of employees' National Insurance contributions was changed in the 1999 budget and from 6 April 1999 is as follows:

Weekly earnings (£)	Monthly earnings (£)	Rate of National Insurance contribution
0–63.99	0–277	Nil
64–575	277–2492	10% † of earnings in excess of £64 per week (or £277 per month)
Above 575	Above 2492	10% of £511 per week (or £2215 per month)

†8.2% for employees who are contracted out.

Example 1

Mr Day's annual salary is £15 978. Calculate his monthly National Insurance contribution if he is contracted out.

Mr Day's monthly salary = £15 978 ÷ 12 = £1331.50

He earns over £277 per month and will pay the rate of 8.2% (because he is contracted out) on his earnings over £277, i.e. on £1331.50 − £277 = £1054.50.

He pays
$$8.2\% \text{ of } £1054.50 = £1054.50 \times 0.082 = £86.47$$
Total monthly contribution   = £86.47

Example 2

Mrs Fielding earns a gross weekly wage of £87.90. Calculate her National Insurance contribution.

Mrs Fielding will pay 10% on her earnings over £64,
i.e. on £87.90 − £64 = £23.90.

She pays
$$10\% \text{ of } £23.90 = £23.90 \times 0.10 = £2.39$$
Total weekly contribution   = £2.39

It is expected that in the budgets in the years 2000 and 2001 the starting point for paying income tax and for paying National Insurance will become the same. Hence in April 2001 the threshold for paying National Insurance will rise to £87 per week. This will be equal to the anticipated personal tax allowance of £4524 per annum.

Exercise 8H

1 Using the table on page 69, calculate the National Insurance contributions for:

  a Alice, who earns £112.70 per week

  b Charlie, who earns £586.40 per month

  c David, who is contracted out and earns £971 per month

  d Anne, who is contracted out and earns £428.64 per month

  e Maria, who earns £4.30 per hour for a 28-hour week.

2 Mrs Ferry earned £52 per week for her part-time job.

  Calculate:

  a her yearly National Insurance contribution

  b her net income for the year.

3 Nadia earns £4.60 per hour for a 40-hour week. Her income tax payment for the week is £29.25.

  Calculate:

  a her gross weekly wage

  b her weekly National Insurance contribution

  c her net pay for the week.

Note. Sections 8.5 and 8.6 and the whole of Unit 9 (following) take into account the changes announced in the Spring Budget for the financial year 1999–2000. Readers using this book during or after the financial year 1999–2000 should make themselves aware of any further changes introduced by the Chancellor of the Exchequer.

9 Value-added Tax

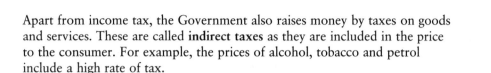

Apart from income tax, the Government also raises money by taxes on goods and services. These are called **indirect taxes** as they are included in the price to the consumer. For example, the prices of alcohol, tobacco and petrol include a high rate of tax.

9.1 Prices exclusive of VAT

The main indirect tax is **value-added tax** or **VAT**. This is charged on most goods that we buy and also on services such as house improvements, garage bills, meals in a restaurant.

Certain items are **zero-rated,** i.e. no VAT is charged, and these include most food (but not luxury foods such as ice cream, confectionery and crisps), children's clothes and books.

The standard rate of VAT is 17.5% (1996–97 value)

Prices are sometimes quoted exclusive of VAT. The price you pay will include the VAT and will be 17.5% more than the price quoted.

Example 1

A garage bill totals £88.70. How much VAT is added to the bill?

$$\text{VAT} = 17.5\% \text{ of } £88.70$$

$$= \frac{17.5}{100} \times £88.70 \quad \text{or} \quad 0.175 \times £88.70$$

$$= £15.52 \text{ (to the nearest 1p)}$$

In the 1994 budget, the rate of VAT on fuel (gas, electricity, etc.) was fixed at 8%.

Example 2

An electricity bill is £112.32 plus 8% VAT. How much is the VAT?

$$\text{VAT} = 8\% \text{ of } £112.32$$

$$= \frac{8}{100} \times £112.32$$

$$= £8.99 \text{ (to the nearest 1p)}$$

Exercise 9A

The prices of the items in questions **1–5** are quoted exclusive of VAT. Calculate the amount of VAT payable in each case, giving your answer to the nearest 1p.

1 a typewriter priced at £152

2 a washing machine priced at £234

3 a calculator at £9.50

4 a box of chocolates at £1.44

5 a necklace at £19.20

6 A garage bill shows the cost of parts to be £30.90 and labour costs of £58.60. VAT is then added to the bill. How much is the final bill?

7 A particular calculator is offered for sale through three different catalogues at the following prices (p&p means postage and packing):

Catalogue A £12.60 exclusive of VAT, post free

Catalogue B £13.80 inclusive of VAT, plus 50p p&p

Catalogue C £12.40 exclusive of VAT, plus 40p p&p.

Calculate the price in each case and hence determine which catalogue is offering the best deal.

8 All prices at a DIY store are shown exclusive of VAT. Mrs Jennings buys eight boxes of tiles, priced at £9.90 per box, a large packet of tile cement at £2.39 and some ready mixed tile grout at £1.38.
Calculate her total bill, including VAT.

9 The cash and carry store shows prices exclusive of VAT. Mr Patel buys the following goods for his grocery shop: £45.30 worth of food which is zero-rated for tax, £27.80 worth of confectionery and £15.25 worth of toiletries.
Calculate the amount of VAT payable and the total amount of his bill, including VAT.

10 The Government decides to change the rate of VAT payable on goods.
Calculate the amount of VAT payable and the total cost of the following goods, including VAT at the rate given:

a a paperback at £4.95, VAT at 12%

b a gas bill of £340, VAT at 8%

c a personal stereo at £28.90, VAT at 9%

d a patio door at £276, VAT at 17%

e two garden gnomes at £6.87 each, VAT at 14%.

9.2   Prices inclusive of VAT

Most shop prices are quoted inclusive of VAT. An item costing £100 *exclusive* of VAT would have VAT of £17.50 added and its price *inclusive* of VAT would be £117.50. Therefore, for all prices which are inclusive of VAT,

every £117.50 of the price includes £17.50 of VAT

That is, if the price exclusive of VAT is 100%, the price inclusive of VAT is 117.5% of the price.

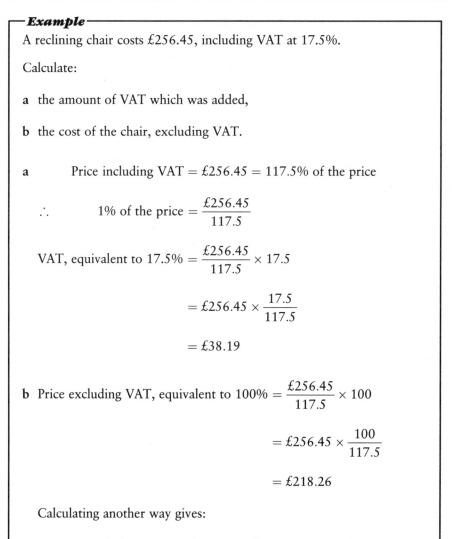

─ Example ──────────────────────────────

A reclining chair costs £256.45, including VAT at 17.5%.

Calculate:

a the amount of VAT which was added,

b the cost of the chair, excluding VAT.

a   Price including VAT = £256.45 = 117.5% of the price

$\therefore$ 1% of the price $= \dfrac{£256.45}{117.5}$

VAT, equivalent to 17.5% $= \dfrac{£256.45}{117.5} \times 17.5$

$= £256.45 \times \dfrac{17.5}{117.5}$

$= £38.19$

b Price excluding VAT, equivalent to 100% $= \dfrac{£256.45}{117.5} \times 100$

$= £256.45 \times \dfrac{100}{117.5}$

$= £218.26$

Calculating another way gives:

Price excluding VAT = £256.45 − £38.19 (VAT) = £218.26

To find the VAT included in a price, divide by 117.5 and multiply by 17.5

To find the price excluding the VAT, divide by 117.5 and multiply by 100.

These calculations only apply to VAT at 17.5%.

For fuel bills where VAT is at 8%:

To find the VAT included in a price, divide by 108 and multiply by 8.

To find the price excluding VAT, divide by 108 and multiply by 100.

If the VAT rate has changed since 1994, can you work out how to change the figures?

Exercise 9B

1 A bottle of whisky costs £8.20.
How much of this price is VAT?

2 Mr Lambert buys a motor-cruiser costing £10 710.
What is the price of the motor-cruiser excluding VAT?

3 How much VAT will a shopkeeper pay to the Government on takings of £76.80 for sweets and £43.95 for ice cream?

4 A garage sells 5000 litres of four-star petrol at 52.9p per litre and 9000 litres of unleaded petrol at 48.4p per litre.
How much VAT does the Government receive from the sale of this petrol?

5 Mrs Jali received a garage bill totalling £125.20 for parts and labour including VAT.
How much did the garage charge for parts and labour?

6 A school buys a computer which costs £899.90, including VAT. Schools are not required to pay VAT on educational equipment.

 a How much VAT can the school reclaim?

 b How much does the school pay for the computer?

7 The items listed below are on sale in a large electrical store. Assume that the Government has suddenly reduced the rate of VAT to 12%. For each item, calculate the amount of VAT paid and the price before VAT was added (give answers to the nearest 1p):

 a steam iron, £28.99

 b hair-dryer, £17.99

 c shaver, £54.99

 d microwave, £199.99

 e washing machine, £329.99

 f vacuum cleaner, £114.99

 g television, £599.99

 h can opener, £11.99

8 The bill for a meal at a restaurant in British Columbia was $63.90, including VAT at the rate of 8%.
How much was the VAT?

9 John's gas bill is £185.76 inclusive of VAT at 8%. How much was the bill before VAT was added?

10 Rebecca's electricity bill is £206.54 inclusive of VAT at 8%. How much VAT was charged?

10 Profit and Loss

10.1   Calculations involving one price

Everyone who is involved with buying and selling goods aims to make a **profit**.

To do so the dealer buys the goods for a certain amount, the **cost price**, and sells the same goods for a higher price, the **selling price**.

The difference between the cost price and the selling price is the profit:

$$\textbf{Profit} = \textbf{Selling price} - \textbf{Cost price}$$

Because it is easier to compare percentages, the **percentage profit** is often required. It is usually calculated on the cost price:

$$\textbf{Profit \%} = \frac{\textbf{Profit}}{\textbf{Cost price}} \times 100\%$$

Sometimes the dealer makes a loss, i.e. the goods are sold for less than the cost price.

In this case, the percentage loss is calculated on the difference between selling price and cost price:

$$\textbf{Loss \%} = \frac{\textbf{Loss}}{\textbf{Cost price}} \times 100\%$$

Example 1

A shoe shop buys a particular style of shoe for £22 per pair and sells these shoes to the customer for £29.99 per pair.

a   What profit does the shop make?

b   What percentage profit does the shop make?

a   Profit = Selling price − Cost price

$$= £29.99 - £22$$

$$= £7.99$$

b   Profit % $= \dfrac{\text{Profit}}{\text{Cost price}} \times 100\%$

$$= \frac{£7.99}{£22} \times 100\%$$

$$= 36.3\%$$

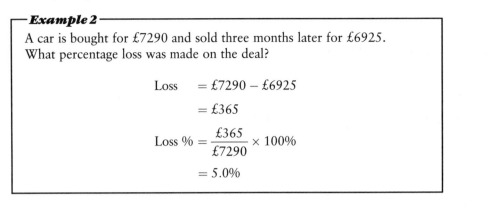

Example 2

A car is bought for £7290 and sold three months later for £6925.
What percentage loss was made on the deal?

$$\text{Loss} = £7290 - £6925$$

$$= £365$$

$$\text{Loss \%} = \frac{£365}{£7290} \times 100\%$$

$$= 5.0\%$$

Example 3

A supermarket buys cheese for 120p per pound. When the cheese
reaches its 'sell by date' the retail price is reduced and the supermarket
makes a loss of 10%.
At what price per pound is the cheese now sold?

$$\text{Cost price (100\%)} = 120\text{p}$$

$$10\% \text{ loss} = \frac{10}{100} \times 120\text{p} = 12\text{p}$$

$$\text{Selling price} = 120\text{p} - 12\text{p}$$

$$= 108\text{p per pound}$$

Exercise 10A

1 For each cost price (CP) and selling price (SP)
given below, find: (i) the profit or loss and
(ii) the percentage profit or loss.

   a CP = £4, SP = £5

   b CP = £20, SP = £27

   c CP = 75p, SP = 66p

   d CP = £90, SP = £108

   e CP = £525, SP = £425

   f CP = £36, SP = £39.99

2 A market trader buys kiwi fruit at £9 for a box
of 96 and sells them at 8 for £1.
What is his percentage profit?

3 A small greengrocer's shop buys crates
containing 24 melons for £36 per crate. The
melons are then sold for £1.68 each.

   a What profit does the shop make on each
   crate of melons?

   b What is this profit as a percentage of the cost
   price?

4 A car was bought for £10 565 and resold at a
loss of 8%. What was the selling price?

5 A greengrocer buys a 10 lb box of grapes for
£7.20. The grapes are sold to customers at
£1.08 per lb.
If 2 lb of grapes are unsaleable, what is the
greengrocer's percentage profit on the cost
price?

6 A retailer buys calculators for £29.20 each plus
17.5% VAT and resells them at £41.98 each.
What is the retailer's profit as a percentage of
his outlay?

7 a A house was bought in 1980 for £82 000
   and sold 10 years later for £160 000.
   What percentage profit was made?

   b If the purchasing power in 1990 of a 1980
   £1 had eroded to 52p, what was the true
   profit as a percentage of £82 000?

10.2 Calculations involving more than one price

> **Example**
>
> A hardware shop buys 18 saws for £8.67 each. Thirteen are sold for £10.85, and the remaining saws are sold at the reduced price of £7.99 in the January sale.
>
> a Calculate the overall profit which the shop makes on the 18 saws.
>
> b Calculate the shop's percentage profit.
>
> a Purchase price for 18 saws $\quad= £8.67 \times 18 \qquad = £156.06$
>
> Amount received for 13 saws $= £10.85 \times 13 \qquad = £141.05$
> Amount received for $\;\;5$ saws $= \;\;£7.99 \times \;\;5 \qquad = \;\;£39.95$
>
> Total amount received $\qquad = £141.05 + £39.95 \; = £181.00$
> Overall profit $\qquad\qquad\;\; = £181.00 - £156.06 = \;\;£24.94$
>
> b Percentage profit $= \dfrac{\text{Profit}}{\text{Purchase price}} \times 100 = \dfrac{£24.94}{£156.06} \times 100\%$
>
> $\qquad\qquad\qquad\qquad\qquad\qquad\quad = 16.0\%$

Exercise 10B

1 A chemist's shop buys 25 bottles of a new brand of perfume for £195. To attract sales, the first ten bottles are sold for £8 each. The remaining bottles are all sold at the normal price of £10 each.
Calculate:

    a the overall profit which the shop makes on this perfume

    b the percentage profit.

2 An electrical store buys fifty personal stereos for £949.50. Thirty five are sold for £24.99 each. The remaining stereos are all sold at the reduced price of £19.99 each.
Calculate:

    a the total amount received for the 50 stereos

    b the overall profit which the store makes on these 50 stereos

    c the store's profit, as a percentage of the purchase price.

3 A grocery store buys 500 tins of tomatoes for 23p per tin. The tins of tomatoes are sold to the public for 29p. Unfortunately, 36 tins have been dented and these tins are sold for 5p less than normal. Assuming all the tins are sold, calculate:

    a the shop's overall profit

    b the shop's percentage profit.

4 A garage sells bunches of fresh flowers which it buys from a local nursery.
The garage pays £1.50 per bunch and sells them for £2.50 per bunch. Any bunches remaining after three days are sold at half price and any remaining after five days are thrown away. On a particular day, the garage buys 12 bunches of flowers. Seven bunches are sold at full price and three are sold at half price.
Calculate:

    a the overall profit on these 12 bunches

    b the percentage profit.

5 The clothing department of a large store purchased 100 matching shorts and T-shirts as a special line for the summer season. The purchase prices were £7 for each pair of shorts and £5 for each T-shirt.
The price to the general public was £9.99 for a pair of shorts and £7.99 for a T-shirt. The items could be bought separately.
Eighty pairs of shorts and ninety T-shirts were sold at these prices.
Calculate:

a the total purchase price of 100 shorts and T-shirts

b the amount the shop received for 80 pairs of shorts and 90 T-shirts.

The remaining shorts and T-shirts were all sold in the 'end of season' sale for £2 each less than the original selling price.

c Calculate the amount the shop received for the shorts and T-shirts during the sale.

d What was the total amount received by the shop for these items?

e Find the overall profit which the shop made on the shorts and T-shirts and express this as a percentage of the purchase price.

6 Barry buys 200 watches and sells 60% of them for £9 each.

a How much does he receive?

Barry then reduces the price of the remaining stock to $\frac{3}{4}$ of the selling price.

b What is the new price of a watch?

Barry sells $\frac{3}{5}$ of the remaining stock at this new price. Unfortunately, the remainder are broken and are thrown away.

c What is the *total* amount of money he received from the sale of the watches?

Barry paid £820 for the 200 watches.

d Express the profit made as a percentage of the cost price. Give your answer correct to 2 decimal places.   (SEG W93)

11 Savings

There are many ways people can choose to invest their money, some of which are discussed later in this unit.

Most people will invest some, if not all, of their money in a building society, post office or bank savings account.

Because they are **lending** their money they will receive **interest** on the money invested.

11.1 Simple interest

The initial sum of money invested is called the **principal**. With **simple** interest the amount of interest to be paid is always calculated on the principal, and therefore the interest remains the same every year.

In practice, this will only happen if the interest is withdrawn each year, or the interest is paid automatically, so that the amount invested is always the same.

Example 1

£3700 is invested at a simple interest rate of 6% for 5 years. Calculate the amount of interest earned.

$$\text{One year's interest} = \frac{6}{100} \times £3700$$

$$= £222$$

$$\text{Five years' interest} = £222 \times 5$$

$$= £1110$$

Simple interest can be calculated using the formula:

$$\textbf{Simple interest} = \frac{\textbf{Rate}}{\textbf{100}} \times \textbf{Principal} \times \textbf{Time}$$

$$I = \frac{R \times P \times T}{100}$$

The formula makes it easier to calculate the rate, the time or the principal.

Example 2

Calculate the length of time taken for £2000 to earn £270 if invested at 9% simple interest.

The simple interest formula is
$$I = \frac{R \times P \times T}{100}$$

Substituting into this gives
$$270 = \frac{9 \times 2000 \times T}{100} = 180 \times T$$

∴
$$180 \times T = 270$$

Dividing by 180 gives
$$T = \frac{270}{180} = 1.5 \text{ years}$$

---

1 Calculate:

  a the simple interest on £1400 for 3 years at 6% per annum

  b the simple interest on £500 at 8.25% per annum for 2 years

  c the length of time for £5000 to earn £1000 if invested at 10% simple interest per annum

  d the length of time for £400 to earn £160 if invested at 8% simple interest per annum.

2 Mr Foyle invests £6750 at 8.5% simple interest per annum.

How much interest has he earned and what is the amount in his account after 4 years?

3 Mr and Mrs Mahon invest £5000 at 9.25% simple interest per annum for 6 months. How much interest will they receive?

4 Mr Allbright invested £10 800 and at the end of each year he withdrew the interest. After 4 years he had withdrawn a total of £3240 in interest. At what annual rate of interest was his money invested?

11.2 Compound interest

Calculation of compound interest

If the interest earned is not withdrawn each year, but is left in the savings account, the amount invested increases as the interest is added. This means that the interest earned the following year will also increase. In five years, with an interest rate of 10% per annum, £1000 'grows' to £1610.51:

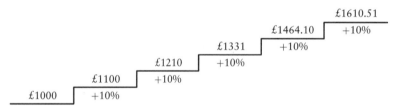

This system of paying interest is called **compound interest**. Building societies usually add compound interest to their accounts every year or every six months.

With some bank accounts, e.g. interest-paying current accounts, the interest is calculated daily and added to the account each month.

Example 1

£3000 is invested in an account paying 12% compound interest per year.

Find the value of the investment after 3 years.

Principal	£3000
Interest at 12% (of £3000)	360
Value of the investment after 1 year	3360
Interest at 12% (of £3360)	403.20
Value of the investment after 2 years	3763.20
Interest at 12% (of £3763.20)	451.58
Value of the investment after 3 years	£4214.78

Example 2

£2000 is invested in a high interest account for one and a half years. The interest rate is 10.5% per annum, and it is paid into the account every six months.

Calculate the value of the investment after this time and the amount of interest earned.

The rate of interest for 1 year   = 10.5%
∴   The rate of interest for 6 months =   5.25%

Principal	£2000
Interest at 5.25% (of £2000)	105
Value of the investment after 6 months	2105
Interest at 5.25% (of £2105)	110.51
Value of the investment after 1 year	2215.51
Interest at 5.25% (of £2215.51)	116.31
Value of the investment after $1\frac{1}{2}$ years	£2331.82

Compound interest earned = £2331.82 − £2000
                         = £331.82

Note that at each stage the amount of interest has been rounded to the nearest penny.

Exercise 11B

1   £6000 is invested at 10% per annum compound interest which is paid annually. How much is in the account after 3 years?

2   £2000 is invested at 10% per annum payable every 6 months.
How much is in the account at the end of $1\frac{1}{2}$ years?

3   Mrs Fletcher invests £6520 at 9.6% per annum for 18 months. The interest is added every 6 months.
Calculate the total amount of money in the account at the end of 18 months and the amount of interest accrued.

4   £12 600 is invested at 11.05% per annum for 3 years.
Calculate the interest earned over this period.
What is the average amount of interest earned per year?

5   A building society offers a rate of 6.8% per annum payable half-yearly.

   a Calculate the interest payable on £1000 at the end of 1 year.

   b What is the equivalent yearly interest rate?

6   One building society offers an interest rate of 8.5% per annum payable half-yearly. A second offers 8.75% payable annually.
Mr Dugan has £2000 to invest. Which savings account should he choose?

   What other factors (apart from higher interest rate) should Mr Dugan take into account when making his choice?

7   Sharon and Tim put £800 into a high interest account which pays compound interest at the rate of 1.1% per month. After it has gained three months' interest, they withdraw all their money. How much do they receive?   (SEG S96)

†11.3 Personal savings and investments

Besides savings accounts, there are many other ways of investing money, e.g. stocks and shares, unit trusts, term shares, Premium Bonds, personal equity plans, collecting fine art or vintage wines.

You may need to do a little research in the library to help with the following exercise.

Exercise 11C

1 Mr and Mrs Askew have just become grandparents. They wish to invest £2000 on behalf of their granddaughter so that she will collect a substantial sum of money on her 18th birthday.
   Should they:

   a buy National Savings Certificates

   b invest the money in a children's savings account

   c buy Premium Bonds

   d buy a valuable piece of jewellery or gold coins

   e invest the money in unit trusts?

   Can you suggest a better investment for this money?

2 a What is the meaning of the term 'taxed at source'?

   b (i) Give an example of an account where the money is *not* taxed at source.
      (ii) Who would benefit by investing in this type of account?

3 Dana and her fiancé wish to save a regular amount each month towards setting up their own home.

   Should they choose:

   a a building society account

   b a unit trust

   c a personal equity plan (PEP)

   d the National Savings 'Yearly Plan'?

4 Give three reasons for investing money in a building society.

5 a For what do the letters TESSA stand?

   b What are the conditions which must be met when investing money in this type of account?

6 Give two factors which determine the interest rate offered for your savings.

12 Banking

12.1 Cheques

All bank current accounts, and some building society accounts, come with a cheque book.

A **cheque** is a written instruction to the bank to pay a sum of money from your account to yourself or to another named person, company or organisation.

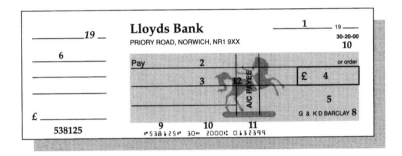

When writing a cheque you must use a pen (**never** a pencil) and fill in correctly:

(1)   the date
(2)   the name of the person, company or organisation to whom you are paying the cheque (the payee)
(3)   the amount of money to be paid, in words (the amount of pence can be written in figures), drawing a line through any unused space
(4)   the amount of money to be paid, in figures, separating the pounds from the pence with a hyphen
(5)   your signature.

The cheque should be completed neatly and any alterations should be initialled.
You should also fill in:

(6) the counterfoil.

The other information on the cheque is to enable the cheque to be cleared and money transferred from and paid into the correct accounts.

The sorting is done by computers at the Clearing Bank.

The additional information is:

(7)   the name and address of the branch of the bank that holds the account
(8)   the name on the account
(9)   the number of the cheque, which will appear on the bank statement
(10)   the branch's sorting code number
(11)   the account number
(12)   the two lines crossing the cheque vertically – all cheques issued by the banks are 'crossed', which means the cheque must be paid into a bank account.

Exercise 12A

1 The cheque shown below was written by R S Langdon to M F Townsend for the amount of £43, on 7 March 1992.

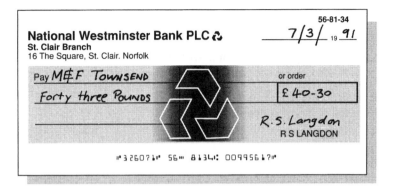

Unfortunately, several mistakes have been made. For each mistake you find, explain what is wrong and what should have been written.

2 Give four reasons why a bank might return a cheque unpaid.

3 What do the words 'refer to drawer' mean, when written on a cheque?

4 Explain the meaning of the phrases 'A/C payee' and 'not negotiable' when written on a cheque.

5 What benefit is there in possessing a cheque guarantee card?

6 Why should a cheque book and a cheque card always be kept separate?

†12.2 Current accounts

For most people, a current account is the main bank account. It is a convenient way of dealing with everyday income and expenditure because the bank offers many useful services in conjunction with a current account. Here are some of them:

- cheque book
- cheque guarantee card
- standing orders
- direct debits
- cash dispenser card
- payment debit card
- monthly statement

Standing orders and direct debits are usually made from a bank account, normally a current account, but the service is also available with some savings accounts.

Standing orders

The account holder instructs the bank to make regular payments, of a specified amount, on his/her behalf.

Standing orders can be used to pay insurance premiums, hire purchase instalments or to transfer regular amounts of money to a savings or investment account.

Because the amount to be paid can only be changed by the **payer**, standing orders are being superseded by the more flexible **direct debit**.

Direct debits

Direct debits can be used to pay regular, fixed amounts in the same way as standing orders, but, because the **payee** can change the amount to be paid, variable payments can be made at variable intervals.

The account holder (payer) is usually given 14 days' notice of any change in the amount to be paid.

Direct debits are used to pay subscriptions or bills such as gas, electricity, telephone, Council Tax, where the amounts are likely to change each year. The payment of a direct debit is at the request of the **payee's bank** and *not* the account holder or the account holder's bank.

Cash dispenser card

As an account holder at a bank or building society, you can apply for a **cash dispenser card**. You will also be given a PIN (personal identity number) which, together with the card, will enable you to obtain cash from a cashpoint machine.

Most of these machines dispense money outside banking hours and many operate during the night.

Payment debit card

A debit card can be used to buy goods from a shop, and it works in a similar way to a cheque. The money is deducted directly from your bank account within a few days.

It can also be used to obtain money from cash dispenser machines or over the counter at a bank.

Bank statements

The bank will send an account holder a **bank statement** each month. This is a concise record of all payments into (credits) and out of (debits) his or her account during the preceding month.

The statement should be checked carefully.

Exercise 12B

1 Study the bank statement and then answer the following questions:

a With which branch does the account holder bank her money?

b What is the account number?

c On what date was the statement produced?

d What was the balance in the account on 13 March?

e By how much did the account become overdrawn?

f How much was withdrawn using cashpoint machines?

g What is the meaning of D/D?

h How much was paid on standing orders?

i What does the symbol * mean?

j What are the numbers 412– – –?

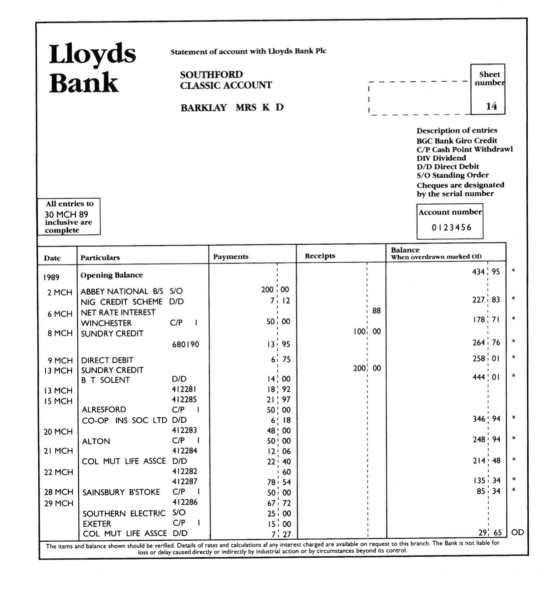

2 Explain briefly how you would check the above bank statement.

3 Name six services or facilities available when you open a bank account.

†12.3   Other accounts

Deposit accounts

A deposit account is used for savings which you wish to leave untouched over a period of time. Interest, from which tax has been deducted, is paid at regular intervals, usually half-yearly.

Withdrawals of any amount can be made from the branch which keeps the account, but notice of withdrawal is usually required; otherwise a penalty is charged based on the amount withdrawn. Statements are usually issued half-yearly.

Savings accounts

Savings accounts have some of the facilities of a current account, such as immediate access to your money, a cash card and regular statements. Some even offer standing orders and direct debits as a service. Interest is generally paid monthly.

Banks and building societies now offer many different types of savings account to suit different needs and these are much more widely used than deposit accounts.

Investment accounts

An investment account is for savings which are intended to be left in the account for relatively long periods of time. A higher rate of interest is paid than with other accounts, but notice is required when money is to be withdrawn, usually 90 days. Interest is lost if money is withdrawn without giving the required notice.

Budget accounts

Budget accounts enable you to plan your spending through the year.

You fill in a form detailing all regular expenses over a year – for example, standing orders, insurance premiums, subscriptions, Council Tax – and estimates are made for electricity, gas, water and telephone bills.

The total amount is averaged over 12 months and you pay this amount into the account each month. The bank pays all the listed bills as they arrive, either directly or by honouring cheques from a special cheque book.

A service charge is made for the administration of a budget account.

†12.4   Banking services

In addition to those services already mentioned, banks will also:

- issue   Eurocheques, traveller's cheques, credit cards, debit cards
- operate   the bank giro system, foreign exchange
- arrange   loans, overdrafts, insurance, mortgages
- advise on   income tax, investments, wills
- manage   family trusts, safe custody services, unit trusts.

Eurocheques

Eurocheques are used when travelling abroad. If you have a Eurocheque book and cheque card, you can use them to draw money from your account at a bank abroad or to pay for purchases from many shops.

Traveller's cheques

Traveller's cheques are a safe way of taking money abroad. They can be bought in various denominations in pounds, francs, dollars, and other currencies.

They must be signed at the bank immediately by the person receiving them. They can be offered in payment for goods or services or exchanged for cash at a bank or bureau de change.

When they are used they must be signed by the same person who signed for them at the bank originally.

You can also use credit cards and debit cards abroad (in the same way as in the UK).

13 Borrowing and Spending

*13.1 True interest rates

Someone who wants to spend money may sometimes need to borrow money or buy on credit.

The cost of borrowing can vary a great deal depending on the method of obtaining credit as well as the amount borrowed.

Usually banks, stores and finance companies quote a flat rate of interest, but this is *not* the true price of the credit.

So that customers can accurately compare credit costs, the **APR (annual percentage rate)** must, by law, be quoted for all credit arrangements.

This is the *true* rate of interest, and it takes into account all the costs involved and the method of repayment. (In general, the lower the APR, the better the deal.)

If you borrow £100 for one year and the 'credit charge' is 8% you repay £108 (the loan plus interest).

But the charge of £8 would only be a true rate of 8% if you had the use of the £100 for the whole year. In fact, you have £100 at the start of the year and nothing at the end.

In practice, you begin repaying the loan almost immediately in twelve equal instalments spread over the year.

The average amount available to you during the year is therefore about half the original amount and the **true rate of interest is nearly double the quoted rate**:

$$\text{APR} \approx 2 \times \text{Flat rate of interest}$$

The APR is, in fact, between 1.8 and 2 times the flat rate of interest as the following example shows:

Suppose you borrow £100 for 1 year at a flat rate of interest of 8% per annum.

The total repayment = £100 + £8 interest = £108.
This is repaid in 12 monthly instalments of £9 (= £108 ÷ 12).

If the interest on the outstanding debt is calculated at a rate of 1.205% per month, the debt will be wiped out by the end of the year, as shown below:

	£
Initial debt	100.00
Interest in first month at 1.205% per month	1.21
Debt at end of first month	101.21
£9 repaid. Debt at beginning of second month	92.21
Interest in second month on a debt of £92.21	1.11
Debt at end of second month	93.32
£9 repaid. Debt at beginning of third month	84.32
Interest in third month	1.02
Debt at end of third month	85.34
£9 repaid. Debt at beginning of fourth month	76.34
Interest in fourth month	0.92
Debt at end of fourth month	77.26
£9 repaid. Debt at beginning of fifth month	68.26
Interest in fifth month	0.82
Debt at end of fifth month	69.08
£9 repaid. Debt at beginning of sixth month	60.08
Interest in sixth month	0.72
Debt at end of sixth month	60.80
£9 repaid. Debt at beginning of seventh month	51.80
Interest in seventh month	0.62
Debt at end of seventh month	52.42
£9 repaid. Debt at beginning of eighth month	43.42
Interest in eighth month	0.52
Debt at end of eighth month	43.94
£9 repaid. Debt at beginning of ninth month	34.94
Interest in ninth month	0.42
Debt at end of ninth month	35.36
£9 repaid. Debt at beginning of tenth month	26.36
Interest in tenth month	0.32
Debt at end of tenth month	26.68
£9 repaid. Debt at beginning of eleventh month	17.68
Interest in eleventh month	0.21
Debt at end of eleventh month	17.89
£9 repaid. Debt at beginning of twelfth month	8.89
Interest in twelfth month	0.11
Debt at end of twelfth month	9.00
£9 repaid. Debt after twelve repayments of £9	0.00

The interest rate of 1.205% per month gives a compound interest rate of 15.46% per year.

Hence a flat rate of 8% per year is equivalent to an APR of 15.46% (almost twice the percentage rate).

(How did we get that figure of 1.205% per month? That involves some tricky arithmetic which you needn't worry about!)

13.2 Buying on credit

If you do not have sufficient funds saved to buy the goods you require, or do not wish to withdraw a large sum of money from your savings, you can buy the goods on credit, i.e. borrow money which will be repaid over a period of time, usually in instalments which include interest.

There are several methods of obtaining credit:

- Hire purchase
- Credit purchase
- Loans
- Monthly accounts
- Credit card/Store card/Charge card.

Hire purchase and credit purchase

These two methods of obtaining credit are similar. A **deposit** is often required and the amount borrowed plus interest is repaid in equal instalments over a set period of time.

The main difference is that with hire purchase you do not own the goods until the last payment has been made and so cannot sell the goods while you are still paying for them. Also, the hire purchase company can repossess the goods if the repayments are not made.

Credit purchase is usually arranged with the store from which you buy the goods. You own the goods from the moment the agreement is signed. If you fail to keep up the repayments, the store will sue through the courts for the amount outstanding.

Example 1

A video recorder can be bought for a cash price of £299.95 or by credit purchase paying 12 monthly instalments of £31.50.
Calculate:

a the credit price

b the interest paid

c the flat rate of interest, as a percentage of the amount borrowed

d the approximate APR.

a Credit price = Total instalments
$\qquad$ = 12 × £31.50
$\qquad$ = £378

b Interest paid = Credit price − Cash price
$\qquad$ = £378.00 − £299.95
$\qquad$ = £78.05

c Flat rate of interest = $\dfrac{\text{Interest}}{\text{Amount borrowed}} \times 100\%$

$\qquad = \dfrac{£78.05}{£299.95} \times 100\%$

$\qquad = 26.0\%$

d Approximate APR = 2 × Flat rate per year
$\qquad$ = 52.0%

Example 2

A washing machine is offered for a cash price of £329.99. It can also be bought on hire purchase for a deposit of £34.99 and 30 monthly instalments of £13.25.
Calculate:

a the total hire purchase price

b the amount of interest paid

c the amount borrowed

d the flat rate of interest

e the approximate APR.

a Total HP price = Total instalments + Deposit
(do not forget to include the deposit)
= 30 × £13.25 + £34.99
= £397.50   + £34.99
= £432.49

b Interest paid = Credit price − Cash price
= £432.49 − £329.99
= £102.50

c Amount borrowed = Cash price − Deposit
= £329.99 − £34.99
= £295.00

d Flat interest rate $= \dfrac{\text{Interest}}{\text{Amount borrowed}} \times 100\%$

$= \dfrac{£102.50}{£295} \times 100\%$

$= 34.7\%$

e Flat rate per year $= \dfrac{\text{Flat rate}}{\text{Time of loan in years}}$

$= \dfrac{34.7\%}{2.5}$

$= 13.9\%$

Approximate APR = 2 × Flat rate per year
= 2 × 13.9%
= 27.8%

Exercise 13A

1 A department store offers a hi-fi system for sale at £429.99. The customer can also buy the system on interest-free credit by paying a deposit of 10% of the cash price and 10 equal monthly instalments.

 a How much is the deposit?

 b How much is each monthly payment?

2 A rival store offers a similar hi-fi system for sale at £349.99 or 12 months' free credit. The credit terms are 10% deposit and 12 equal monthly instalments.
 How much is each instalment?

3 A television set can be bought for £189.99 cash or by credit purchase.

The credit purchase terms are: no deposit and 24 monthly payments of £9.31.
Calculate:

 a the total credit price

 b the amount of interest paid

 c the flat rate of interest

 d the approximate APR.

4 A BMW motorbike has a cash price of £4195. It can be bought on hire purchase for a deposit of 10% and 36 monthly payments of £135.84.
Calculate:

 a the total hire purchase cost

 b the approximate APR.

5   Mr Weston is buying a new car. The cash price of the car is £7399. He is offered a £1500 trade-in price on his old car. The hire purchase terms on the new car are a deposit of 25% of the cash price and 24 monthly payments of £290.
    Calculate:

    a   the deposit

    b   the amount borrowed by Mr Weston from the HP company

    c   the total hire purchase cost of the car

    d   the interest paid, as a percentage of the loan

    e   the approximate APR.

6   Mrs Sayeed wishes to buy a new car. The cash price of the car is £9098.

    The hire purchase terms offered by the garage are: a deposit of 15% and *either* 36 monthly repayments of £271.50 *or* 60 monthly repayments of £179.12.

    Find, for each deal:

    a   the total hire purchase price of the car

    b   the approximate APR.

7   The same camera is available in two different shops at the same price of £98.60, but with different credit terms.

Shop A requires a deposit of 10% and twelve monthly repayments of £8.43.

Shop B requires a deposit of 15% and ten monthly repayments of £9.60.

By calculating the true rates of interest (APR), decide which shop is offering the better deal.

8   A car is offered for sale in a garage with a price of £8400 on the windscreen. John can either pay by cash or credit.
    If John buys the car on credit, he pays a deposit of 20% and 36 monthly instalments of £210.

    a   (i)   Find the deposit he would have to pay.
        (ii) Find the total credit price.

    If John pays cash he will pay the *cash price* which will give John a discount of $12\frac{1}{2}\%$ off the windscreen price.

    b   What is the extra amount paid for credit compared with cash?
    c   If John buys the car on credit he requires a loan. The loan will be the difference between the cash price and the deposit.

        (i)   Express the extra amount paid for credit compared with cash as a percentage of the loan.
        (ii) Hence make an estimate of the APR which John is charged for credit.
                                            (SEG Spec98)

Loans

If a large amount of money is required, for example, for a car or holiday or home improvements, a common form of credit which is applied for is a **loan**.

A **bank loan** or **personal loan** can be arranged with the bank manager or with a finance company. The interest rates can be lower than for other forms of credit.

It is important to choose a reputable company and to compare interest rates by looking at the quoted APR.

A bank loan is applied for through the bank manager. The amount borrowed is added to the borrower's account as a single payment and is then repaid over a fixed period of time in instalments. One disadvantage of this type of loan is that the interest is fixed in advance so that the amount to be repaid is known exactly.

An **overdraft** is a temporary, and sometimes expensive, method of obtaining credit used for a short-term loan.

The bank will charge interest on the amount of the overdraft plus 'transaction charges' on all cheques, standing orders, direct debits and cashpoint withdrawals during this period.

Exercise 13B

1 The table below is published by a finance company. It shows the monthly repayments required for a given personal loan and the APR charged.

Months	12 APR 23.9%	24 APR 23.9%	36 APR 23.9%
Amount of loan	Monthly repayments	Monthly repayments	Monthly repayments
£300	28.02	15.50	11.39
£400	37.36	20.67	15.19
£500	46.70	25.84	18.99
£600	56.04	31.01	22.79
£700	65.38	36.18	26.59
£800	74.73	41.35	30.39
£900	84.07	46.52	34.19
Amount loan	APR 21.9%	APR 21.9%	APR 21.9%
£1000	92.61	50.87	37.14
£1100	101.88	55.96	40.86
£1200	111.14	61.05	44.57
£1300	120.40	66.14	48.29
£1400	129.66	71.23	52.00
£1500	138.92	76.31	55.72
£1600	148.19	81.40	59.73
£1700	157.45	86.49	63.14
£1800	166.71	91.58	66.86
£1900	175.97	96.67	70.57
£2000	185.23	101.75	74.29
£2100	194.49	106.84	78.00
£2200	203.76	111.93	81.72
£2300	213.02	117.02	85.43
£2400	222.28	122.11	89.15
£2500	231.54	127.19	92.86

For each of the following loans, calculate:

(i)   the total amount to be repaid
(ii) the total interest
(iii) the flat rate of interest.

Compare the flat rate of interest with the given APR:

a £1000 for 12 months   d £300 for 24 months

b £1000 for 24 months   e £2400 for 36 months

c £1000 for 36 months   f £1700 for 24 months.

2 Explain what is meant by the phrases:

a 'security for a loan'

b 'a protected loan'.

3 Mr and Mrs D'Angelo wish to borrow £5000 to be repaid over a period of 5 years. After visiting several banks and building societies, they find that the best terms are:

a monthly repayments of £123.83 for a 'secured' loan

b monthly repayments of £147.85 for an 'insured' loan

c monthly repayments of £132.56 for a straightforward repayment
   loan.

   (i) Calculate the total cost of each of the above loans.
   (ii) What are the advantages and disadvantages of each type of loan?

4 Jamal's bank manager agrees to an overdraft of up to £100 on his
   account each month. There is an arrangement fee of 1% of the amount
   allowed for this service.

   Because Jamal is a student, the interest charged is only 1.2% (APR
   15.3%) and there are no transaction charges. At the end of January his
   account is overdrawn by £80 and at the end of February by £74.20.

   a How much was the arrangement fee?

   b How much does he pay each month for his overdraft?

5 Mr and Mrs Madden buy their clothing through a mail order catalogue
   using their 12 month credit scheme.
   Their costs are:

   - the catalogue price for the goods, *plus*
   - handling and delivery charges:
     £1.99 for 1 item, £2.80 for 2 items, £3.30 for 3 or more items.

   A service charge of 2.4p in the pound is added each month to the
   outstanding balance.

   a At the end of June their account is clear. During July they buy a dress
   and a pair of jeans priced at £29.99 and £34.99 respectively.

   (i) What is the total price charged?
   (ii) They make a first repayment of £6.
        How much interest is added to the account they receive one
        month later?

   b The bill is repaid by means of a further ten repayments of £6 (eleven
   altogether) and one of £13.74.

   (i) What was the total cost of the two items?
   (ii) What was the credit charge?
   (iii) Find an approximate value for the APR.

13.3   Credit cards

Credit cards, store cards and charge cards can be used to pay for a wide range
of goods and services.

Credit cards

Access cards and *some* Visa cards are credit cards. (Other Visa cards are debit cards or Automatic Telling Machine cards – they are not considered here.) With credit cards you receive a statement every month which lists the purchases made in the previous month.

The bill can be settled in full, within 25 days, in which case there are no interest charges to be paid.

Alternatively a minimum payment of, usually, £5 or 5% of the total bill (whichever is the greater) can be made, or any larger repayment. Interest will then be charged each month on the outstanding amount.

Typical interest charges are 1.8% per month (APR 23.9%) or 1.55% per month (APR 20.3%) plus an annual charge.

Many cards offer different rates and terms, and some offer free insurance for goods bought using them.

Store cards

A store card operates in a similar way to a credit card, but it is only accepted in the store or group of stores in which it was issued.

The interest rates are usually higher than for credit cards. An APR above 30% is not uncommon.

To make matters more complicated, some store cards (e.g. Marks and Spencer's cards) are credit cards, some are charge cards and some are budget accounts.

Charge cards

An annual fee is usually paid for a charge card (e.g. American Express and Diner's) and the bill must be settled in full each month.

Example

Mr Buckley borrows £450, on average, each month on his credit card and interest is charged at 2.0% per month.

If he changed credit card companies, he would pay an annual fee of £12 and interest charges of 1.8% per month.

Would he benefit by changing?

Comparing the monthly charges for the two companies:

2.0% interest on £450 = £9.00

1.8% interest on £450 = £8.10
average monthly fee   = £1.00

Total   = £9.10

Conclusion:
Mr Buckley would not benefit by changing his credit card.

Exercise 13C

1 What advantages does the possession of a credit card offer the card holder?

2 What additional 'perks' are available to credit card holders?

3 What is meant by the term 'credit limit'.

4 What should a credit card holder do in the event of a card being lost or stolen?

5 What is the minimum age for a credit card holder?

6 What disadvantages are there in possessing a credit card?

7 Mr Johnstone receives a bill for £250 from his credit card company. He makes the minimum repayment and is charged interest of 2.0% per month on the balance.

   a What repayment does he make?

   b How much is the balance?

   c What interest charges will be added to his next statement?

8 Mrs Magee receives her credit card statement which shows a total bill of £262, including her annual £12 fee. She pays £24.50 by the due date and is charged interest of 1.8% per month on the balance.

   a What is the balance?

   b What interest charges are added to the next statement?

   c Should either Mrs Magee or Mr Johnstone (see question 7) change credit card companies?

9 a Mr and Mrs Collier see the following advertisement in their paper:

CD COMPACT MIDI
£399.99
FREE HEADPHONES

They make some inquiries and discover that they will need to make 35 repayments of £17 and a final payment of £8.49 in order to pay for the CD system. Calculate:

   (i) the credit charge
   (ii) the approximate APR.

   b The Colliers then find the same equipment for sale though a mail order catalogue with weekly payments of £4.85 over 2 years (i.e. 104 repayments). Calculate:

   (i) the credit charge
   (ii) the approximate APR.

   c Compare the two deals.

13.4 Discrimination in spending

It pays to be discriminating when spending money. Taking time to make comparisons means getting the best value for your money.

You can save money by looking for special offers, buying during the sales (if you are sure you are getting a bargain) or buying in bulk.

Example

Mrs Newman is shopping for soap powder and finds a new brand on the supermarket shelves.

The powder is sold in three packet sizes:

a starter pack: weight 145 g, price 39p
a medium size: weight 800 g, price £1.55
a large size: weight 2 kg, price £3.79

Which size should Mrs Newman buy to give her the best value for her money?

There are two methods which can be used to compare value for money on the basis of cost.

Method 1
Prices are compared by calculating the cost of 1 g of powder for each size (given here to 3 significant figures).

1 g costs: (i) Starter pack $\dfrac{39p}{145} = 0.269p$

(ii) Medium size $\dfrac{£1.55}{800} = 0.194p$

(iii) Large size $\dfrac{£3.79}{2000} = 0.190p$

Packet type:	*Starter*	*Medium*	*Large*
Price per g:	0.269p	0.194p	0.190p

This shows that the large size gives the best value for price.

Method 2
Value is compared by calculating the amount of powder which can be bought for the same amount of money (e.g. £1) in each case (again to 3 significant figures).

Amount of powder for £1: (i) Starter pack $\dfrac{145}{0.39} = 372g$

(ii) Medium size $\dfrac{800}{1.55} = 516g$

(iii) Large size $\dfrac{2000}{3.79} = 528g$

Packet type:	*Starter*	*Medium*	*Large*
Price per £1:	372 g	516 g	528 g

This also shows that the large size will give Mrs Newman more for her money.

Of course, price is not the only thing to be taken into account when making a purchase. It is also important to consider, where appropriate:

- the quality of the product
- the length of time for which non-durables can be stored

- personal preferences of taste, colour, style etc.
- the manufacturer's reputation for reliability
- the shop's reputation for service.

For example, Mrs Newman might consider that the starter pack will be the best value at the moment. If she finds that she prefers her old powder, she has not spent much money on a product she does not want.

The medium size packet costs very little more per kg than the large size and is lighter to carry. It also takes up less storage space. For some people, this size may be the best buy.

Exercise 13D

1 a A well-known brand of margarine sells at 38p for 250 g and 67p for 500 g.

     (i) For how much per kg is each size sold?
     (ii) Which size is the better value for money?

  b The price of the smaller size of margarine is reduced by 5p on 'special offer'. Which size is the best buy?

2 Mayonnaise costs 99p for a 400 g jar and £1.39 for a 600 g jar. Mrs Kaplan has a voucher for '10p OFF' a jar of mayonnaise.

  a If she uses the voucher, how many grams of mayonnaise will cost 1p:

     (i) with the 400 g jar
     (ii) with the 600 g jar?

  b Which jar gives the best value for money?

  c What other factor(s) should Mrs Kaplan take into consideration before making her purchase?

3 The village shop stocks soap powder in three sizes:

E3 contains 1.05 kg of powder and costs £1.22
E10 contains 3.50 kg of powder and costs £3.79
E15 contains 5.25 kg of powder and costs £5.49.

  a Calculate comparative costs for each size, based on the E15 size.

  b Which size would you recommend as the best buy for the following people, giving at least one reason for each choice?

     (i) Mr and Mrs Donovan and their three children
     (ii) Mrs Kyle, a 66-year-old widow
     (iii) Mr and Mrs Groves, a newly married couple with no children, but a large mortgage
     (iv) Mr Hambledon, a bachelor living in a small flat.

4 Compare the following advertisements and decide which is the best value for money, giving clear reasons for your choice.

£339.99
• Free 5 year guarantee
• 24 wash programmes
• 11 lb wash load
• 1300 rpm spin speed

£329.95
• 5 year guarantee only £90
• 21 programmes
• 10 lb wash load
• 300/1000 rpm spin speed

5 Mr and Mrs Palfrey decided to find out which nearby supermarket gives the best value for money. They made a shopping list of items which they buy regularly and then visited the four nearest supermarkets in turn making a note of all the prices.

The results of their survey are shown below:

Supermarket	A	B	C	D
Apples (2 lb)	58p	70p	72p	64p
Bread (large)	46p	47p	41p	45p
Butter (250 g)	59p	62p	62p	63p
Cereal (pkt)	72p	75p	72p	71p
Chicken (3 lb)	£2.37	£2.94	£2.85	£2.79
Coffee (100 g)	£1.92	£1.91	£1.93	£1.89
Fish (1 lb)	£2.60	£2.19	£2.82	£2.39
Meat (1 lb steak)	£3.40	£3.46	£3.69	£3.66
Potatoes (3 lb)	60p	66p	66p	75p
Tea bags (80)	£1.27	£1.36	£1.29	£1.30

a (i) If they had bought all the groceries listed above in the same supermarket, at which supermarket would they have paid the least amount?

(ii) Assuming each grocery item on the list above was bought at the most economical price, calculate the lowest price they could have paid for the above items, buying from these supermarkets.

b The distance from the Palfreys' home to each of the supermarkets is:

A 20 miles

B 3 miles

C 1 miles

D 10 miles.

Supermarket D also sells petrol which is, on average, 5p per gallon cheaper than any local petrol station.

Which supermarket should Mr and Mrs Palfrey use:

(i) if they shop weekly

(ii) if they shop monthly

(iii) if they need a few extra items?

6 Krunchi Kornflakes are sold in large and family size packets:

The manufacturer is trying to increase sales.
A *new* large packet is sold with an 'extra 15% free'.

a What is the weight of this *new* large packet?

b What is the price **per 100 grams** of Krunchi Kornflakes in this *new* packet?

The manufacturer reduces the price of the family packet so that it costs 5% less per 100 grams than the *new* large packet.

c (i) Using your answer to part **b**, calculate the new price **per 100 grams** for the family packet.
   (ii) By how much must the manufacturer reduce the price of the family packet?   (SEG S96)

14 Travel

14.1   Foreign currency

Currency exchange

If you plan a trip abroad for business or holiday, you must work out the money you will need and the form in which you will take it.

Most people take a limited amount of cash in the currency of each country to be visited. They take the remainder in the more secure form of traveller's cheques. Alternatively, they use Eurocheques, credit cards or debit cards to obtain cash.

In the UK British currency (pounds sterling) is usually exchanged for foreign currency at a bank or travel agency. Exchange rates between one currency and other currencies change frequently and are published daily in newspapers and displayed where money is exchanged.

By consulting the **selling price** you can calculate the amount of foreign currency you will be sold for your pounds and from the **buying price** you can calculate the amount of pounds you will receive in return for your foreign currency.

Example

Mr Öztürk is to travel on a business trip to Turkey. He changes £350 into Turkish lira on a day when the bank selling price is 189 000.

On returning home he changes his remaining 13 093 000 lira into sterling. The bank buying price is 207 000 to £1.

Calculate:

a   the amount of lira he receives

b   the amount of pounds he receives on his return.

a   The bank pays 189 000 lira for every £1 it buys.
For £350 he will receive 189 000 × 350 lira

$$= 66\,150\,000 \text{ lira}$$

b   The bank charges 207 000 lira for every £1 it sells

For 13 093 000 lira he will receive $£\dfrac{13\,093\,000}{207\,000}$

$$= £63.25$$

Exercise 14A

1   Using the bank selling price, change:

   a £12 to Austrian schillings

   b £140 to Spanish pesetas

   c £96.50 to American dollars.

2   Using the bank buying price, calculate the sterling equivalent of:

   a 200 French francs

   b 11 100 Italian lire

   c 650 Japanese yen.

EXCHANGE RATES		
	Bank Buys	Bank Sells
Australia $	2.33	2.18
Austria Sch	20.85	19.65
Belgium Fr	62.40	58.50
Canada $	2.08	1.98
Denmark Kr	11.38	10.78
Finland Mkk	7.00	6.60
France Fr	10.05	9.45
Germany Dm	2.978	2.798
Greece Dr	292	266
Italy Lira	2205	2075
Japan Yen	266	250
Netherlands Gld	3.33	3.15
Norway Kr	11.48	10.82
Portugal Esc	262.5	246.50
Spain Pta	190	178
Sweden Kr	10.86	10.20
Switzerland Fr	2.618	2.458
USA $	1.77	1.67

3   Geraldine and Peter ate a meal in a restaurant while on holiday in Rhodes.
   The meal for two cost 4256 drachma. Use the bank selling price to calculate the cost of the meal in pounds.

4   The Andersons spent a holiday touring in Yugoslavia. While travelling they used 200 litres of petrol which cost 1236 per litre. The exchange rate was 4330 dinar to £1.

   a How much did the petrol cost them in pounds?

   b What was the price per litre of the petrol in pence?

5   Before going on holiday to Germany and Austria, the Williams family changed £600 into Deutschmarks. While in Germany they spent 824 DM and then changed their remaining marks into Austrian schillings as they crossed the border. The exchange rate was 7.047 Sch to 1 DM. Calculate:

   a the number of marks they received

   b the number of schillings they bought.

6   If Mr Oztürk (in the example on page 102) had postponed his trip to Turkey until the following week the bank selling price would have been 195 000 lira to the pound.
   On his return, the buying price would have been 215 000 lira to the pound.

   a How many lira would he have received?

   b How many lira would he have had left on his return to Britain (assuming that he would have spent the same amount)?

   c How many pounds would he have received on his return?

   d How much money would he have saved by travelling the following week?

7   **a** Mr Elton changed £100 into French francs for a day trip to France. How many francs did he receive?

   b Unfortunately the excursion was cancelled and so he changed all his francs back to pounds. How many pounds did he receive?

   c How much money did he lose because of the cancellation?

Commission

In practice, banks also charge commission for each currency exchange.

The commission is £3 on currency exchanges up to the value of £150 and 2% of the value above £200.

On traveller's cheques the commission is £2 for up to £200 in value and 1% of the value above £200.

Exercise 14B

Questions **1** and **2** refer to Exercise 14A.

1 **a** How much commission did Mr Elton pay when changing pounds to francs?

    b How much commission did he pay when changing the francs back to pounds?

    c How much did his cancelled trip cost him, including commission charges?

2 **a** How much commission did the Williams family pay for their Deutschmarks?

    b How much commission would they have paid if they had taken the £600 in traveller's cheques?

3 **a** For a three-week holiday, touring Germany, Switzerland and Austria, a group of four friends decided to take the equivalent of £900 abroad.

At the bank they changed £110 into Deutschmarks, £100 into Swiss francs and £120 into Austrian schillings. How much of each currency did they receive (after commission was deducted)?

    b After buying the foreign currencies, they changed as much as possible of the remaining money into traveller's cheques.

The smallest value of traveller's cheque which can be bought is £10.

      (i) How much did they exchange for traveller's cheques?

     (ii) How much commission did they pay for the cheques?

    c What was the total cost per person of the foreign currencies and cheques?

14.2   Time

There are two methods in general use for showing the time:

(i) the 12-hour clock   (ii) the 24-hour clock.

With the 12-hour clock the day is divided into two periods, from midnight to noon (am) and noon to midnight (pm).

The 24-hour clock uses a single period of 24 hours, starting at midnight. The time is written as a four-digit number, without a decimal point, and there is no need to specify morning (am) or afternoon (pm).

This method is always used for timetables and is in common use on video recorders and digital clocks.

Example

Write the times

a ten to nine in the evening,

b twenty past seven in the morning, using both the 12-hour and 24-hour clocks.

a The 12-hour clock time is 8.50 pm.
The 24-hour clock time is 2050 (i.e. 8.50 + 12 hours).

b The 12-hour clock time is 7.20 am.
The 24-hour clock time is 0720 (note the zero at the beginning to make a four-digit number).

Exercise 14C

In the following exercise the time is written using words, the 12-hour clock or the 24-hour clock.
For each question, give the time using the other two methods.

1   Six fifteen in the morning

2   11.10 pm

3   0930

4   1.40 am

5   1350

6   2220

7   Ten to ten in the evening

8   10.45 am

9   Twenty-five past midnight

10   1656

Example

The 0745 train from Newcastle is scheduled to arrive in Southampton at 1516.
How long is the journey?

Method 1

From 0745 to 0800 = 15 mins
From 0800 to 1500 = 7 hours
From 1500 to 1516 = 16 mins
Total journey time = 7 hours + 15 mins + 16 mins
= 7 hours 31 mins

Method 2

	Hours	Mins		Hours	Mins	
Train arrives	15	16	=	14	76	(76 = 16 + 60)
Train departs	7	45		− 7	45	
Time taken			=	7	31	

The journey time is 7 hours 31 minutes.

Exercise 14D

1   Mr Little catches the 0654 train to London.
If he arrives at the station at twenty to seven, how long does he wait?

2   Miss Brothers travels on the 0704 train. She arrives at the station nine minutes before the train is due.
At what time does she arrive at the station?

3   Mr Ghosh arrives at the airport at 8.35 am and his plane takes off 55 minutes later.
At what time does the plane take off?

4   At what time does the 0929 train at Coventry arrive if it is 13 minutes late?

5   a   Mrs Richardson's afternoon train is scheduled to arrive in Birmingham at eight minutes past four, but it is 15 minutes late.
At what time, on the 24-hour clock, does she arrive?

b   Her connecting train leaves at 1651.
How long does she have to wait?

6 Chris has an appointment in London at 2.15 pm. The train journey from his local station takes 1 hour and 10 minutes. He allows a further 20 minutes to travel on the underground to his place of appointment. In case there are any delays on the journey, he allows an extra 15 minutes travelling time. What length of time should Chris allow from his local station?

7 a The boat from Dover to Calais departs at 1715. The crossing takes $1\frac{1}{2}$ hours and French time is 1 hour ahead.
At what time does the boat dock in France, French time?

 b The same boat leaves Calais for Dover at 2000. At what time does it arrive in England, British time?

14.3 Timetables

All timetables use the 24-hour clock.

Exercise 14E

Mondays to Saturdays

BASINGSTOKE (Bus Station) ⇌		0730			0840	0935	1035	1135	1235	1335	1435	1535
Basingstoke (Winton Square)		0735			0845	0940	1040	1140	1240	1340	1440	1540
Worting Road (South Ham)		0739			0849	0944	1044	1144	1244	1344	1444	1544
Worting (White Hart)		0743			0853	0948	1048	1148	1248	1348	1448	1548
Newfound (Fox Inn)		0747			0857	0952	1052	1152	1252	1352	1452	1552
Deane Gate		0753			0903	0958	1058	1158	1258	1358	1458	1558
Overton (Post Office) ⇌		0758	0756		0908	1003	1103	1203	1303	1403	1503	1603
Laverstoke (Mill)		0802	0802		0912	1007	1107	1207	1307	1407	1507	1607
Whitchurch (Square)		0809	0809		0919	1014	1114	1214	1314	1414	1514	1614
Whitchurch (Bere Hill Estate)	0650		0813		0923		1118		1318		1518	
Whitchurch (Square)	0654	0809	0817		0927	1014	1122	1214	1322	1414	1522	1614
Hurstbourne Priors (Portsmouth Arms)	0659	0814	0822		0932	1019	1127	1219	1327	1419	1527	1619
The Middleway			0821			1026		1226		1426		1626
Longparish (Plough Inn)	0708			0831	0941		1136		1336		1536	
Longparish (Station Hill)	0712			0835	0945		1140		1340		1540	
London Road (Admirals Way)		0723*	0825	0843	0953	1030	1148	1230	1348	1430	1548	1630
ANDOVER (Bridge Street) arr.		0727	0829	0847	0957	1034	1152	1234	1352	1434	1552	1634
ANDOVER (West Street) ⇌ arr.		0729C			0959C		1154C		1354C			

ANDOVER (West Street) ⇌ dep.	0624	0734			0911		1111		1311		1511		
ANDOVER (Bridge Street) dep.	0626	0736	0831		0913	1031	1113	1231	1313	1431	1513	1631	1735
London Road (Admirals Way)	0630	0740	0835		0917	1035	1117	1235	1317	1435	1517	1635	1739
Longparish (Station Hill)					0925		1125		1325		1525		1747
Longparish (Plough Inn)					0929		1129		1329		1529		1751
The Middleway	0634	0744	0839			1039		1239		1439		1639	
Hurstbourne Priors (Portsmouth Arms)	0641	0751	0846		0938	1046	1138	1246	1338	1446	1538	1646	1800
Whitchurch (Square)	0646	0756	0851		0943	1051	1143	1251	1343	1451	1543	1651	1805
Whitchurch (Bere Hill Estate) ⇌					0947		1147		1347		1547		1809
Whitchurch (Square)	0646	0756	0851	0851	0951	1051	1151	1251	1351	1451	1551	1651	1813
Laverstoke (Mill)	0653	0803	0858	0858	0958	1058	1158	1258	1358	1458	1558	1658	1820
Overton (Post Office) ⇌	0657	0807	0902	0902	1002	1102	1202	1302	1402	1502	1602	1702	1824
Deane Gate	0702	0812	0907	0907	1007	1107	1207	1307	1407	1507	1607	1707	1829
Newfound (Fox Inn)	0708	0818	0913	0913	1013	1113	1213	1313	1413	1513	1613	1713	1835
Worting (White Hart)	0712	0822	0917	0917	1017	1117	1217	1317	1417	1517	1617	1717	1839
Worting Road (South Ham)	0716	0826	0921	0921	1021	1121	1221	1321	1421	1521	1621	1721	1843
Basingstoke (Winton Square)	0720	0830	0925	0925	1025	1125	1225	1325	1425	1525	1625	1725	1847
BASINGSTOKE (Bus Station) ⇌	0725	0835	0930	0930	1030	1130	1230	1330	1430	1530	1630	1730	1852

1 Use the bus timetable to answer the following questions.

 a Mr Tully arrives at Basingstoke bus station at 10.15 am. How long does he have to wait for a bus?

 b How many buses from Basingstoke stop at Longparish?

 c Miss Dawes catches the 0908 bus at Overton. To get to Salisbury she must change buses at Andover. The Salisbury bus leaves Andover at 1038.

 How long does she have to wait at Andover?

 d Mrs Goff lives in South Ham and visits her mother in Witchurch for at least 3 hours every Wednesday. If she catches the 1044 bus from Worting Road, which buses can she catch from The Square in order to be home before 5 o'clock?

2 Assume that the time in France and Belgium is 1 hour ahead of British time.

Dover/Boulogne [$1\frac{3}{4}$ hours]	0030		0330		0630		0930		1230		1530		1830		2130		
Boulogne/Dover		0130		0430		0730		1030		1330		1630		1930		2230	2359

Dover/Calais [$1\frac{1}{4}$ hours]		0200	0400	0600	0730	0900	1030	1200	1330	1500	1630	1800	1930	2100	2230
Calais/Dover	0015	0200	0400	0600	0730	0915	1045	1215	1345	1515	1645	1815	1945	2115	2245

Dover/Ostend [4 hours]	0015		0430				1000			1400		1630		1930	2130	
Ostend/Dover				0600			0945	1145	1345				1745		2100	2345

a How many ferries leave Dover in the evening between 7 o'clock and 10 o'clock?

b At what time does the 7.30 pm boat arrive in Ostend?

c The 1330 ferry from Dover arrives in Calais at 1610.
How many minutes late is it?

d The Carmichael family have hired a chalet in Boulogne, but it is not available until after 12 noon.
Assuming it takes 45 minutes to pass through customs and drive to the chalet, what is the earliest ferry they should catch from Dover?

e Mr and Mrs Davenport plan to return to England on the 0130 ferry from Boulogne but arrive just as the boat is leaving.
How much time will they save if they catch the next available ferry from Calais?
(Assume that it will take more than half an hour, but less than 2 hours, to drive from Boulogne to Calais.)

f The 1215 ferry from Calais to Dover leaves 20 minutes late.
Because of heavy seas, the crossing takes 45 minutes longer than usual.
At what time does the ferry arrive in Dover?

15 Household Costs

15.1 Mortgages

Loans based on value of property

A **mortgage** is a loan from a building society, bank or other financial institution to buy a house.

The size of the mortgage obtained depends on:

- the value of the property to be purchased,
- the earnings of the prospective purchaser(s).

The loan is normally up to 80% of the value of the property (which is usually less than the purchase price), but can be as much as 100%.

The remainder of the purchase price must be paid before the buyer can occupy the house. It is called the **deposit**.

Example

This semi-detached, four-bedroom house is sold for £160 000. The building society surveyor values the property at £150 000. The building society agrees to a mortgage of 90%.

Calculate the amount of the loan and the deposit.

Loan = 90% of the valuation

$$= \frac{90}{100} \times £150\,000$$

$$= £135\,000$$

Deposit = Purchase price − Loan

$$= £160\,000 − £135\,000$$

$$= £25\,000$$

Exercise 15A

A bank offers mortgages of 95% to buyers of properties valued at £70 000 or less, 90% for properties not exceeding £120 000 in value and 80% if the value is more than £120 000.
For each of the following houses, calculate the maximum loan available from this bank and the deposit required:

1 a semi-detached house sold for £78 500, valued at £75 000

2 a four-bedroom family house sold for £110 000, valued at £106 000

3 a three-bedroom bungalow sold for £61 000, valued at £62 500

4 a five-bedroom house sold for £147 000, valued at £142 000

5 a town house sold for £260 000, valued at £200 000.

Loans based on income

A bank will usually give a loan of up to three times the borrower's gross basic annual income. For joint borrowers, the maximum loan is *either* three times the larger income plus the smaller income *or* 2.25 times the joint income.

---**Example**---

A couple require a mortgage to buy a house. Their basic incomes are £12 500 and £14 000. What is the maximum loan they can obtain from this bank?

Arrangement 1.   Loan $= 3 \times £14\,000 + £12\,500$

$= £54\,500$

Arrangement 2.   Loan $= 2.25 \times (£14\,000 + £12\,500)$

$= £59\,625$

The maximum loan is therefore £59 625.

Exercise 15B

1   Find the maximum loan obtained by someone whose basic income is £17 000 per annum.

2   A man earns £14 000 per annum and his wife earns £21 000 per annum.
What is the maximum loan they can obtain?

3   Two friends decide to become joint owners of a property. Their basic annual incomes are £15 000 and £10 000.
What is the maximum amount of money they can borrow?

4   A couple require a mortgage from the bank. If the smaller basic income is £10 000, determine (by drawing a graph, or otherwise) the values of the larger income for which it is better to take 2.25 times the joint incomes.

5   A couple's basic annual incomes are £22 000 and £15 000. They wish to purchase a £97 000 house which has been valued at £92 000.
By considering the three different arrangements they could make with the bank, find the largest possible loan they could obtain.

Mortgage repayments

Once the mortgage is arranged and the property has been bought the mortgage, plus interest, must then be repaid over a number of years. The payments are made monthly and will depend on:

- the size of the loan
- the number of years over which the loan is to be repaid
- the current rate of interest.

There are many different types of mortgage available, the two most common being a **repayment** and an **endowment** mortgage.

†**A repayment mortgage** is the original type of mortgage whereby the repayments are made over a set number of years. A single payment is made each month, part of which is interest on the loan and part repayment of the capital.

This is the most flexible type of mortgage, but as time passes the repayment of interest becomes smaller and the repayment of capital is then a larger part of the monthly premium. As tax relief is given on the interest, this means that the monthly premiums increase over the period of the loan.

It is also necessary to take out a mortgage protection policy.

With an **endowment mortgage** two payments are made each month. One is to the lender to repay the interest on the loan and the other is to an insurance company. When the insurance policy matures the loan is repaid. The insurance policy also includes full life cover.

The monthly repayments vary during the period of the loan according to the current interest rate. Initially the premiums for an endowment mortgage are higher than for a repayment mortgage.

Tax relief is allowed on the interest paid on the loan. For mortgages of £30 000 or less this tax relief is deducted at source, i.e. it is deducted from the monthly premium. This scheme is called MIRAS (Mortgage Interest Relief At Source).

Which mortgage you choose depends on long-term inflation and your personal circumstances. It is difficult to obtain impartial advice from specialists because many of them are paid by commission or tend to deal with one particular insurance company.

The following table gives the monthly repayments, after tax relief, for a loan of £1000.

No. of years	10	15	20	25	30
Monthly repayment	£12.07	£10.14	£9.26	£8.84	£8.64

(In real life the monthly repayments may vary depending on the current interest rate.)

Example

A loan of £25 200 is made on a house over 25 years. Find:

a the monthly repayment,

b the total amount paid.

a
$$\text{Monthly repayment} = \frac{25\,200}{1000} \times £8.84$$
$$= 25.2 \times £8.84$$
$$= £222.77$$

b Total repayments in 1 year $= 12 \times £222.77 = £2673.24$

Total repayments in 25 years $= 25 \times £2673.24 = £66\,831$

Exercise 15C

1 Using the repayment table on page 110 find, for each of the following loans:

  (i) the monthly repayment
  (ii) the total amount repaid.

  a £15 000 over 20 years

  b £12 000 over 30 years

  c £29 500 over 25 years

  d £30 000 over 15 years

2 Mr and Mrs Yardley pay a deposit of 20% on a house costing £34 000. They obtain a 25-year mortgage on the remainder.

  a What is their monthly premium?

  b How much will they have paid for the house at the end of 25 years?

3 A young couple purchase a flat for £31 000 which was valued at £29 500. They obtain a 95% mortgage over 30 years. Calculate:

  a the amount of the loan

  b the monthly repayments

  c the total repayments

  d the amount of the deposit

  e the total cost of the flat.

4 The following table shows the monthly repayments for a loan of £1000 before tax relief has been deducted:

No. of years	15	20	25	30
Monthly repayment	£12.91	£12.04	£11.66	£11.45

For the following loans, calculate:

  (i) the monthly premium
  (ii) the total amount repaid.

  a £45 000 over 25 years

  b £36 000 over 20 years

  c £32 500 over 15 years

  d £51 000 over 30 years.

5 Anna and Colin are buying a house for £120 000. They pay a deposit of 40% and take out a mortgage to cover the remainder of the cost. The Lee Towers Building Society informs them that the monthly repayment is £7.06 per £1000 borrowed per month.
What is their monthly repayment?

15.2 The cost of renting

Many people rent accommodation rather than buy their own property. For some this is a necessity because they cannot afford the high cost of a mortgage; others choose to rent because there are fewer responsibilities, and sometimes because it is more economical if a job requires frequent moves to new areas. The cost of renting often includes a charge for gas, electricity and water rates. Repairs and redecoration are usually the responsibility of the landlord.

Exercise 15D

1 Mr and Mrs Evans rent a furnished flat at a cost of £120 per week.
How much is this per year?

2 Daniel rents furnished accommodation at £370 per month.
What is the annual cost of renting?

3 Four students share a flat. The monthly rent is £576.
How much per week should each student contribute to cover the cost of the rent?

4   Anwen is a student looking for accommodation for two years. She sees the following advertisements in a local newspaper:

Which arrangement will cost her the least amount?

5   a   The Jacksons live in a council house and pay £470 per month in rent. How much do they pay per year?

b   They have the opportunity to buy their council house for £61 000. They can obtain a 90% mortgage and the repayment rate is £8.80 per month per £1000.
  (i)   How much would they need for the deposit?
  (ii) How much would they pay per year in mortgage repayments?

c   Assuming they have sufficient money saved for the deposit, should the Jacksons buy their council house? Give reasons for your decision.

6   a   What advantages are there in renting accommodation?

b   What advantages are there in buying a house?

*15.3   The Council Tax

In 1991 the Government announced its intention to replace the Community Charge with the Council Tax. The Council Tax was introduced in 1993. Each household was placed in one of eight bands according to the value of the property.

Households in each of the bands will be charged a certain proportion of that levied on those in band D.

The bands (in England) and the amount charged are given below.

Band	Value of property	Amount charged (as a proportion of band D)
A	up to £40 000	$\frac{2}{3}$
B	£40 001–£52 000	$\frac{7}{9}$
C	£52 001–£68 000	$\frac{8}{9}$
D	£68 001–£88 000	1
E	£88 001–£120 000	$\frac{11}{9}$
F	£120 001–£160 000	$\frac{13}{9}$
G	£160 001–£320 000	$\frac{15}{9}$
H	Over £320 000	2

There is a 25% reduction for a household consisting of only one adult; an adult is anyone aged 18 or over.

Owners of unoccupied properties are charged 50% of the normal rate. When calculating the number of occupants in a household certain categories of people are disregarded. These include students, student nurses, YTS trainees, and apprentices.

Council Tax payments can be made in one lump sum or in 10 equal monthly instalments.

┌─ *Example* ───

Mrs Metcalf lives with her student son in her house valued at £135 000. The Council Tax charged in her town is £361 in band D.

What Council Tax does Mrs Metcalf pay?

Her house is in band F.
The Council Tax for band F is £361 $\times \frac{13}{9}$ = £521.44
The student son is disregarded for Council Tax purposes.
∴　Mrs Metcalf receives a 25% discount for only one adult in the household.
the discount is £521.44 $\times \frac{25}{100}$ = £130.36
She pays £521.44 − £130.36 = £391.08

Exercise 15E

1　Find the Council Tax paid by the households identified below:

　a　Mr and Mrs Wolski who live in a house valued at £95 000 where the band D rate is levied at £450

　b　Mr and Mrs Yaqub who have two children over 18 years of age
　　Their house is valued at £249 000 and the Council Tax is levied at £643 for houses in band D

　c　Mr and Mrs Armstrong who live with their 7-year-old child in a house valued at £340 000. The band D rate is £630.

2　Find the monthly Council Tax paid by John who lives on his own in a flat valued at £37 000. The band D rate is £495.

3　Find the monthly Council Tax paid by Pam who lives with her two children in a bungalow valued at £67 000. The children are Jane who is a 19-year-old student and Angela who is a 23-year-old student nurse. The band D rate is £450.

15.4　Household bills

Apart from the Council Tax, substantial bills which a householder can expect to receive each year are:

- Water rates
- Electricity bill
- Gas bill
- Telephone bill

The electricity, gas and telephone bills arrive quarterly; the water rates bill arrives annually.

Water rates

Water rates are still levied by some local councils at so much in the pound of rateable value plus a standing charge.

The rateable value is an amount assigned by the local district valuer to all properties in the council's area.

The charges for Winchester in 1996–97 are shown below:

Service	Annual standing charge	Charge per £ of rateable value	Minimum charge
Water supply	£22	36.3p	£38.00
Wastewater	£33.40	48.8p	£38.00

Example

The average customer in this area has a rateable value of £210 and receives all water services.

Calculate:

a the average annual water rates bill

b the approximate daily cost for water.

a Total standing charge $\qquad\qquad = $ £55.40
Water supply $\qquad = 36.3p \times 210 = $ £76.23
Wastewater (sewerage) $\quad = 48.8p \times 210 = $ £102.48

Total annual charge $\qquad\qquad = $ £234.11

b Approximate daily cost $\quad = \dfrac{£234.11}{365} = 64.1p$

Exercise 15F

1 Calculate the yearly water rates for a bungalow with rateable value £425.

2 The occupier of a house with rateable value £234 pays the water rates in two half-yearly instalments.
How much is each payment?

3 A cottage has a septic tank for sewerage, which is emptied by private contract.
Calculate the occupier's yearly water rates if the rateable value is £194.
How much will the owner pay monthly if she opts to pay in 10 instalments?

4 Two years ago, the standing charge for water services was £22. The owner of the cottage (see question 3) paid £93.20 for water.
How much (to the nearest 0.1p) was the charge per pound of rateable value for water supply?

The following information refers to questions 5, 6 and 7.

Many people now pay for water according to the amount they actually use. The amount of water used is measured by a water meter. Typical charges are shown below:

Service	Annual standing charge	Pence per cubic metre
Water supply	£24.00	55.1p
Wastewater	£28.00	80.3p

5 Calculate the total water bill, including all services, for the following amounts of water used:

  a 50 cubic metres   **c** 31.6 cubic metres

  b 10.9 cubic metres   **d** 9000 gallons

  (Assume 1000 gallons = 4.5 m³.)

6 The owner of the bungalow (see question 1) estimates that the quarterly water consumption is about 38 cubic metres.
How much would the water bill have been if the water supply had been metered?

7 The owner of the cottage (see question 3) estimates the annual water consumption to be 52 cubic metres.
Should she have a water meter installed?

Electricity bills

The amount of electricity or gas which a household uses is recorded by a meter.

There are two types of meter in use.

The digital meter is a line of figures:

The clock meter has dials which are read from left to right. Each dial is numbered from 0 to 9 and the hands move clockwise and anticlockwise alternately.

When a hand is between numbers the lower number is recorded. (When a hand is between 9 and 0, record 9.)

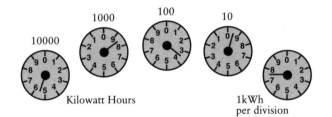

To calculate an electricity bill the price per unit is multiplied by the number of units used and a standing charge is added. VAT is charged at 8% on electricity.

Example 1

What is the reading on the set of dials above?

The reading is 58397.

Example 2

In February, the reading on Mr Spencer's meter is 28279 and the previous reading, in November, was 26814.

The cost of electricity is 5.44p per unit and the quarterly standing charge is £6.27.

Calculate:

a the number of units of electricity used

b the cost of electricity used

c the total electricity bill inclusive of 8% VAT.

a Number of units used $= 28\,279 - 26\,814 = 1465$

b Cost of electricity $= 1465 \times 5.44p = £79.70$

c Total bill (excluding VAT) $= £79.70 + £6.27 = £85.97$

VAT at 8% $= £6.88$

Total bill (including VAT) $= £92.85$

---

1   What is the reading on the following sets of dials?

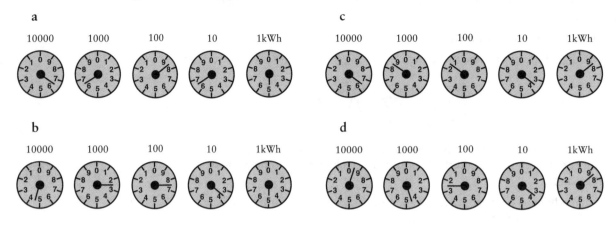

a

| 10000 | 1000 | 100 | 10 | 1kWh |

c

| 10000 | 1000 | 100 | 10 | 1kWh |

b

| 10000 | 1000 | 100 | 10 | 1kWh |

d

| 10000 | 1000 | 100 | 10 | 1kWh |

2   Calculate the total electricity bills, including VAT at 8%, for the following amounts:

	Present reading	Previous reading	Cost per unit	Standing charge
a	37162	35841	5.90p	£9.70
b	86719	85017	6.83p	£9.90
c	26341	24863	5.70p	£7.67
d	70905	69438	6.04p	£8.45
e	42186	41902	7.63p	£9.20

In questions **3** and **4**, assume all prices given are inclusive of VAT.

3   Mr Sinclair receives the following electricity bill based on an estimated reading:

Previous reading	Present reading	Tariff	Units	Price per unit	Amount £	p
77945	78616E	DOMESTIC	671	6.04p	40	53
QUARTERLY CHARGE					8	45
		TOTAL THIS ACCOUNT			48	98

The true reading was 78262.

a By how much has Mr Sinclair been overcharged?

b Draw a set of dials and put in arrows to show the true meter reading.

4   Mrs Yamoto receives an electricity bill for £59.08.
The standing charge is £7.67 and the cost per unit is 5.70p.

a How many units of electricity were used?

b If the present reading is 66112, what was the previous reading?

Gas bills

Gas bills are calculated in a similar way to electricity bills, but the number of units used is the number of cubic feet of gas used.

This number has then to be converted into therms and multiplied by the price per therm.

A standing charge is added.

Typically 105 therms = 100 cubic feet of gas.

Example

In one quarter Mr and Mrs Blake used 260 hundred cubic feet of gas. The cost of the gas was 36p per therm and the standing charge was £8.98 per quarter.

Calculate the amount paid for gas.

Amount of gas used   = 260 hundred cubic feet of gas

$$= \frac{260}{100} \times 105 \text{ terms}$$

$$= 273 \text{ therms}$$

Cost of gas   = 36p × 273

      = £98.28

Standing charge   = £8.98

Amount paid   = £107.26

Exercise 15H

1   In the following questions, the cost of gas is 39.8p per therm and the standing charge is £8.70 per quarter.

Calculate the quarterly gas bills for the following people:

a   Mr Hudson, who used 401 therms

b   Mrs Snell, who used 275 therms

c   Miss Dawe, who used 392 therms

d   Mr Ghosh, who used 456 therms.

2   Mrs Kirk used 370 hundred cubic feet of gas in one quarter. How many therms did she use?

3   Mr McBain used 95 therms of gas at a cost of 34.2p per therm. The standing charge was £9.00 per quarter.
How much was his gas bill?

4   Miss Hennigan's gas meter reads 3469 on 10 December and 3617 on 10 March. The charge per therm is 43p and the quarterly standing charge is £9.40.

Calculate:

a   the number of units of gas consumed

b   the number of therms of gas consumed

c   the cost of the gas

d   the total gas bill for the quarter.

5   a   The reading on Stephen Bond's gas meter at the beginning of the quarter was 5628.

At the end of the quarter the reading is as shown below:

(i)   What reading is shown on the meter?
(ii) How many units (i.e. hundreds of cubic feet of gas) have been used during the quarter?

b   The cost of gas is 39.8p per therm and there is a standing charge of £8.70 per quarter.

(i)   How many therms have been consumed?
(ii)   What is the cost of the gas consumed?
(iii) What is the total gas bill?

Telephone bills

Telephone bills are sent three or four times per year. You are charged for the use of the system, the hire of the telephone, and the calls which have been made.

Some people buy their own telephone, and their bills will not include the charge for the hire of the telephone.

VAT is added to telephone bills as a service charge.

Exercise 151

1   Mrs Kingsley has her own telephone. She pays the system charge of £21.86 and the cost of 415 units at 4.20p per unit. Calculate her total bill (excluding VAT).

2   A meter reading on 1 March was 27258, and on 31 May it was 27623. The cost per unit was 4.20p and the standing charges totalled £21.86. Calculate the total bill (excluding VAT).

3   a   Mr Stears' telephone bill is made up of the following charges:

- quarterly rental, £21.86
- 196 metered units @4.20p per unit
- operator calls, £1.68

Calculate the total bill (excluding VAT).

b   VAT at 17.5% is added to the bill. Calculate the final bill which Mr Stears receives.

4   Complete the following telephone bill:

Quarterly rate		£
SYSTEM		21.86
APPARATUS		3.10
TOTAL		a

Date	Meter reading	Units used
13 MAR	024367	
12 JUN	024536	b

UNITS AT 4.20p	c
TOTAL (EXCLUSIVE OF VAT)	d
VAT AT 17.5%	e
TOTAL PAYABLE	f

16 Insurance

Is insurance really necessary?

In the UK:

- every 10 minutes a house catches fire
- every 2 minutes a road accident occurs
- every 40 seconds a car is stolen
- every 30 seconds a house is burgled

so you may begin to understand why insurance *is* necessary.

These misfortunes don't always happen to someone else, and so some form of insurance cover is necessary for practically everyone.

Most people will need insurance for:

- *Their home*. House buildings insurance will cover the house against possible damage from fire, flood, storm or subsidence.
- *Their possessions*. House contents insurance will cover such items as furniture, carpets, clothes, books, jewellery against loss or damage.
- *Their life*. Life insurance will protect dependents from financial hardship in the event of the death of the insured.
- *Their car or motorbike*. Car or motorbike insurance is required by law.
- *Their holiday*. Holiday insurance is strongly recommended when booking any holiday.

16.1   House buildings and contents insurance

A buildings policy covers the following against damage: the fabric of your home (i.e. the foundations, roof, walls, floors, doors and windows), the fixtures (kitchen units, fitted wardrobes, light fittings, etc.), and any outbuildings.

The contents can be insured against loss or damage on either a new-for-old basis or at their true value, i.e. the new price less a deduction for wear and tear.

A new-for-old policy is a sensible choice for most people as the full costs of repairs are met by the insurance company. For example, if you burn a hole in a two-year-old carpet which might have been expected to last for ten years, a new-for-old policy will give you the money to buy a new carpet. A basic policy will give you only eight-tenths of the new price.

The cost of contents insurance depends on the type of policy (new-for-old costs more) and the area in which the house is situated.

Many insurance companies put each area of the country into a low, high or top risk category depending on the likelihood of a housebreaking or burglary occurring. The premium payable is usually decided by the post code and will be quoted as a fixed rate per £1000 (or £100) of cover.

The policy might cost extra if:

- there are lots of easily stolen, valuable items
- there have been claims made in the past.

There are optional extras which will increase the cover provided by your policy and increase the premium you pay. For example:

- An 'all-risks' extension will cover loss or damage to jewellery and such items as spectacles, contact lenses, sports equipment, cameras, which may be lost or damaged outside the home.
- Freezer contents cover will insure you in case of a power cut or breakdown.
- Bicycles can be covered for loss or damage anywhere in the UK, provided that, if a bike is stolen, it was securely locked at the time.

Policies should be index-linked so that the amount of cover increases with the rate of inflation. There is then less danger of being underinsured.

Insurance based on risk

Below are examples of the prices offered by several companies. Buildings and contents can be insured with the same company or with two separate companies.

Company	Buildings (Cost per £1000 of cover)	House contents					
		Basic			New-for-old		
		Top risk	High risk	Low risk	Top risk	High risk	Low risk
Company A	2.10	10.00	4.45	2.50	14.00	6.00	3.00
Company B	1.90	7.00	4.00	3.00	8.50	4.50	3.50
Company C	2.10	7.50	4.40	2.40	10.00	5.20	3.00
Company D	1.80	8.20	2.50	2.50	10.00	3.00	3.00
Company E	2.04	7.40	3.33	2.70	9.96	4.44	3.48

Example

Mr and Mrs Perry own a house in a top-risk area. They wish to insure the house for £75 000 and the contents for £14 000 (on a new-for-old basis) with Company B. Calculate the annual premium payable.

$$\text{Premium for buildings} = \frac{75\,000}{1000} \times £1.90 = £142.50$$

$$\text{Premium for contents} = \frac{14\,000}{1000} \times £8.50 = £119.00$$

Total annual premium = £142.50 + £119.00 = £261.50

1 What is the yearly premium for £14 000 of basic cover for house contents for someone living in a high-risk area and insuring with Company B?

2 A couple insure their house with Company C for £40 000.
What is their yearly premium?

3 a How much does it cost per year to insure contents (basic cover) for £15 000 with Company C, in a low-risk area?

   b How much extra would a couple living in a top-risk area pay if the contents were insured for £15 000 and were covered by a new-for-old policy?

4 Miss Morgan lives in a top-risk area. She decides to insure the house with Company D and the contents with Company B. The house is to be covered for £95 000 and the contents (on a new-for-old basis) for £24 000.
How much will she pay for insurance:

   a yearly   b monthly?

5 Mr and Mrs Samra live in a top-risk area. They take out new-for-old contents insurance with Company E for £18 000.

   a Calculate their annual premium.

   b They decide to pay the premium monthly. What is their monthly premium?

Insurance zones

Many insurance companies divide the country into zones, which also depend on the post code. The premium payable is then decided by the zone number.

Below is a table used by one insurance company.
(All rates are per £100 sum insured.)

BUILDINGS			ALL ZONES 13.5p			

CONTENTS	ZONE 1	ZONE 2	ZONE 3	ZONE 4	ZONE 5	ZONE 6	ZONE 7
	25.5p	28.5p	32.5p	40.5p	52.5p	65.0p	81.0p

Rates can be reduced by the following discounts:
   $7\frac{1}{2}$% if you agree to pay the first £50 of any claim, or
$17\frac{1}{2}$% if you agree to pay the first £100 of any claim.

Example

Mr and Mrs Perkins live in Sunderland (Zone 4) and wish to insure their house for £45 000 and its contents for £11 500. They agree to pay the first £50 of any claim.
Calculate their annual premium.

Premium for buildings $= \dfrac{45\,000}{100} \times 13.5\text{p} = $ £60.75

Premium for contents $= \dfrac{11\,500}{100} \times 40.5\text{p} = $ £46.58

Total premium $=$ £60.75 + £46.58 = £107.33

$7\frac{1}{2}$% discount $=$ £107.33 $\times \dfrac{7\frac{1}{2}}{100} = $ £8.05

Total premium payable = £107.33 − £8.05 = £99.28

Exercise 16B

1 Ms Robinson rents a house in Oxford (Zone 4) and she wishes to insure the contents for £16 700.

Calculate the annual premium payable.

2 Miss Lewis owns a house in Penzance (Zone 1). She decides to insure the house for £75 500 and the contents for £18 500. In addition she agrees to pay the first £50 of any claim.

Calculate the annual premium payable.

3 Mr Jackson insures his house in Kingston (Zone 3) for £94 800 and its contents for £18 400. He agrees to pay the first £100 of any claim on the contents insurance.

Calculate his monthly premium to the nearest penny.

4 Mrs Burgess moves from Walsall (Zone 3) to Glasgow (Zone 5). She insured her old house for £72 000 and its contents for £18 000. She intends to insure her new house and its contents for the same amounts.

Calculate the increase in her annual premium and express this as a percentage of the original premium.

5 Mr and Mrs Carr live near London (Zone 6). Their house is insured for £95 000 and the contents for £25 000. What insurance premium do they pay?

The following year they increase the contents insurance by £1000, as they have just bought a word processor, and the insurance company increases its rates by 8%.
By how much will their premium increase?

6 Noreen insures her house for £91 000 and its contents for £18 000. The annual premiums for the insurance are

Buildings 21p per £100 of cover
Contents 98p per £100 of cover.

Noreen calculates the total annual cost of the premiums to be £3675.
Use estimation to check Noreen's calculation to see if she is correct.
You must show all your working.   (SEG S96)

16.2   Car insurance

The law requires that every car driver is insured against damage to another person's property or vehicle or injury to any person.

Your basic premium for car insurance will depend, not only on the **insurance company** you choose, but also on:

- the type of cover you take out
- the make and model of car you own
  Insurance companies classify cars into groups which depend on the original cost, the performance, the cost of repairs and spare parts. The lower the group number the cheaper the insurance, which reflects the insurers' belief that these cars are less likely to be involved in accidents and/or are cheaper to repair.
- the area in which you live
  Accidents are more likely to occur in areas where traffic is heavy. Some areas have a higher rate of theft or vandalism.
- your age, sex and driving experience
  There is usually an extra premium to pay for drivers under 25 years of age as this age group has a higher accident rate.
  Women drivers are considered a better insurance risk than men drivers and some insurance companies will give a discount to women drivers or will use a separate table of premiums.
  Convicted drink-drivers will have to pay a much larger premium.
- whether you use the car for business or pleasure.

The type of cover can be:

Third party is the minimum insurance allowed by law. It covers injury or damage to other people and their property caused by your car. It is the cheapest form of insurance.

Third party, fire and theft also covers damage to your own car from fire, explosion, lightning, theft or attempted theft and loss of your car by theft. It is more expensive than third party insurance.

Comprehensive insurance is the most expensive type of car insurance. This also covers you for damage to your own car in an accident, whether or not it was your fault, and often provides additional cover for loss of contents.

Table I shows, for a sample of cars, the groups to which cars are assigned.

Table I

Group 3	Citroen AX, Fiat Punto, Nissan Micra, Mini, Renault Clio, Vauxhall Corsa, VW Polo.
Group 4	Nissan Almera, Peugeot 106, Peugeot 306, Renault Megane, Rover 214
Group 5	Ford Fiesta, Ford Escort, Lada Samara
Group 6	Citroen ZX, Peugeot 205, Vauxhall Astra, VW Golf
Group 7	Ford Mondeo, Nissan Primera, Renault Laguna
Group 8	Citroen Xantia, Toyota Corolla, Vauxhall Vectra
Group 9	Honda Civic, Mazda 323, Toyota Carina
Group 10	BMW 1.8, Ford Maverick, Honda Accord, Rover 416, Vauxhall Frontera
Group 11	Ford Galaxy, Mercedes 1.8, Rover 620, Rover 820, VW Sharon
Group 12	Audi A4, Ford Scorpio, MGF, Vauxhall Omega.

Table II shows the areas, for insurance purposes, to which some of the counties and cities of mainland Britain are assigned.

Table II

Area 1	Berkshire, Central Scotland, Cornwall, Dorset, Gwent, Lincolnshire, Norfolk, Powys
Area 2	Avon, Cumbria, Dyfed, Hampshire, Gloucestershire, Northumberland, Shropshire, Wiltshire, Yorkshire
Area 3	Durham, Lancashire, Strathclyde, Tyne and Wear
Area 4	Buckinghamshire, Merseyside, Oxfordshire, Tayside
Area 5	Birmingham, Outer Glasgow, Manchester
Area 6	Inner Glasgow, Liverpool, Greater London
Area 7	Inner London

Table III shows a typical basic comprehensive premium payable, depending on age, make of car and place of residence. The same table is used for men and women drivers.

Table III

Car group	Age or above	Area 1	Area 2	Area 3	Area 4	Area 5	Area 6	Area 7
3	20	536	574	614	657	702	752	804
	25	367	395	427	461	498	537	580
	30	337	364	393	424	458	495	535
	35	306	330	357	385	416	450	486
4	20	611	654	700	749	800	857	917
	25	419	457	487	525	568	612	661
	30	384	415	448	483	522	564	610
	35	349	376	407	439	474	513	554
5	20	730	785	840	899	960	1028	1100
	25	502	556	584	630	682	734	793
	30	461	498	538	580	626	677	732
	35	419	451	488	527	569	616	665
6	20	869	931	996	1066	1138	1219	1305
	25	595	659	693	747	809	870	940
	30	547	591	638	688	742	803	868
	35	497	535	579	625	672	731	789
7	20	1030	1104	1181	1264	1350	1446	1548
	25	706	782	821	886	959	1032	1115
	30	649	701	757	816	881	952	1030
	35	589	635	686	741	798	866	935
8	20	1222	1310	1410	1500	1601	1715	1836
	25	837	928	974	1051	1138	1224	1323
	30	769	831	898	968	1044	1129	1221
	35	649	755	814	879	946	1028	1109
9	20	1450	1553	1662	1779	1899	2033	2177
	25	993	1099	1155	1246	1349	1452	1569
	30	912	986	1064	1147	1239	1339	1448
	35	829	892	966	1043	1122	1219	1316
10	20	1719	1842	1971	2110	2253	2412	2582
	25	1177	1304	1370	1478	1600	1722	1861
	30	1082	1169	1262	1361	1469	1589	1718
	35	983	1059	1145	1237	1381	1495	1560

Discounts

You can reduce the cost of your insurance by obtaining discounts on the basic premium.

Most insurance companies will give discounts for the following:

- **limiting the number of drivers** to yourself or self and spouse.
- **a voluntary excess**, that is agreeing to pay the first part of any claim.
- **no-claims bonus.** If you have not made a claim on your policy over a period of time the insurers will award you a 'no-claims' discount. The longer the claim-free period, the larger the discount, usually up to a maximum of 60%.

---Example---

Megan Jones is 27 and the proud owner of a Peugeot 306. She lives in Gwent and has been driving for 5 years without having an accident. She agrees to limit the driving to herself and for this she is allowed a 15% discount on the basic premium.
Calculate her comprehensive car insurance premium.

From Table I: a Peugeot 306 is in Group 4
From Table II: Gwent is in Area 1
From Table III: a 27-year-old, with a car in Group 4, living in
                      Area 1 pays a basic premium of £419

Discount for limiting to one driver = 15% of £419 $= 0.15 \times £419$
                                                 $= £62.85$
Premium now $= £419 - £62.85 = £356.15$

As Miss Jones has made no claim on her insurance for more than 4 years, she is entitled to the maximum no claims bonus of 60%.
(The no-claims bonus is nearly always deducted last.)

No claims bonus = 60% of £356.15 $= 0.6 \times 356.15 = £213.69$
Comprehensive premium = £356.15 − £213.69 $= £142.46$

(*Note*. **Discounts are made consecutively. Do *not* add the percentages.**)

Exercise 16C

1   Use Tables I, II and III to find the basic comprehensive premium for:

   a the 25-year-old owner of a Ford Fiesta living in Hampshire.

   b the 22-year-old owner of a Mini living in Birmingham.

   c the 40-year-old owner of a BMW 1.8 living in Wiltshire.

   d the 34-year-old owner of a Vauxhall Vectra living in Greater London.

2   **a** Because the owner of the Mini (see question **1b**) limits driving to himself and his wife, he is allowed a 10% discount on the basic premium.
      What is his premium now?

   b He agrees to pay the first £50 of any claim for damage and is allowed a further 10% discount.
      Calculate the new premium.

   c Because he has driven for three years without making a claim he receives a no claims bonus according to the scale:

      • 30% after 1 claim-free year
      • 40% after 2 claim-free years
      • 50% after 3 claim-free years
      • 60% after 4 claim-free years.

      How much does he now pay?

3   The owner of the Ford Fiesta (see question **1a**) decides to limit the driving to herself, for which she is allowed a 15% discount. She has six claim-free years of driving.
   Calculate her insurance premium.

4 a Mr Flynn is 21 and has only been driving for one year. He lives in Manchester and drives a Peugeot 106.
What is his basic premium?

   b He agrees to a voluntary excess of £100, for which he receives a discount of 10%.
What is his premium now?

   c He has made no claims since he began driving. Find his actual car insurance premium.

In questions 5 and 6, calculate the net insurance premium.

5

Policy holder's name and address	Car details:	
S W REID	Registration	**D174 BEF**
27 HILL ROAD	Type of Body:	**HATCHBACK**
GATESHEAD	Make and Model:	**VAUXHALL**
TYNE AND WEAR		**ASTRA**

Date of Birth: **16/4/61**   Cover: **Comprehensive**

Discounts:
Named Driver: **10%**
Voluntary excess: **15%**
No-claims Bonus: **50%**

6

Policy holder's name and address	Car details:	
M A ROBBINS	Registration	**G541 LJH**
BRIAR COTTAGE	Type of Body:	**4 DOOR SALOON**
OVERTON	Make and Model:	**FORD**
HANTS		**MONDEO**

Date of Birth: **27/7/46**   Cover: **Comprehensive**

Discounts:
Named Driver: **–**
Voluntary excess: **10%**
No-claims Bonus: **60%**

17 Further Percentages

17.1 Inverse percentages

Inverse percentages were used in Unit 9 to find the original price of an item when the price inclusive of VAT was known. In this unit the same method is used in a more general context.

Example

Andy bought a guitar which he later sold to a friend for £57.40, making a loss of 18% on the amount he paid.

How much did the guitar originally cost?

The **original price** is always represented by **100%**

Andy's selling price is represented by $100\% - 18\% = 82\%$

So 82% represents the selling price of   £57.40

1% represents $\dfrac{£57.40}{82}$

100% represents the cost price of $\dfrac{£57.40}{82} \times 100 = £70$

Therefore the guitar originally cost £70.

If preferred, the information can be represented on a table instead:

	%	£
Cost price	100	?
Loss	18	?
Selling price	82	57.40

$$\text{Cost price} = \frac{£57.40}{82} \times 100$$

$$= £70$$

If required, the amount of money lost on the deal could be found by a similar method:

$$\text{Loss} = \frac{£57.40}{82} \times 18$$

$$= £12.60$$

Exercise 17A

1 In the following questions you are given the percentage profit and the selling price. For each case find the original cost price.

a 20%, £30   d 110%, £466.20

b 28%, £69.12   e 28.7%, £57.92

c $12\frac{1}{2}$%, £1662

2 In the following questions you are given the percentage loss and the selling price. For each case, find the original cost price.

a 10%, £54   d 28%, £864

b $7\frac{1}{2}$%, £31.45   e 8.25%, £77.99

c 5%, 96p

3 Doris is a pensioner and does not pay income tax. Her bank account earns interest of £540 from which tax at a rate of 25% has been deducted.
How much tax can she reclaim?

4 Mario's wage is increased by 9.2%. He now earns £235 per week.
How much did he earn previously?

5 After one year Mrs Brennan's car is valued for insurance at £5600, a depreciation on the price when new of 22%.
How much, to the nearest £10, did Mrs Brennan pay for the car?

6 A garden centre buys plants which it resells at a profit of 28%.
How much was the original price of a rose bush which is sold for £3.40?

7 A car is sold for £5225 after a depreciation of 45% of the original purchase price.
Calculate the original purchase price of the car.
(SEG Spec 98)

8 In a sale, everything is reduced by 15% of the marked price. Kerry was given a reduction of £7.49 on a cassette player. What was its marked price?
(SEG W93)

*17.2 Compound percentages

---Example---

A wholesaler adds 20% profit to the price of goods when he sells to a retailer. The retailer then adds 15% profit to the same goods before selling to the customer. What percentage above the original price does the customer pay?

The answer is NOT 35%!

The original price is equivalent to 100%
The wholesale price is equivalent to 100 + 20% of 100 = 120%
The retail price is equivalent to 120 + 15% of 120 = 138%

The customer pays 38% above the original price.

Exercise 17B

1 The wholesale price of a certain item is 25% more than the manufacturing cost. The retail price is 20% more than the wholesale price. What is the percentage difference between the retail price and the manufacturing cost?

2 Mr Miles invests a sum of money for 2 years at an interest rate of 10% per annum.

By what percentage has his original sum of money increased after two years?

3 The menswear department is having a sale. The normal selling price of a particular style of shirt includes a profit of 30%. During the sale the normal price is reduced by 15%.
What percentage profit does the shop make on these shirts during the sale?

4 Last year when Mrs Berry had her car serviced, 80% of the cost of servicing was for labour and 20% was for parts. In one year the labour costs increased by 9% and the cost of parts by 12%. What is the percentage increase in Mrs Berry's bill compared to the previous year?

5 The manufacturing cost of a child's toy is made up of 50% labour, 30% materials and 20% overheads.
The factory improves the machinery which increases the overheads by 80%, but decreases the labour costs by 25%. At the same time the cost of materials increases by 10%.
What is the overall percentage increase in the manufacturing cost?

6 A motor trader predicts that his sales will increase in 1997 and buys 45% more vehicles. His predictions were incorrect and he is unable to sell 28% of the total vehicles bought. Calculate the percentage increase in sales for 1997. (SEG Spec 98)

17.3 Depreciation

Most possessions, such as cars, caravans, electrical equipment, depreciate in value as time passes.

For example, at the end of each year the value of a car will be less than its value at the beginning of the year.

Depreciation is calculated by a similar method to compound interest.

Example

A car was bought for £6500 in 1990. During the first year of ownership its value depreciated by 20% and during each subsequent year by 15%.

Calculate the value of the car three years later.

	£
Cost of car	6500
Depreciation of 20% (of £6500)	1300
Value after 1 year	5200
Depreciation of 15% (of £5200)	780
Value after 2 years	4420
Depreciation of 15% (of £4420)	663
Value after 3 years	3757

The value of the car after 3 years is £3757.

Exercise 17C

1   A small business buys a computer costing £4500. The rate of depreciation is 20% per annum. What is its value after three years?

2   A motorbike is bought second hand for £795. Its price depreciates by 11% per year. For how much could it be sold two years later? (Give your answer to the nearest pound.)

3   Mr and Mrs Parsons' carpet is accidentally damaged by fire. It was bought only three years ago for £450. The insurance investigator (loss-adjuster) decides that it will have depreciated in value by 10% each year. What was the value of the carpet just before the accident?

4   Mr Carey invests £3000 in unit trusts in the hope that they will appreciate in value. Unfortunately, although they appreciate in value by 8% during the first year, they depreciate in value during the second year by 8% and during the following two years depreciate by 10% of the value at the beginning of each year.

How much are the unit trusts worth after four years?

18 Trial and Improvement

Although Trial and Improvement is in the SEG Money Management and Number syllabus, the work is based upon algebraic techniques. Students should study Section 30.1 before working through this topic.

Unless an equation is simple, it is rare to be able to solve it exactly.
The solution of an equation can be found by trial and improvement methods, or by graphical methods.
The trial and improvement method is shown in the worked examples below.

Example

By trial and improvement, solve the equation $x^3 = 30$ to two decimal places.

If $x = 1$ then $x^3 = 1$
  $x = 2$ then $x^3 = 8$
  $x = 3$ then $x^3 = 27$
  $x = 4$ then $x^3 = 64$.

$\therefore$ We have found two consecutive integers, 3 and 4, one giving x^3 below 30, and one giving x^3 above 30.

$\therefore$ There is a solution of $x^3 = 30$ between $x = 3$ and $x = 4$.

To find a closer approximation to its value, repeat this trial and improvement method for values between 3 and 4.

If $x = 3.1$, then $x^3 = 29.791$
  $x = 3.2$, then $x^3 = 32.768$

$\therefore$ There is a solution between $x = 3.1$ and $x = 3.2$.

Repeating this produces:

If $x = 3.10$ then $x^3 = 29.791$
  $x = 3.11$ then $x^3 = 30.080$

$\therefore$ The solution is between $x = 3.10$ and $x = 3.11$.

If $x = 3.105$ then $x^3 = 29.935$.

$\therefore$ The solution is above 3.105.

Hence the solution of $x^3 = 30$ is 3.11 (to 2 decimal places).

Exercise 18A

By trial and improvement, solve the equations below to the number of decimal places requested.

1 $x^2 = 17$ to 1 d.p.

2 $x^2 = 27$ to 1 d.p.

3 $x^3 = 34$ to 1 d.p.

4 $x^2 = 22$ to 2 d.p.

5 $x^3 = 37$ to 2 d.p.

6 $x^3 = 56$ to 1 d.p.

7 $x^2 = 19$ to 2 d.p.

8 $x^3 = 31$ to 3 d.p.

9 $x^2 = 29$ to 3 d.p.

19 Proportion or Variation

Although Proportion is in the SEG Money Management and Number syllabus, the work is based upon algebraic techniques. Students should study Unit 30 before working through this topic.

When one quantity is directly related to another we say that the first quantity is **proportional to** or **varies as** the second quantity.

For example, the amount spent on pints of beer is proportional to the number of pints bought; if you buy twice as many pints as the next person, you will spend twice as much. If you travel at a constant speed, then the distance you travel is proportional to, or varies as, the time you spend travelling.

19.1 Linear proportion

The algebraic expression $\qquad y \propto x$

means $\qquad y$ is directly proportional to x

or $\qquad y$ varies directly as x

This can be rewritten as $y = kx$ where k is a constant amount and k is called the **constant of proportion**.

Example

If y varies directly as x and $y = 2$ when $x = 3$, find y when $x = 9$.

i.e.
$$y \propto x$$
$$y = kx$$

When $y = 2$ and $x = 3$: $\qquad 2 = k \times 3$

$\therefore \qquad k = \dfrac{2}{3}$

When $x = 9$: $\qquad y = \dfrac{2}{3} \times 9$

$$y = 6$$

Note. You use the values given (in this case $y = 2$ when $x = 3$) to find k. Then you use this value of k to find the unknown quantity.

Exercise 19A

1   Copy and complete the table so that $y \propto x$.

x	2	4	8	—
y	20	40	—	95

What is the equation connecting x and y?

2   y varies directly as x and $y = 21$ when $x = 7$. Find y when $x = 3$.

3   $y \propto x$ and when $y = 24$, $x = 6$.
Find:

a y when $x = 1$

b x when $y = 12$.

4 s varies directly as t and $s = 35$ when $t = 7$.

Find:

a s when $t = 3$

b t when $s = 49$.

5 P is directly proportional to Q.
   $P = 15$ when $Q = 60$.

Find P when Q is 84.

Find Q when P is 12.5.

6 The distance (s) travelled by a car is directly proportional to the time (t) it takes to travel this distance. The car travels 180 miles in 4 hours.

a Find the equation connecting s and t.

b How many miles has the car travelled after 3 hours?

7 The amount by which a spring is stretched is its extension, (E). This is proportional to the force (F) required to stretch it, so that $E \propto F$.
   A spring is stretched by a distance of 24 mm by a force of 2 newtons.

Find:

a how much the spring is stretched by a force of 5.5 newtons

b the force needed to stretch the spring by 60 mm.

*19.2 Proportion (not linear)

Example 1

The area, $A\,\text{m}^2$ of a logo varies as the square of the height of the sign, h m. When the height is 6 m, the area is 48 m^2.
Find:

a the area when the height is 4 m

b the height when the area is 32 m^2.

$$A \propto h^2$$
$$A = kh^2$$

When

$\therefore$
$$h = 6, \quad A = 48$$
$$48 = k \times 6^2$$
$$\frac{48}{36} = k$$
$$k = \frac{4}{3}$$

a When
$$h = 4, \quad A = \frac{4}{3} \times 4^2$$
$$= \frac{64}{3}$$

$\therefore$
$$\text{Area} = 21\tfrac{1}{3}\,\text{m}^2$$

b When
$$A = 32, \quad 32 = \frac{4}{3}h^2$$
$$h^2 = 32 \times \frac{3}{4}$$
$$h^2 = 24$$
$$h = \sqrt{24}$$
$$h = 4.899$$

$\therefore$ height is 4.90 m.

Example 2

The pressure of a gas, p N/m^2, varies inversely as its volume, v m^3.

When the pressure is 250 N/m^2 its volume is 0.2 m^3.
By first forming an equation connecting p and v, find

a the pressure when the volume is 0.5 m^3, and

b the volume when the pressure is 400 N/m^2.

p varies inversely as its volume means that p varies as $\dfrac{1}{v}$

i.e. $$p \propto \frac{1}{v}$$

$\therefore$ $$p = k \times \frac{1}{v} \quad \text{or} \quad \frac{k}{v}$$

When $$p = 250, \quad v = 0.2$$

$\therefore$ $$250 = \frac{k}{0.2}$$

$$k = 0.2 \times 250$$
$$= 50$$

$\therefore$ $$p = \frac{50}{v}$$

a When $$v = 0.5$$

$$p = \frac{50}{0.5}$$

$$p = 100$$

$\therefore$ Pressure is 100 N/m^2

b When $$p = 400$$

$$400 = \frac{50}{v}$$

$$v = \frac{50}{400}$$

$$v = 0.125$$

$\therefore$ Volume is 0.125 m^2

Exercise 19B

1 $P \propto Q^2$ and $P = 12$ and $Q = 4$.
Find P when $Q = 12$.

2 y varies as the square of x and $y = 48$ when $x = 4$.

  a Show that $y = 3x^2$.

  b Find y:
   (i) when x is 2.
   (ii) when x is $\frac{1}{2}$.

3 The cost of buying a rectangular carpet varies as the square of its longer side. A carpet with a longer side of 3 m costs £180.

Find the cost of a carpet with a longer side of 4 m.

4 The braking distance, d feet, of a car is proportional to the square of its speed, v mph.
When the speed is 30 mph, the braking distance is 50 feet.
Find

   a the braking distance when the car is travelling at 70 mph, and

   b the speed when the braking distance is 220 yards.

5 The time taken to travel between two cities is inversely proportional to the speed. When a plane is travelling at 500 mph, it takes 0.3 hours.
Find

   a the time taken when travelling at 340 mph

   b the speed of Concorde which takes 5 minutes.

6 The gravitational force on a body, f newtons, varies inversely as the square of its distance, r miles, from the centre of the earth. When a body is 400 miles from the centre, i.e. on the Earth's surface, f is 320.
Find

   a the force when the body is 40 000 miles above the Earth

   b the distance from the centre of the Earth when the force is 2 newtons.

7 John estimates that the value of a car, £v, is inversely proportional to its age, y years.
After 2 years the car has a value of £7000.
By first forming an equation involving v and y, find

   a the value of the car after 6 years

   b the age of the car when it is worth £3000.
   (SEG Spec 98)

20 Statistical Terms

In 1834, the Royal Statistical Society was founded, and defined statistics as 'using figures and tabular exhibitions to illustrate the conditions and prospects of society'. Statistics is now used to deal with the collection, classification, tabulation and analysis of information and opinions.

Data. Data is the information which you have obtained (or have been given). The word 'data' is a plural, really, and the singular is 'datum' (a single piece of information). These days, though, nearly everyone says 'The data *is*...' rather than 'The data *are*...', so we shall do the same.

Variables. Something which can change from one item to the next is a *variable*. A variable can either be **quantitative** (i.e. numerical like the number of people on a bus), or **qualitative** (i.e. non-numerical, like colour).

There are two types of quantitative variables:

 (i) *Continuous*. A continuous variable is a variable which could take all possible values within a given range, e.g. the height of a tree.

 (ii) *Discrete*. A discrete variable is a variable which increases in steps (often whole numbers), e.g. the number of rooms in a building.

   A discrete variable does not have to consist only of whole numbers. For example, the size of shoes is also a discrete variable, and the sizes go up in steps of halves (5, $5\frac{1}{2}$, 6, $6\frac{1}{2}$, etc.)

In real life, you can measure continuous variables to a certain degree of accuracy. For example, the height of a tower can be measured to the nearest metre. Although the measurements recorded produce discrete data, the height is still a continuous variable.

Population. The term '*population*' means everything (or everybody) in the category you are considering. For example, if you were studying the length of life of light bulbs from a factory, the population would be all such light bulbs.

Exercise 20A

For each population below, state whether the variable given is qualitative or quantitative. If it is quantitative, state whether it is discrete or continuous.

1 the number of people in a room

2 the height of a person

3 the age of a person

4 the number of people on a train

5 the colour of the shoes of people on a train

6 the speed of a car

7 the number of dresses sold in a shop

8 the number of cars in a car park

9 the number of children in a family

10 the newspaper which a person reads on a Sunday

11 the weight of a person

12 types of shrub sold in a garden centre.

STATISTICS AND PROBABILITY

21 Sampling, Surveys, Questionnaires

One of the problems with statistical surveys involving people is, that whatever your opinion, there are likely to be many other people with the same opinion. If you ask only these people, your opinion will be seen to be that of the whole population. If you ask only people with the opposite opinion, you will be seen to be in a minority. Therefore, you must ask a variety of people, so that you have a true picture of the population.

Remember, however, that in statistics, the term **population** does not necessarily refer to people. If you wished to survey the ages of cars on the road, your population might be all the cars in Britain.

21.1 Surveys

When you record any information – for example, about other people's opinions or numbers of surviving African elephants or types of road accidents – you are carrying out a **survey**. The survey results may be obtained by asking questions, by observation or by research.

To obtain completely accurate information, you would have to ask *everybody* (in your town or country or whatever), and receive answers from everybody, or observe *all* the elephants in Africa.

21.2 Censuses

When information is gathered about all the members of a population, the survey is called a **census**.

A national census is carried out every ten years. The last one was in 1991. Every adult in Britain is asked a large number of questions on mainly factual matters, for example the number of rooms in their house, their age, and the number of cars they possess.

A national census is a very large undertaking, and the results, though accurate, take a substantial length of time to be produced. Apart from the vast number of people to be asked, and the placing of their answers in computers, it is very difficult to ensure that every adult has in fact replied. It costs the country a great deal of money to complete a national census.

21.3 Samples

It is usually impossible for firms, newspapers, biologists, medical researchers, etc., to obtain information about the whole population, because the survey:

- may be expensive
- may take a long time
- may involve testing to destruction – e.g. if you wish to find out how long batteries last, you test them until they run out
- may be impossible to carry out for every member of the population – e.g. a survey to find the weights of trout in Scottish rivers.

A small part of the population is chosen for the survey and this is called a **sample**.

The statistician then assumes that the results for the sample are representative of the population as a whole. The larger the number of people asked, the more likely their response is to be a valid result for the whole population.

Clearly it is vital that for the survey to be accurate the sample you choose must be representative of the whole population.

To achieve this, every member of the population must have an equal chance of being chosen.

*21.4   Sampling methods

Random sampling

A random sample is one in which every member has an equal chance of being selected.

Campaign groups for or against a particular issue (such as the possible siting of a new supermarket near a park) can often obtain a large majority for their point of view simply by selecting which passers-by to question (perhaps the people living near the park who will be worried about the possibility of noise). By careful selection, majorities as high as 70% can easily be obtained both for and against the same issue! (Some people may well want a supermarket behind their back garden.)

The simplest way to obtain a random sample is to give every member a number, and to select numbers from tickets in a box (as in a raffle), or (if there are too many for this method), select numbers by computer. (Random numbers can also be obtained by using the RAN button on some calculators.)

It is common to use the electoral roll of a suitably sized area (on which every adult is listed) to obtain a numbered list from which to select a sample.

Periodic sampling

With periodic or systematic sampling, a regular pattern is used to pick the sample, for example, every hundredth firework on a production line. This can give an unrepresentative sample if there is a pattern to the list which is echoed by the sample.

Stratified random sampling

A stratified sample is more accurate than a random sample, and is used in opinion polls, when 1 or 2% accuracy is important. A stratified sample (or **strata sample**) is one in which the population is divided into categories. The sample should then be constructed to have the same categories in the same proportions.

Random sampling is then used to select the required numbers in each category.

For example, if you wished to find out about the earnings of students in a sixth-form college, it would be sensible to have both lower sixth and upper sixth students represented. You may also wish to make sure that one-year students, males and females, are fairly represented. Suppose there are 1000 students in college, of whom 220 are lower sixth one-year students, 420 are lower sixth two-year students and 360 are upper sixth students.

A sample of 50 would contain the following numbers:

$$\text{LVI one-year students} = \frac{220}{1000} \times 50 = 11$$

$$\text{LVI two-year students} = \frac{420}{1000} \times 50 = 21$$

$$\text{UVI students} \qquad = \frac{360}{1000} \times 50 = 18$$

The eleven LVI one-year students would be randomly chosen from the 220 students in college. The other two strata would be chosen in the same way.

Quota sampling

For a quota sample, a manufacturer may determine the proportions of each group to interview.

For example, if a manufacturer wishes to launch a new chocolate bar on the market, it may be more important to canvass the opinions of children and those who do the shopping than any other sector of the market.

A market researcher paid to survey a sample of 100 people could be instructed to ask, say, 20 people under the age of 18, 30 in the age range 19–40 who do the family shopping, 10 in the same age range who don't, 30 in the age range over 40 who do the family shopping, and 10 in this age range who don't. The researcher will probably use convenience sampling (see below) to choose who to ask, but once one of the quotas is filled, no more people in that category may be asked. The researcher will continue to ask people in the other categories until the sample of 100 has been surveyed.

This is a common method used for market research, but inexperienced (or lazy!) researchers may choose an unrepresentative sample.

Convenience sampling

The most convenient sample is chosen, which, for a sample of size fifty, usually means the first fifty people you meet. There is obviously no guarantee that this sample will be representative. In fact it is highly likely that it won't be.

21.5   Bias

The results of a survey are biased if the sample is not representative of the whole population. Bias can be introduced if:

- the sample is unrepresentative. Even when using random sampling an unusual sample may be chosen, and this is just bad luck.
- an incorrect sampling method is used. Sampling methods, other than random, or stratified random sampling, are very likely to produce biased samples.

If you wanted to know people's views on drinking, a survey held outside a public house at closing time would clearly produce a different response from one held outside the office of the 'Teetotallers' League'! Neither would be representative of the complete population. Both of these samples would be biased.

Bias can also be introduced if questions asked in the survey are not clear or are leading questions (see Section 21.6).

Exercise 21A

In questions 1–8:

a identify the population

b criticise the method of obtaining the sample

c recommend an alternative way of obtaining a sample.

1   A journalist at a local newspaper wants to canvass popular opinion about plans for a new shopping centre in town. He goes into the High Street, and asks people, until he has asked 50.

2   Stephanie wishes to find out the earnings of college students. She goes into a college common room, and asks 40 girls.

3   The police wish to ascertain how many cars have a valid tax disc. One day, they set up a survey point on a road out of a town, between 5pm and 6pm. They stop a car, and check its tax disc. As soon as it has left, they stop the next car.

4   A geography student needs to collect five soil samples from his garden for a project. He stands in the middle, and throws a coin in the air. Where it lands, he takes a sample.

5   For a survey into the smoking habits of teenagers, Carol went to a tobacconist's near a school at 3.30pm, which was when the school day ended. She asked everyone entering the shop how much they spent on cigarettes in a week.

6   To find out how many homes in a telephone area have central heating, a salesgirl telephones 100 people, picked at random from a telephone directory.

7   To find out the make of car that people in an area of town use, Peter went out after lunch and knocked on doors until he had one hundred responses. He was pleased with his efficiency, as he had finished by 4pm.

8   To investigate what influenced people in their decision on mode of transport to work, John went to the station just before the 8.15 train departed, and asked as many people as he could.

9   The table shows the number of people working in different sections of a paint manufacturing company.

Work Section	Number of men	Number of women
Manufacturing workshop	500	100
Storage and Distribution	200	100
Purchasing	30	50
Marketing and management	10	10

The owner wants to question 100 employees on how to improve production.
He proposes to allocate each person a number and then select 100 numbers at random.

a State **one** disadvantage of such a selection process.

It was suggested that a Stratified Sample would be more representative of the workers in the company.

b (i) Calculate the number of people in the Stratified Sample of 200 who would represent the purchasing section.
   (ii) How many women should be included within the people chosen from the purchasing section?   (SEG W94)

21.6 Questionnaires

If your questions are written down and given to people to complete, the list of questions is called a **questionnaire**. The questions you ask must be chosen with care. They must:

(i) **not give offence.** Some people do not wish to give their precise age, or social class, so you *either*: (a) find an alternative question, (e.g. 'Which of these age ranges applies to you?'), *or* (b) fill in the information by using your own judgement.

(ii) **not be leading.** 'What do you think of the superb new facilities at . . .' will *lead* most people to agree they are better than the old facilities. People do not usually want to contradict the questioner. However, the point of the survey should not be to obtain agreement with your view, but to obtain other people's opinions.

(iii) **be able to be answered quickly.** The person answering the questions will often have only a small amount of time to spare and will not want you to write long sentences on their point of view. To obtain information easily from the survey it is helpful to have Yes/No answers or 'boxes' for the answers which are ticked. Here is an example:

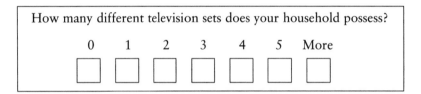

A questionnaire must also be easy for anyone to understand. The questions themselves must also be designed carefully. The question 'How much do you watch TV?' could result in the following types of response:

'A lot', 'Not much'
'Every night', 'twice a week'
'For two hours a night'
'Whenever there's sport, a film, . . .'

A better question is 'How many hours do you spend watching TV?', but this may encourage wild guesses because of poor memory.

An even better question to ask is 'How many hours did you watch TV *last* night?'. You can then offer a range of possible answers such as:

'Not at all'

'Up to $\frac{1}{2}$ hour'

'$\frac{1}{2}$ to 1 hour'

'1 to 2 hours'

and so on.

If you suspect different times are spent on different days, it is up to you, as a statistician, to ask a few people each day over a period of a week.

All surveys are open to error. The larger the sample, the more accurate the result.

21.7 Pilot surveys

It is common for companies to carry out an initial survey on a small area of the country in order to identify potential problems with the questions and to identify typical responses. This limits the errors in expensive large-scale surveys.

Exercise 21B

Criticise the questions asked in this exercise and suggest questions which should be asked to find the information required.

1 What do you think of the improved checkout facilities?

2 Do you agree that BBC2 programmes are the best on TV?

3 What is your date of birth?

4 Sheepskin coats are made from sheep. Do you wear a sheepskin coat?

5 Dolphins are wild animals. Do you enjoy watching dolphins perform?

6 Sunbathing causes skin cancer. Do you sunbathe?

7 Vitamin D is obtained from sunlight. Do you sunbathe?

8 Is the new decor a major improvement on the old?

9 Would you rather use your local shops than a major supermarket miles away?

10 Ioshi works for a company that employs about 100 people at each of 10 factories. She is designing a questionnaire to survey the employees on the number of working days lost through stress related illness.

a One of the questions that she thought of is as follows.

How many days did you take off last year?

Less than 10	**10 to 20**	**Over 20**

Suggest **two** improvements to this question.

b Describe a method that could be used to select a representative sample for the survey.
(SEG S95)

11 A bank wishes to carry out a survey.

It intends to ask its customers the following questions. In each case the appropriate response is to be circled.

To help the bank improve the questionnaire show how you would change **either** the responses **or** the wording of the question **or** both.

a How often do you come into the bank?

Very often   Often   Occasionally

b What is your age?

Under 18   18 to 25   25 to 30   Over 30

c What do you think of the new improved facilities?

Very good   Good   Average   Poor
(SEG Sp95)

21.8 Hypothesis testing

A statistical survey should have a purpose. It may be used to find out people's opinions (e.g. an opinion poll) or to discover what the population requires of a new product (consumer research), but often it is used to test a theory. A statement is made about a population, or populations, which can be tested statistically. This statement is called a **hypothesis**.

For example:

1 'Women live longer than men.'
2 'You can't tell margarine from butter'.
3 'A new leisure centre would benefit the town.'
4 'The most popular colour of car is red.'

are four statements which are **hypotheses** (plural of 'hypothesis'). Each of these hypotheses can be investigated statistically.

To test the truth of the hypothesis someone must devise an appropriate method to collect data. This must then be analysed before a conclusion can be made, based on the results of the analysis.

The hypothesis can be tested by carrying out a survey or experiment, by observation, or by using published data (which is the result of someone else's survey).

For the four examples above:

1 To test the hypothesis that women live longer than men, government statistics published over several years could be used. Averages and measures of spread could be calculated (see Units 25 and 27) and compared.

2 An experiment could test whether or not it was possible to tell the difference in taste between margarine and butter, perhaps by blindfolding people and seeing whether they can identify which is which.

3 A method for deciding if a new leisure centre is needed could be to devise a suitable questionnaire and survey a sample of the town population.

   Among other things, the questionnaire would need to find out what facilities people required for sports, how their needs were being met at the present time, and whether they would consider using a new local leisure centre.

4 One method of testing the hypothesis that 'the most popular colour of car is red' would be to carry out a survey of cars on a busy stretch of road and record results taken at different times on different days on a survey sheet.

You will understand more about analysing data when you have worked through the Units 25 (Averages), 26 (Cumulative Frequency), and 27 (Dispersion).

Exercise 21C

1 State an appropriate method which could be used to test the following hypotheses:

   a If it rains on St Swithin's day, it will rain for the next forty days and forty nights.

   b There is insufficient parking space provided at the college for students and staff.

   c Television does not (or does) cater for the viewing needs of teenagers.

   d Consumers prefer ....
      (choose any product or set of products which interests you).

2 Devise an experiment to test the hypothesis 'Teenagers have quicker reactions than adults over 30.'

3 Explain carefully how you would test the germination rate of seeds (or the growth rate of plants) under different conditions.

22 Classification and Tabulation of Data

Data can be classified in many different ways; the most common are quantitative and qualitative (as defined in Unit 20).

22.1 Tabulation

The purpose of tabulation is to arrange information, after collection and classification, into a compact space so that it can be read easily and quickly. It then may be represented pictorially to enable relevant facts to be seen readily, as explained in the next unit.

Tabulation consists of entering the data found in columns or rows.

Example

The numbers of students living in certain villages were:

Village	Number of students
Ashurst	31
Botleigh	15
Crow	28
Downton	24
Eaglecliffe	19
Fillingdales	33
Total	150

The data can be made more detailed by subdividing the rows and/or columns to give more precise information.

An example would be:

Village	Number of students	
	Male	Female
Ashurst	15	16
Botleigh	10	5
		etc.

It is important that the tables produced are neat, all rows and columns are clearly identified, and that units (where appropriate) are given.

22.2 Tally charts

It is common to record the data by means of a **tally chart**. Suppose you were noting the speeds of cars at a particular point. You would draw up a list of possible speeds and as each car went by, you would record a |. To enable you to total your results quickly, you would mark every fifth | in a row diagonally to produce a block of five ⊥⊢⊦.

A section of your results for this exercise would look like this:

Speed of car (mph)	Tally	Total
61	⊞⊞ ⊞⊞ |||	13
62	⊞⊞ ⊞⊞		11
63						4
64	⊞⊞ ⊞⊞ ⊞⊞ |||	18

22.3 Frequency tables

The above table (with or without the tally) shows the raw data obtained. It is called a **frequency table**.

The grouping of the data to identify how many are in each category produces a **frequency distribution**.

It is often found useful to combine this data into a more compact form by grouping the particular values, as shown below:

Speed of cars (mph)	Frequency
0–39	0
40–49	7
50–59	54
60–69	75
70–79	25
80–89	2

In order to complete such a table, it is necessary to collect the raw data, fix the magnitude of each class interval, and group the data accordingly.

Exercise 22A

1   In a board game a die was thrown several times. Here is a sequence of the scores:

3, 4, 1, 5, 6, 1, 2, 3, 2, 4, 5, 4, 3, 1, 2, 5, 6, 3, 1, 4, 2, 5, 6, 4, 5, 1, 6, 5.

Use a tally chart to obtain a frequency table for this data.

2   The number of people on the 87 bus was counted each time it passed the City Centre. Here is the data:

11, 25, 60, 16, 23, 2, 44, 26, 49, 58, 29,
 8, 14, 24, 7, 16, 47, 5, 30, 34, 9, 12,
33, 10, 55, 21, 56, 32, 19, 6, 1, 21, 21,
42, 9, 35, 25, 55, 37, 46, 32, 14, 59.

With intervals 0–10, 11–20, 21–30, 31–40, 41–50, 51–60, use a tally chart to obtain the frequency distribution.

3   The numbers of seats *not* occupied on 80 transatlantic flights in one day were:

34, 8, 9, 6, 12, 30, 9, 11, 5, 39, 6, 25,
26, 42, 33, 16, 13, 30, 5, 29, 43, 34, 11, 26,
 2, 39, 35, 19, 20, 40, 15, 11, 20, 34, 31, 17,
23, 2, 17, 15, 32, 3, 44, 6, 1, 7, 26, 35,
18, 25, 37, 4, 39, 37, 34, 26, 33, 7, 21, 16,
18, 15, 29, 35, 21, 6, 40, 39, 13, 12, 4, 4,
38, 39, 12, 0, 4, 33, 34, 18.

Summarise the information into class intervals 0–4, 5–9, 10–14, 15–19, 20–24, 25–29, 30–34, 35–39, 40–44.

4 A specialist in 'vowel research' counted the number of times each vowel was used on the page of a book. This is what she found:

A ⅢⅢ ⅢⅢ ⅢⅢ Ⅰ   O ⅢⅢ ⅢⅢ ⅢⅢ ⅠⅠ
E ⅢⅢ ⅢⅢ ⅢⅢ ⅢⅢ ⅠⅠ   U ⅠⅠⅠ
I ⅢⅢ ⅢⅢ ⅠⅠ

Then she realised that she'd missed the last paragraph on the page. Here it is:

'Look out, Danny! There's some broken floorboards here! Come back!'

Continue the tally, and hence obtain the frequencies of the use of each vowel.

5 Serge collected the following data giving the heights of male students in metres.

1.82	1.64	1.71	1.86	1.57	1.67
1.73	1.76	1.84	1.52	1.79	1.65
1.80	1.67	1.71	1.64	1.58	1.81
1.67	1.74	1.69	1.56	1.63	1.74
1.68	1.83	1.69	1.58	1.64	1.73

An observation sheet has been started using equal class intervals for the heights.

a Complete the first column.

Height (m)	Tally	Frequency
1.50 and less than 1.55		
1.55 and less than 1.60		
⋮		

b Enter Serge's data onto the observation sheet.   (SEG S94)

23 Pictorial Representation of Data

The presentation of data in the form of tables has been considered in Unit 22. However, most people find that the presentation of data is more effective, and easier to understand, if the data is presented in pictorial or diagrammatic form.

The pictorial presentation used must enable the data to be more effectively displayed and more easily understood. The diagrams must be fully labelled, clear and should not be capable of visual misrepresentation (see Section 24.3). Types of pictorial representation in common use are the pictogram, bar chart, frequency polygon, and histogram.

Statistical packages may also be used to present data in a variety of ways. You cannot, however, rely completely on a computer to produce your pie charts, pictographs, histograms, etc. In an examination you must also be able to carry out the necessary calculations yourself and draw the most appropriate diagrams for the given data.

23.1 Pictograms

In a **pictogram** data is represented by the repeated use of a pictorial symbol. The example below shows how a pictogram works.

Example

A survey of 1000 people living in Freeton was taken, to see what colour of cars they owned.

Represent this data in the form of a pictogram.
The results of the survey were:

Colour of cars	Number of cars
Red	60
White	200
Blue	100
Grey	50
Gold	80
Black	30

Here is one possibility. A full car symbol represents 20 cars; half a car represents 10 cars. It is not possible to show small fractions of a symbol accurately, and the detail required should not normally be to more than half of a symbol (but certain symbols may allow for a quarter).

Colour of cars   Key: 🚗 = 20 cars

Red   🚗 🚗 🚗

White 🚗 🚗 🚗 🚗 🚗 🚗 🚗 🚗 🚗 🚗

Blue   🚗 🚗 🚗 🚗 🚗

Grey   🚗 🚗 🚗

Gold   🚗 🚗 🚗 🚗

Black 🚗 🚗

Exercise 23A

For each question illustrate the data given by means of a pictogram.

1 The numbers of bottles of champagne sold in five villages in 1991 were:

Abbotshurst	60	Tobbenham	80
East Lynne	120	Westering	100
Marlinsby	50		

2 The makes of a number of cars passing a junction were:

Ford	35	Citroën	20
Rover	30	Renault	10
BMW	5	Vauxhall	25

3 The number of flights for each airline out of Gatwick in a one-hour period was:

British Airways	8	Aer Lingus	2
Swissair	1	Britannia	6
Virgin Atlantic	1	Monarch	3

4 The workforce of a factory was asked by which mode of transport they came to work. The results were:

Car	35	Motorcycle	15
Train	25	Walk	20
Bus	10		

5 The contents of a fruit bowl were:

Apples	7	Bananas	3
Pears	5	Peaches	7
Kiwi fruit	6	Oranges	2

6 Students in a department of a college were asked about the type of accommodation in which they lived. The data was:

Flat	25	Semi-detached house	40
Maisonnette	5	Detached house	30

7 Forty children were asked their favourite type of bread.
The answers were:

White (sliced)	16	Brown	14
White (unsliced)	6	French stick	4

23.2 Bar charts

A **bar chart** is a diagram consisting of columns (i.e. bars), the heights of which indicate the frequencies. Bar charts may be used to display discrete or qualitative data.

Example

Fifty households were surveyed, and the number of children in each family was recorded as follows:

Children in family	Frequency
0	8
1	11
2	17
3	8
4	5
5	1

Represent this data by means of a bar chart.

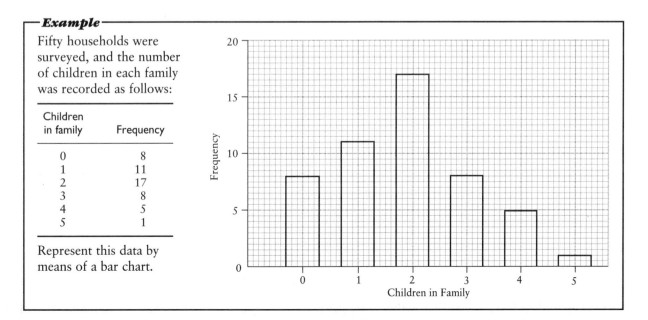

Dual bar charts

Dual bar charts are used when two different sets of information are given on connected topics.

Example

The number of people over 17 years old, and the number of people holding driving licences in a particular street were found over a period of years. These are shown on the right.

Year	1986	1987	1988	1989	1990	1991
No. of people over 17	32	27	29	31	33	39
No. of people with driving licence	12	17	19	11	24	28

Represent this data by means of a dual bar chart.

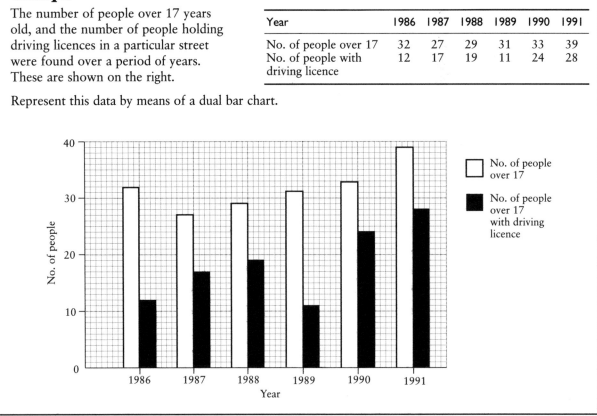

Sectional bar charts

Sectional bar charts, or **component bar charts**, are used when two, or more, different sets of information are given on the same topics. They are particularly useful when the *total* of the two or more bars is also of interest.

Example

The numbers of saloons and hatchbacks sold by a garage were recorded:

Month	Jan	Feb	Mar	Apr	May	Jun
Saloons	18	7	8	12	10	13
Hatchbacks	16	12	9	7	9	8

Represent this data by means of a sectional bar chart.

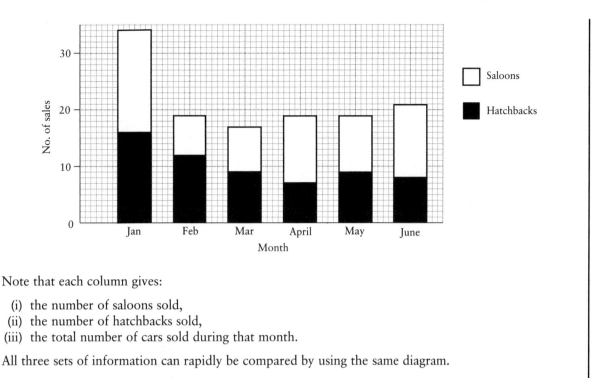

Note that each column gives:

(i) the number of saloons sold,
(ii) the number of hatchbacks sold,
(iii) the total number of cars sold during that month.

All three sets of information can rapidly be compared by using the same diagram.

Exercise 23B

1 The goals scored in 38 football league matches on Saturday 24 February 1990 were:

Number of goals in match	0	1	2	3	4	5
Number of matches	3	6	10	12	5	2

Illustrate this information by means of a bar chart.

2 200 people were asked their favourite television channel.
The results were:

BBC1	78
BBC2	24
ITV	75
Channel 4	23

Illustrate this information by means of a bar chart.

3 On a walk through a forest the following types of tree were seen:

Oak	80	Beech	35
Elm	6	Conifer	70
Chestnut	18	Cedar	21

Illustrate this data by means of a bar chart.

4 The holiday destinations of 100 people entering an airport were:

	France	Spain	Greece	Italy	Morocco	USA
Male passengers	3	18	9	8	5	9
Female passengers	4	11	15	2	6	10

Draw a sectional bar chart to illustrate this data.

5 Twenty people noted the television channel they were watching at 8.15pm on two successive nights.

The results were:

	First night	Second night
BBC 1	6	7
BBC 2	4	1
ITV	6	6
Channel 4	3	4
Satellite	1	2

Draw a suitable bar chart to illustrate this data.

6 The papers sold in one day were:

	Male buyer	Female buyer
Sun	8	4
Daily Mail	3	7
Daily Mirror	4	2
Daily Telegraph	3	1
The Times	2	3

Draw a sectional bar chart to illustrate this data.

7 The days with more than one hour of sun, and the number of days with rain, were recorded as:

	Sunny days	Rainy days
January	1	27
February	11	21
March	18	12
April	14	14
May	20	7
June	27	4
July	28	8

Draw a dual bar chart to illustrate this data.

23.3   Pie charts

A **pie chart** is another type of diagram for displaying information. It is particularly suitable if you want to illustrate how a population is divided up into different parts and what proportion of the whole each part represents. The bigger the proportion, the bigger the slice (or 'sector').

Example

The mode of transport of 90 students into college was found to be:

Walking	12
Cycling	8
Bus	26
Train	33
Car	11
Total	**90**

Represent this data by means of a pie chart.

A circle has 360°. Divide this by 90 to give 4°. This is then the angle of the pie chart that represents each individual person.

Since 12 people walk to college, they will be represented by $12 \times 4° = 48°$.

Similarly for the others:

	Angle in pie chart
Walking	$12 \times 4 = 48°$
Cycling	$8 \times 4 = 32°$
Bus	$26 \times 4 = 104°$
Train	$33 \times 4 = 132°$
Car	$11 \times 4 = 44°$
	Total = 360°

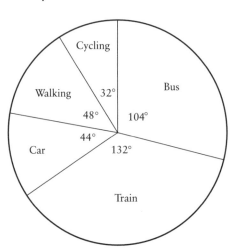

Method for calculating the angles on a pie chart

Here is a summary of how to work out the size of each bit of 'pie'.

(i) Add up the frequencies. This will give you the total population (call it *p*) to be represented by the pie.

(ii) Divide this number into 360.

(iii) Multiply each individual frequency by this result. This will give you the angle for each section of the pie chart.

Exercise 23C

By means of pie charts, illustrate the data given below:

1 The types of central heating used by households in a village were:

Solid fuel	14
Gas	105
Electricity	41
None	20

2 The holiday destinations of 60 people were:

France	Spain	Greece	Tunisia	USA	Caribbean	Portugal
21	15	6	3	8	5	2

3 240 students were asked what they were intending to do during the next year. The results were:

80 going to university
86 staying at college
64 going into employment
10 no firm intention

4 The numbers of bedrooms in 720 houses were recorded as:

1 bedroom 80
2 bedrooms 235
3 bedrooms 364
4 bedrooms 39
5 bedrooms 2

5 At a sports centre, the ages of 100 people were recorded as follows:

Under 20 years 30
20–29 years 15
30–39 years 12
40–59 years 14
60 years and over 29

6 Each pound spent at the Winchester Theatre Royal box-office is used to meet the theatre's expenses as follows:

Performance fees 60p
Salaries 17p
Premises and depreciation 10p
Administration 5p
Publicity 5p
Equipment 3p

23.4 Line graphs

A bar chart can be replaced by a line graph, provided that the quantity on the horizontal axis is a continuous variable, e.g. age, temperature or time.

In this case the data is plotted as a series of points which are joined by straight lines.

Line graphs associated with time are called **time-series graphs**.

They are used, for example, by geographers to illustrate monthly rainfall or yearly crop yield, etc., and by businesses to display information about profits or production over a period of time.

They show trends and have the advantage that they can be easily extended.

Example

The temperatures, recorded every six hours, of a patient in a hospital ward are given on the table:

Time (hours)	Mon 06	Mon 12	18	00	06	Tues 12	18	00	06	Wed 12
Temperature (°F)	99.0	99.12	99.12	99.2	99.2	98.88	98.68	98.6	98.6	

Represent this data by means of a line graph.

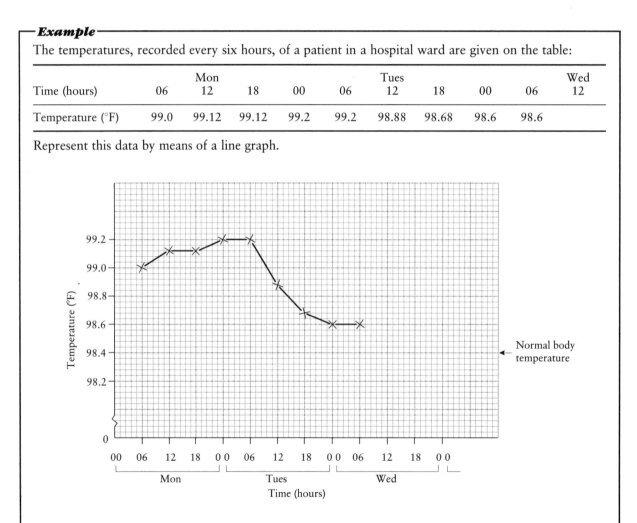

Small gaps are usually left near to the crosses to emphasise that the graph is a general trend line and that values between the points have no real meaning.

(In the given situation, the temperature will, in reality, have changed smoothly as time increased. Joining the points with a smooth curve would be more representative.

However, if air temperature had been recorded, there could have been several fluctuations of temperature between readings.)

Illustrate the data given using a line graph.

1 The rainfall during a period of six months was:

Month	Jan	Feb	Mar	Apr	May	Jun
Rainfall (mm)	75	192	86	89	25	19

2 The numbers of cars sold per month at a garage were:

Month	Jul	Aug	Sep	Oct	Nov	Dec
Number of cars	15	92	27	18	21	11

Comment on any trends you notice.

3 The numbers of passengers carried by an airline (in thousands) were as follows (Sp = spring, Su = summer, etc.):

1988				1989				1990			
Sp	Su	Au	Wi	Sp	Su	Au	Wi	Sp	Su	Au	Wi
21	48	31	17	22	49	29	18	23	41	24	25

Comment on any trends you notice.

4 The hours of sunshine per month on Bliss Island were:

Month	Mar	Apr	May	Jun	Jul	Aug	Sep	Oct
Hours of sunshine	145	195	241	304	310	307	261	175

5 The numbers of ferries per day sailing from Dover by a certain shipping company were:

Month	Jan	Feb	Mar	Apr	May	Jun
Number of ferries	22	23	26	27	31	35

6 The maximum temperatures for six successive months at Sunbourne were:

Month	Apr	May	Jun	Jul	Aug	Sep
Temperature (°C)	16	23	22	33	29	13

23.5 Histograms

In Section 23.2 we used bar charts to show the frequencies of qualitative data and also of quantitative data which had only a few discrete values (i.e. the number of children in a family).

When discrete data has a greater range of values, or when data is continuous, it is usually helpful to group the data into classes and show the frequencies of the classes.

When we do this with discrete data, we are actually treating it as if it were continuous. For example, the discrete examination marks out of 60 gained by 100 students may be grouped in tens as follows:

Marks	0–	10–	20–	30–	40–	50–60
Frequency	3	11	21	44	15	6

For discrete data, we should not write the groups as 0–10, 10–20, 20–30, 30–40, 40–50, and 50–60 as it is then not possible to know to which groups the numbers 10, 20, 30, 40 and 50 have been assigned. With continuous data, this does not cause a problem as the probability of obtaining a value of exactly 10 is zero (measurements will always be slightly above or below).

The diagram which shows frequencies of data grouped like this is called a **histogram**.

Since the data is continuous, or treated as such:

(i) the histogram must have a continuous horizontal scale;
(ii) each column will have, as its base, its class interval;
(iii) there will be no spaces between the columns, unless the frequency relating to a class interval is zero.

A histogram constructed from discrete data

> **Example**
>
> The students' examination marks were grouped into tens:
>
> | Marks | 0– | 10– | 20– | 30– | 40– | 50–60 |
> |---|---|---|---|---|---|---|
> | Frequency | 3 | 11 | 21 | 44 | 15 | 6 |
>
> The histogram is as shown below:
>
>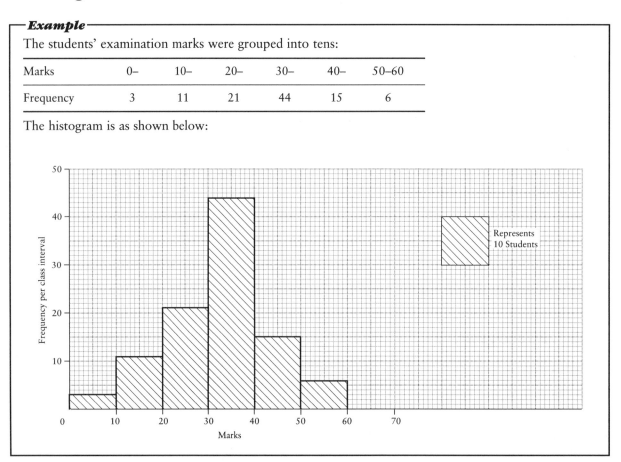

When the class intervals are all equal, the heights of the columns (being proportional to the areas) show the frequencies directly.

Because this is often called a bar chart or a frequency diagram, as well as the correct name, histogram, it is common to label the vertical axis simply 'frequency' when the class intervals are all equal. More precisely, the vertical axis should be marked 'frequency per class interval'.

A histogram constructed from continuous data

---Example---

The diameters of 140 apples in a box were measured and the results were recorded as follows:

Diameter (cm)	4–	5–	6–	7–	8–9
Frequency	12	20	56	40	12

This frequency table tells us that the apples have been put into groups or classes of 1 cm width.

In the first class there are 12 apples whose diameters might have any measurement from 4 cm up to, but not including, 5 cm.

If the apples were being measured on the millimetre scale and an apple was measured and found to have a diameter of exactly 5.0 cm, it would belong to the second class.

The histogram representing this data is as shown below.

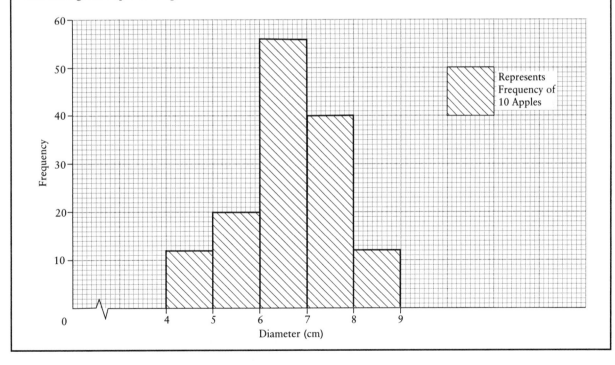

Class boundaries

If the class intervals are defined by rounded numbers as in 0–9, 10–19, 20–29, 30–39 etc. and the data is continuous, we need to be able to allocate values such as 9.3, 29.5 and 29.8 into the groups.

We define boundaries to the classes at the half-way points of 9.5, 19.5, 29.5, 39.5, etc. Values below the boundaries belong to the class below, and values at and above the boundaries belong to the class above.

The **upper class boundary** is the maximum possible value which would be in that class.

The **lower class boundary** is the minimum value which would be in that class.

For example, 9.3 belongs in the class 0–9 and 29.5 and 29.8 belong in the class 30–39.

The class boundaries will depend on the degree of rounding. For example:

  (i) The heights of trees may be measured to the nearest metre. The class
       4–7 m contains trees of height 3.5 m up to (but not including) 7.5 m
       and the class boundaries are 3.5 and 7.5 m.
 (ii) The heights of buildings may be measured more accurately, to the
       nearest 0.1 m. The class 4.0–7.0 m contains buildings of height 3.95 m up
       to (but not including) 7.05 m and has class boundaries 3.95 m and
       7.05 m.

Note. Age boundaries are different from those for other measures. The class of student ages 17–19 years has class boundaries of 17 and 20 years. This is because students aged 19 are 19 until their 20th birthdays.

Sometimes we are able to show the class boundaries clearly when we draw the columns of the histogram. If possible, we do so. With other data the scale does not allow such fine detail. In this case, it is understood that the difference between the class boundary and the rounded value in the 'frequency table' is too small to show on the graph paper.

A histogram constructed from continuous data showing class boundaries

───**Example**───

The frequency table for the heights of 108 conifers each measured to the nearest metre is as follows:

Height of conifer (m)	1–3	4–6	7–9	10–12	13–15
Frequency	12	24	27	30	15

Class boundaries are 0.5, 3.5, 6.5, 9.5, 12.5 and 15.5 (m). Each class width is 3 m, and this is the width of the base of each column in the histogram.

In this case, we are able to show the class boundaries clearly on the histogram.

*23.6 Histograms with bars of unequal widths

Histograms often have class intervals of different widths. In these cases the **area** of the bar or column represents the frequency, whereas in a bar chart the **height** of the bar represents frequency.

Thus: **in histograms with columns of unequal widths, the heights of the columns are found by dividing the frequency by the width of the class interval.**

The scale on the vertical axis is 'Frequency per unit class interval' or 'Frequency density' where

$$\text{Frequency density} = \frac{\text{Frequency of class interval}}{\text{Width of class interval}}$$

The data in the example on page 158 could be grouped as follows:

Height of conifer (m)	1–2	3–6	7–9	10–11	12–14	15
No. of conifers	6	30	27	28	12	5

Group	1–2	now has class interval	0.5–2.5,	class width = 2
	3–6	now has class interval	2.5–6.5,	class width = 4
	7–9	now has class interval	6.5–9.5,	class width = 3
	10–11	now has class interval	9.5–11.5,	class width = 2
	12–14	now has class interval	11.5–14.5,	class width = 3
	15	now has class interval	14.5–15.5,	class width = 1

The data to be plotted therefore is:

Height of conifer (m)	1–2	3–6	7–9	10–11	12–14	15
Frequency density	3	7.5	9	14	4	5

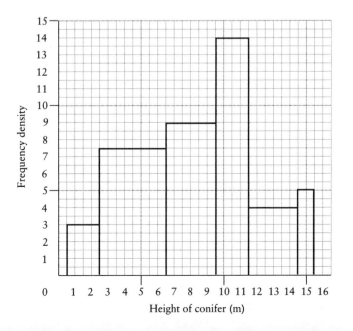

Exercise 23E

1   The speeds of 100 cars on a motorway were recorded. The data found was:

Speed (mph)	30–40	40–50	50–60	60–70	70–80	80–90
No. of cars	2	11	35	42	9	1

Represent this data by means of a histogram.

2   The heights of 80 students were recorded. The data was:

Height (cm)	150–160	160–170	170–180	180–190	190–200	200–210
No. of students	4	7	15	47	6	1

Represent this data by means of a histogram.

3   Below is a table showing the number of seats not booked on an airline's daily flight between London and Miami over 10 weeks.

19	1	8	11	15	19	21	17	1	23	19	11	12	15
21	11	8	4	15	27	21	20	14	18	7	11	23	21
8	17	1	19	12	16	21	25	28	29	17	15	11	8
16	8	2	8	6	10	11	15	9	8	6	2	3	8
21	18	27	32	37	4	11	19	21	34	21	15	12	11

By means of a tally chart, find the frequencies in the class intervals 0–4, 5–9, 10–14, ..., 30–34, 35–39. Represent this grouped data by means of a histogram.

*4   The number of accidents in High Town was recorded over two years, and the information was grouped in weekly periods.

Number of accidents	0–1	2–3	4–5	6–9	10–14	15–17	18–20
Number of weeks	1	7	12	36	28	19	1

Represent this data by means of a histogram.

*5   The IQs of 100 students were measured and the results were as follows:

IQ	100-109	110–119	120–129	130–139	140–159
No. of students	12	34	38	13	3

Represent this data by means of a histogram.

*6   The wages of 50 workers were as below. Represent this data using a histogram.

Weekly wage (£)	70–80	80–100	100–150	150–175	175–200	200–300
No. of workers	2	8	18	14	5	3

23.7 Frequency polygons

A method of presenting data which is an alternative to a histogram is the **frequency polygon**. They are often used to compare frequency distributions, i.e. to compare the 'shapes' of the histograms, as it is possible to draw more than one frequency polygon on the same graph. It is easier to make comparisons using frequency polygons than using histograms.

For ungrouped data, the frequencies are plotted as points. For grouped data, which is more usual, the frequencies are plotted against the mid-point of the class interval. In both cases the points are joined with straight lines.

Example 1

Represent the data of question 2, Exercise 23E by means of a frequency polygon.

The new table is:

Mid-point of class	155	165	175	185	195	205
Frequency	4	7	15	47	6	1

Frequency polygon:

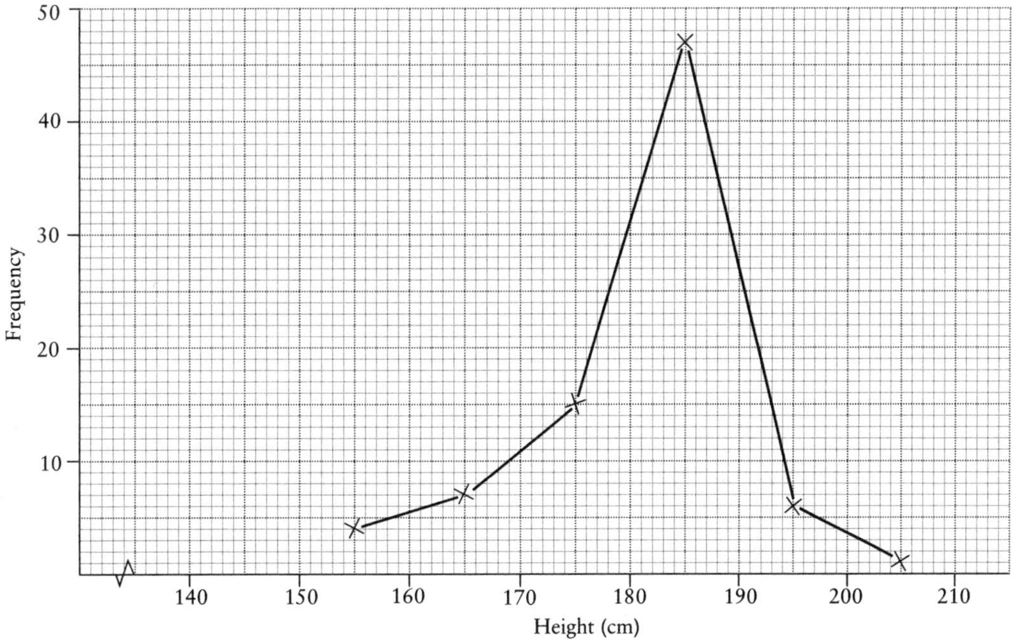

When frequency polygons are used to compare two or more sets of data, it is the shapes of the distributions which are important. For this reason, the mid-points of the classes are often plotted against the actual frequencies rather than the frequency densities which would be used for a histogram.

Example 2

The mock examinations results in Mathematics for two successive GCSE groups are recorded on the table below.

Mark	1–20	21–40	41–60	61–80	81–100
Group 1 % frequency	5	12	35	28	20
Group 2 % frequency	7	26	48	9	10

a Draw the frequency polygon for each group.

b Assuming the ability of the pupils was the same in each year, comment on the mock examination papers.

a In this example, the percentage frequencies are plotted against the class mid-points, which are 10.5, 30.5, 50.5, 70.5 and 90.5.

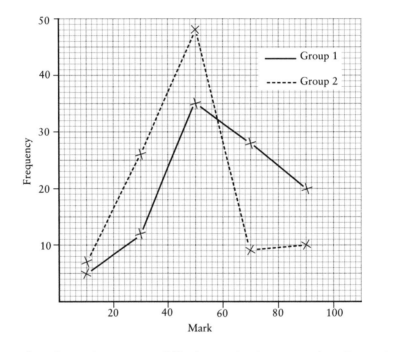

b Group 2 appear to have been given a more difficult examination paper than Group 1.

Exercise 23F

1 Draw the frequency polygon for the data given in question **1** of Exercise 23E.

2 A teacher noted the absence rates of her maths class on Mondays and Fridays.

The results are given on the table below.

No. absent from class	0	1	2	3	4	5	6	7	8	9	10	11
Monday Frequency	3	6	6	7	4	4	3	0	0	0	0	0
Friday Frequency	0	2	2	3	4	5	2	0	6	3	2	1

a Draw the frequency polygon for each day, using the same axes.

b Comment on the absence rates of the two days.

3 An experiment was carried out using eight dice to find the number of sixes in each throw of the dice. The objective was to see if the experimental results matched the expected theoretical calculations. The frequency polygon shows the expected results.

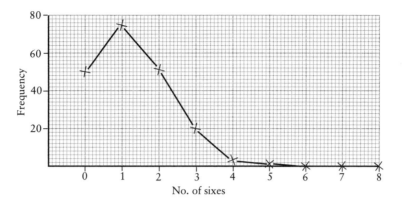

The experimental results were:

Number of sixes	0	1	2	3	4	5	6	7	8
Frequency	50	75	51	20	3	1	0	0	0

a Copy the given graph and plot the frequency polygon for the experimental results on the same axes.

b Comment on the two sets of results.

4 The table below shows the number of letters per word in 100 words of two books.

No. of letters	1	2	3	4	5	6	7	8	9	10	11	12
Frequency, Book 1	3	12	28	7	14	11	6	6	7	4	1	1
Frequency, Book 2	6	8	37	22	7	11	6	0	1	2	0	0

a Draw the frequency polygons using the same axes.

b One extract was taken from a child's story and the other from an adult science fiction story.

State which is which, giving reasons for your decision.

5 Choose two daily newspapers, e.g. a broadsheet and a tabloid. Compare them by drawing frequency polygons of the number of words per sentence in one hundred sentences taken from similar sections in each paper.

(If you keep these results, they could be used in future work to calculate means, medians and standard deviations.)

*6 The table below shows the adult population of males and females in 1986.
(Source: The Office of Population Censuses and Surveys.)

Age	16–24	25–34	35–44	45–59	60–64	65–74	75 and over
Male population (millions)	4.1	3.9	3.8	4.5	1.4	2.1	1.2
Female population (millions)	4.0	3.9	3.8	4.5	1.6	2.7	2.4

Draw the frequency polygons and comment on your results.

*23.8   Shapes of distributions

Histograms and frequency polygons are also used to illustrate the shape of a frequency distribution.

Many frequency distributions have a recognisable shape which can easily be described.

1   A **symmetrical** distribution has the same shape either side of a central vertical line.
    Example: IQ scores in the population.

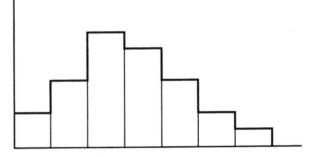

2   A **positively skewed** distribution has a 'tail' on the right of the graph.
    Example: Results on a difficult exam paper.

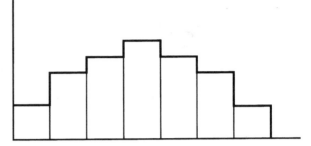

3   A **negatively skewed** distribution has a 'tail' on the left of the graph.
    Example: Results on a relatively easy exam paper.

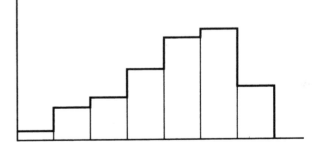

4   A **bimodal** distribution has two 'peaks'.
    Example: Lengths of strides of female and male students.

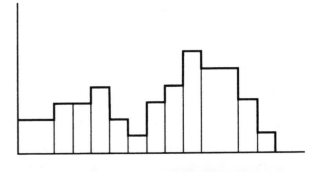

†23.9 Normal distribution

The pulse rates (beats per minute) of one hundred adults were found to be:

72.4	68.2	69.3	71.1	66.8	67.2	65.4	66.4	74.0	70.2
68.0	78.9	76.0	70.8	64.3	82.3	70.2	73.4	75.8	76.1
74.2	76.7	65.5	71.6	74.1	68.7	66.8	71.4	71.3	77.5
83.3	74.9	71.5	75.7	71.6	73.2	82.5	73.6	81.8	76.2
73.8	78.2	65.6	76.9	76.8	81.5	77.2	69.9	78.2	68.1
75.8	75.4	80.3	78.0	68.3	76.0	78.5	77.4	79.0	76.9
78.8	71.7	74.4	69.8	77.6	73.4	77.3	86.0	79.3	71.3
74.9	72.4	66.9	73.7	74.4	68.8	82.6	71.4	72.9	71.9
73.7	79.8	74.0	71.8	73.4	76.0	79.2	79.4	75.2	69.8
79.4	69.1	76.7	74.9	84.3	69.2	74.7	69.9	81.9	76.0

When this data is organised as a frequency distribution with classes 61–65, 66–70, etc., the frequency distribution is

Pulse rate	61–65	66–70	71–75	76–80	81–85	86–90
Frequency	4	19	36	32	8	1

The associated histogram is given below:

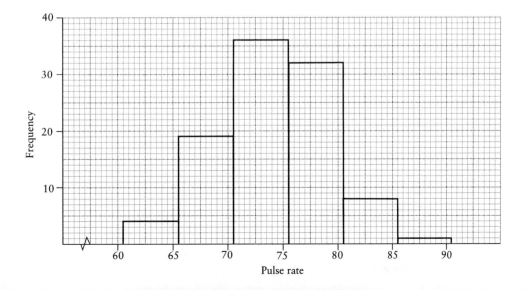

If the data were regrouped into smaller classes, 61–62, 63–64, etc., the relevant histogram would become:

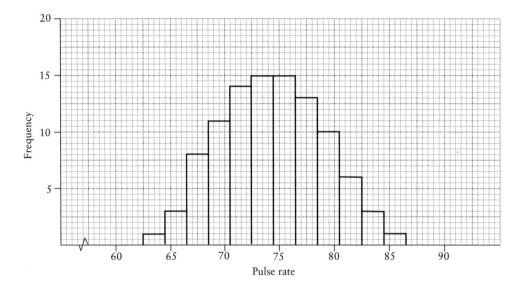

With continuous variables, and accurate measurements, the class widths could be made smaller and smaller so that the mid-points of the bars became closer and closer. Eventually, if the mid-points were close enough, joining them would produce a **frequency curve**:

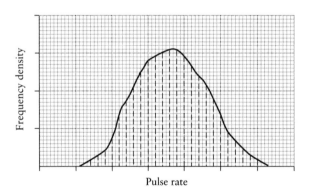

All continuous variables may be treated in this way, and many variables fit a distribution with a distinctive **bell shape**.

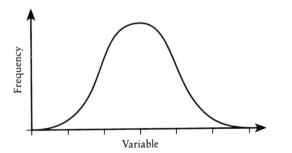

This is called a **normal distribution**.

Whether represented by a histogram or a **normal curve**, all normal distributions have several features in common:

- They are symmetrical about a central line through the highest point.
- The majority of the values lie close to this central line.
- Very few values lie in the lower and upper 'tails'.
- They are bell-shaped.

Many events which take place in the real world fit a normal distribution.

Some examples are:

1 The weights of packets of crisps which are marked 'Contents weight 75 g'.

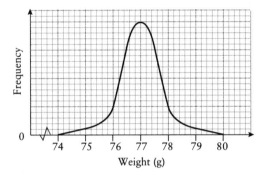

2 Intelligence quotient (IQ) of the population.

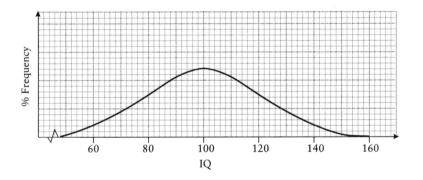

3 The time taken for a garage to change a tyre on a car.

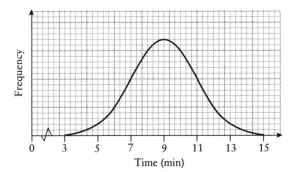

Exercise 23G

1 a Illustrate the following frequency distribution by a histogram.

Hand span (cm)	14–15	15–16	16–17	17–18	18–19	19–20
Frequency	1	1	3	5	16	21
Hand span (cm)	20–21	21–22	22–23	23–24	24–25	25–26
Frequency	33	19	28	22	8	3

  b Describe the distribution.

  c Suggest a reason for the shape of this distribution.

2 An experiment was carried out which involved throwing two dice one hundred times and recording the score.

   The results were:

Score	2	3	4	5	6	7	8	9	10	11	12
Frequency	2	6	8	12	13	17	16	13	8	4	1

   Draw a histogram and describe the distribution.

3 The scores given when a single dice was thrown fifty times are shown on the histogram below:

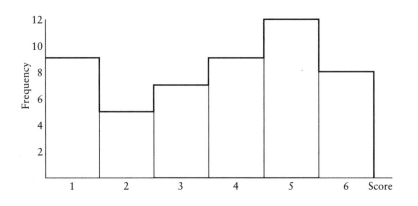

  a What shape would you expect the distribution of scores on a die to have?

  b Give a reason for the shape of the given distribution.

4 In an experiment, ten drawing pins were tossed and the number of drawing pins landing with the pin vertical was recorded.

No. with pin up	0	1	2	3	4	5	6	7	8	9	10
Frequency	0	4	17	33	47	43	30	18	7	1	0

  a Draw a graph of the frequency distribution and comment on the shape.

  b Carry out the experiment yourself and draw a frequency polygon of the results.

   (The drawing pins used in the given experiment were lightweight pins with coloured tops. If you use a different type of drawing pin, such as the large brass pins, you will obtain different results.)

†5 State which of the following distributions are normal:

    a the weights of the female adult population of Britain

    b the weights of racing boat crews
    (The crews consist of eight rowers and a cox, who steers the boat. The cox is usually considerably lighter than the others.)

    c the number of faulty matches in a box of matches

    d the lengths of mature leaves on a shrub

    e the number of sweets in a packet of Smarties

    f the number of children in a British family

    g the body temperature of healthy people.

6 Which of the sketches of distributions would best fit the following data?

    a the waiting time of passengers on the London Underground

    b the total score when the scores on two dice are added together

    c the foot sizes of students in a tertiary college

    d the reaction times of students in further education

    e the concentration of insulin in the blood of a diabetic, measured between injections

    f the test results on a relatively easy paper.

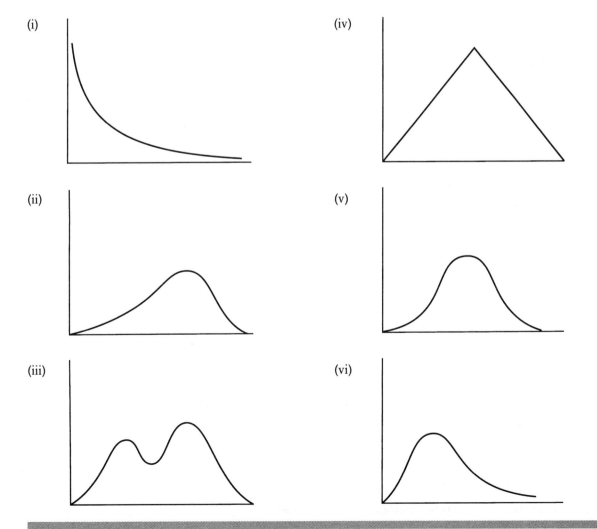

(i)

(ii)

(iii)

(iv)

(v)

(vi)

24 Interpretation of Statistical Diagrams

24.1 Reading and interpreting diagrams

We have seen in Unit 23 that there are many ways of presenting data in pictorial form. It is clearly necessary to be able to interpret correctly any diagrams given. Examples of pie chart and histogram interpretation are given below.

Interpreting pie charts

The initial interpretation is the fact that the largest portion of a pie chart relates to the largest group, and the smallest portion to the smallest group. However, if any of the data is known, the rest of the data can be calculated.

Example 1

The pie chart below shows the number of students in different sections of a college. 220 students are in the Construction department.

a How many students are there in the college?

b How many students are there in Catering?

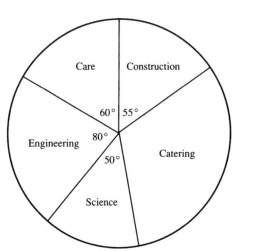

a $55°$ represents 220 students.

∴ $1°$ represents $\dfrac{220}{55} = 4$ students.

The complete circle ($360°$) represents $4 \times 360 = 1440$ students.

∴ There are 1440 students in the college.

b The angle representing Catering is

$360° - (80° + 55° + 60° + 50°) = 115°$.

∴ The number of students in Catering is $4 \times 115 = 460$.

*To interpret histograms

Example 2

The histogram shows the heights (in metres) of trees in a plantation. There are 40 trees with heights between 10 m and 12 m.

a How many trees are there with heights between 4 and 6 m?

b How many trees are there in the plantation?

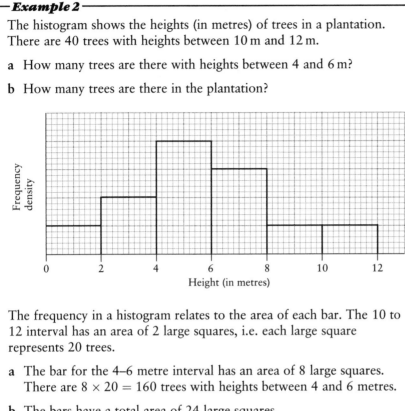

The frequency in a histogram relates to the area of each bar. The 10 to 12 interval has an area of 2 large squares, i.e. each large square represents 20 trees.

a The bar for the 4–6 metre interval has an area of 8 large squares. There are $8 \times 20 = 160$ trees with heights between 4 and 6 metres.

b The bars have a total area of 24 large squares. There are $24 \times 20 = 480$ trees in the plantation.

Example 3

The histogram represents the speeds of 200 cars along a motorway.

a How many cars are exceeding the 70 mph speed limit?

b What percentage of cars is travelling within 10 mph of the speed limit?

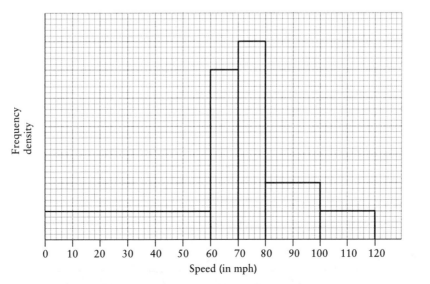

When the histogram has bars of unequal widths, we still use the method described in Example 3.

The total number of large squares is 25.

This represents 200 cars.

$\therefore$ 1 large square represents $\dfrac{200}{25} = 8$ cars.

a The cars exceeding 70 mph are represented by 13 large squares.

$\therefore$ $13 \times 8 = 104$ cars are exceeding the 70 mph speed limit.

b Cars travelling between 60 and 80 mph are within 10 mph of the speed limit.

These cars are represented by 13 large squares.

$\therefore$ There are $13 \times 8 = 104$ cars within 10 mph of the speed limit.

$\therefore$ The percentage of cars within 10 mph of the speed limit $= \dfrac{104}{200} \times 100 = 52\%$

Exercise 24A

1 The pie chart shows the different drinks sold at lunchtime in college. 720 drinks were sold in total.

Find the number of

a Coke

b orange

c coffee

d chocolate drinks sold during lunchtime.

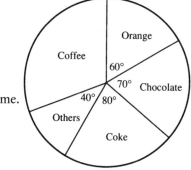

2 The pie chart shows the number of passengers flying from London to Miami on one afternoon. 1800 passengers in total flew this route on that afternoon.

Find the number flying:

a Virgin

b American Airlines

c British Airways.

The plane used by Virgin is a Boeing 747 seating 370. What percentage of the Virgin seats were occupied?

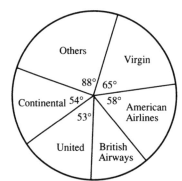

3 A bookshop sold 1080 books, and noted the types of book sold.

Find the number of books
sold which were classed as:

a Thriller

b Hobby

c Travel

d Science.

A pie chart showing: Hobby 70°, Travel 70°, Science 80° (Thriller), 70°, 100° (Romance), Thriller 80°. Labels: Hobby, Science, Travel, Thriller, Romance with angles 70°, 70°, 80°, 100°.

4 The pie chart shows the results of an election
in a constituency. There are 72 000 voters
of whom 80% voted. Which party won,
and what was the winning margin?

A pie chart with: Labour 148°, Conservative 154°, SLD 23°, Others.

5 The pie chart shows the different petrols sold in one week.

The garage sold 25 000 gallons of diesel.

a How much unleaded petrol
was sold?

b How much 4-star petrol
was sold?

c What were the total sales?

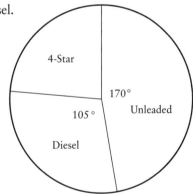

A pie chart with: 4-Star, Unleaded 170°, Diesel 105°.

6 The pie chart shows the types of dwelling in which people in a village
live. There are 720 dwellings in the village. By measuring the angles, find
how many are:

a detached houses

b bungalows

c semi-detached houses.

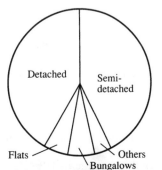

A pie chart with: Detached, Semi-detached, Flats, Bungalows, Others.

7   In 1994 a survey was conducted by asking sixty students how they
    travelled to college. This pie chart shows the results of the survey.

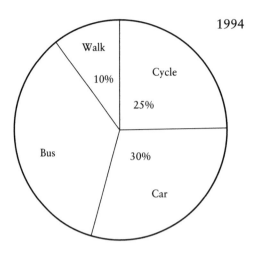

a   (i) What percentage of students travel to college by bus?

    (ii) Calculate the number of students who travel to college by bus.

In a similar survey 30 years ago the following pie chart was obtained.

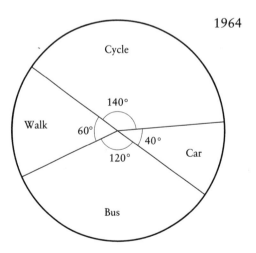

b   Which method of transport was most common in 1964?

c   The pie charts indicate a change in the method of transport over the
    last 30 years.
    Describe **two** changes that are shown.

In 1964, fourteen students travelled to college by car.

d   How many students cycled to college in 1964?   (SEG Sp94)

8 The pie chart shows the use of agricultural land in South Australia.

   a Find the percentage of land used for wheat.

   b The total acreage is 3 000 000 acres. What area is used for hay?

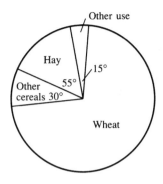

9 The pie chart shows details of the 200 pets kept by people living in a village.

   How many

   a dogs

   b cats

   c rabbits

   were there?

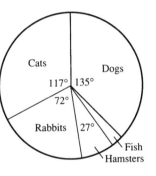

*10 The histogram shows the number of lunches of various prices sold in a restaurant.
Reconstruct the frequency table.

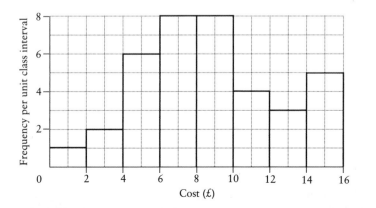

*11 The histogram shows the distance travelled by 200 lobsters in one week.
Reconstruct the frequency table.

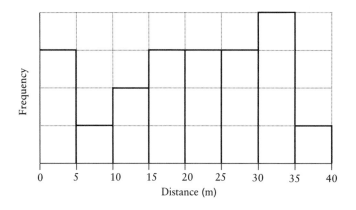

*12 At a fête, the number of peas in a jar was guessed. The histogram
represents the guesses made.
Reconstruct the frequency table.

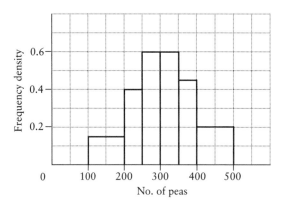

*13 The histogram below represents the distribution of ages in a small village.
Find the number of people who live in the village.

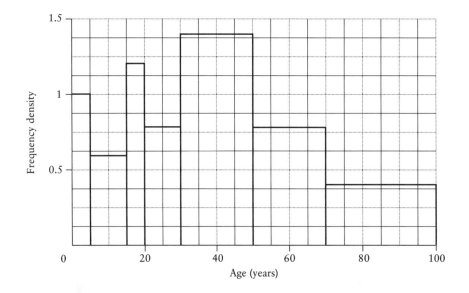

*14 The histogram shows the number of cars arriving at a ferry terminal in the last hour before departure. 250 cars arrived in this hour. The latest official check-in time is half an hour before departure.
How many cars arrived after the official check-in time?
What percentage of the cars arrived within 10 minutes of the check-in time?

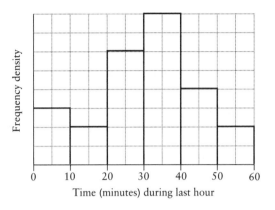

*15 The following histogram shows the number of metal rods, of different lengths, made in a factory in one week.
Find the total number of rods.

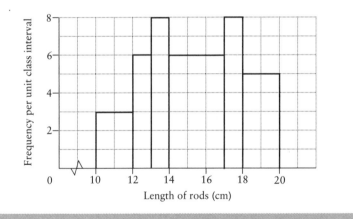

24.2 Drawing inferences from diagrams

When you can readily transfer data to pictorial form, and can convert pictorial representation back into numerical data, you can start to draw inferences from the data given in either form.

Drawing inferences from statistical data is not an exact science. There is rarely a correct, precise answer. If some of the data does not fit the overall pattern, this should be noted. Then, the reasons for the apparent contradiction should be considered.

We can illustrate this by examples.

Example 1

The number of people each day visiting an open-air swimming pool in August were:

341, 352, 347, 355, 361, 341, 344, 352, 344, 360, 371, 347, 329, 351, 621, 357, 348, 359, 354, 372,....

What can be inferred from this data?

The 15 August figure of 621 is clearly exceptional and should be easily identified.

The reasons why this figure is exceptional would be unlikely to be found without further questioning – for example:

(i) was an alternative swimming facility closed for the day?
(ii) was the weather on the 15th substantially hotter than on the other days?
(iii) does the data refer to a country with a Bank Holiday on 15 August (e.g. France)?

These are all possible reasons and you may be able to suggest others.

Some data gives results from which suitable inferences can be fairly quickly drawn.

Example 2

A men's clothes shop has two branches: one in a city centre, and one at an out-of-town shopping complex. The weekly sales of the two shops over a period of three months are given below:

Date (week commencing)		December				January				February			March	
		7	14	21	28	5	12	19	26	2	9	16	23	2	9
Sales in	City shop	18	21	23	14	29	31	28	7	5	8	6	4	5	8
thousands of pounds	Out-of-town shop	22	27	52	15	4	5	6	8	9	8	10	9	7	8

What can be inferred from this data?

From the above data, the following inferences could be made:

(i) Sales immediately prior to Christmas are higher than at other times in the three-month period.

(ii) Sales in the out-of-town complex are generally higher than those in the city shop.

(iii) The city shop had a 'sale' during the first three weeks of January.

(iv) During the 'sale' the trade at the out-of-town shop was reduced.

In reality, the firm could well investigate the types of garments sold over the years to decide whether or not to promote the same articles in both shops at the same or different times. Computerisation of sales enables shops to keep far better checks on stock sold. This enables them to react more quickly to consumer demand and to supply each individual shop with the goods which its specific customers require.

You may find it helpful to study Unit 25 (Averages) and Unit 27 (Dispersion) before doing this exercise.

1 In an Ian Fleming novel the number of words per sentence in the first 100 sentences are as follows.

Number of words	0–4	5–9	10–14	15–19	20–24	25–29	30–34	35–39
Number of sentences	14	37	28	12	5	1	1	2

a Copy the axes and draw a frequency polygon to illustrate this information.

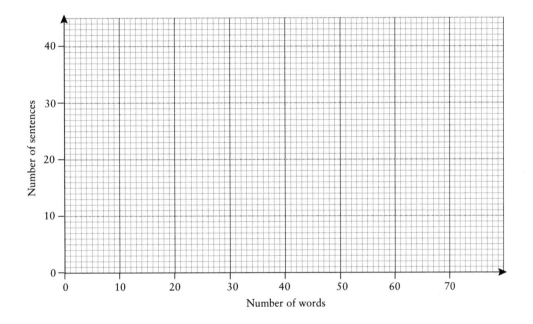

The frequency polygon of the number of words per sentence in the first 100 sentences of a novel by a different author is drawn below.

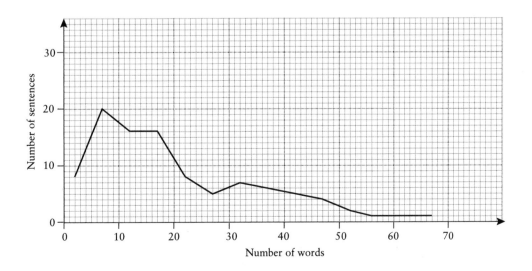

b Use the frequency polygons to compare the sentence length of the two novels. (SEG Sp94)

†2 A plant has two varieties **A** and **B**.
A study was made of the height of the two varieties.
The histograms represent the frequency distributions obtained.

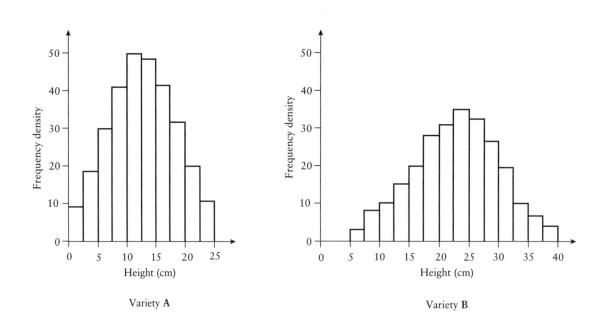

Variety **A**   Variety **B**

a Name the type of distribution shown by the height of these plants.

b Make **two** statistical comments about the height distributions of these plants.   (SEG W94)

3 The height, to the nearest inch, of 500 adult females is given in the following table.

Height	58	59	60	61	62	63	64	65	66	67	68	69	70	71	72
Frequency	1	0	4	3	9	32	67	84	102	135	41	18	3	0	1

a Complete the grouped frequency table.

Height (inches)	Class mid-point	Frequency
54.5–59.5	57	1
59.5–64.5		
64.5–69.5		
69.5–74.5	72	

b Use your grouped frequency distribution from **a** to calculate an estimate of the mean height of the 500 females.

c The frequency polygon showing the distribution for the height of 500 adult males is shown below. Copy it and on the same axes, draw the frequency polygon for the grouped distribution of the female height.

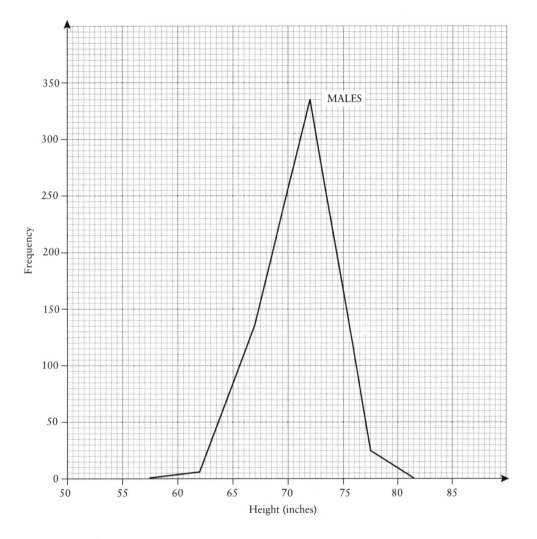

*d Use the frequency polygons to compare the means and dispersions of the two distributions.

(SEG Sp94)

24.3 Dangers of visual misrepresentation

Statistics can very easily be presented in a form which, although correct, is misleading. The simplest way in which this occurs is by the use of the 'false origin'.

For example, when a rail line is electrified, the number of cars per day on the parallel road reduces from 38 500 per day to 38 200.

A correct bar chart would show:

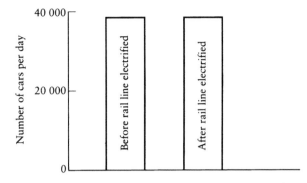

From this, readers would draw the conclusion that there has been little change.

Carefully deleting most of the vertical scale, and starting at 38 000 produces the bar chart below:

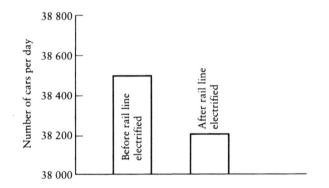

To a casual reader, the conclusion would be quite clear: there has been a significant reduction in the number of cars using the road.

Although this would rarely be spotted, the scale can be more subtly altered as below:

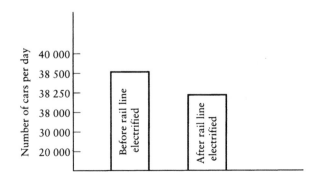

This over-expansion of the relevant part of the vertical scale can be justified as giving prominence to that part of the graph in which we are interested. However, its effect is to mislead readers.

If a section of a scale is to be deleted, it should be clearly identified – usually with a squiggle on the axis as shown on the right:

Bars of different widths can also mislead:

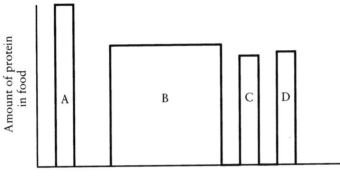

Although B does not have the highest amount, a casual readers' eye is drawn to it as its 'area' is much greater than any other bar.

The phrase 'amount of protein in food' is also open to misrepresentation, as it does not specify how much of food A is compared with how much of food B. Are they proportions, or amount per penny, or amount per gram, or something else?

False impressions can also be given by careful selection of which figures to show. In a time-series, you have to have a start and end date. In the presentation of information demanding an increase in wages, salaries or subsidies, it is to be expected that the data will start at a peak year. In other cases the data will stop at a favourable moment.

Example

The diagram shows the Japanese share index for 1981–90. The diagram on the right shows the same index stopping in Summer 1987, when it would show an unbroken climb.

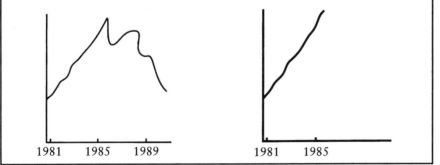

All diagrams should:

(i) be clearly labelled and titled

(ii) have the scales clearly identified

(iii) have the units given

(iv) should be drawn as outlined in Unit 23.

Any diagram which is not so drawn may, intentionally or not, mislead.

Exercise 24C

Criticise the diagrammatical representations shown in the following questions:

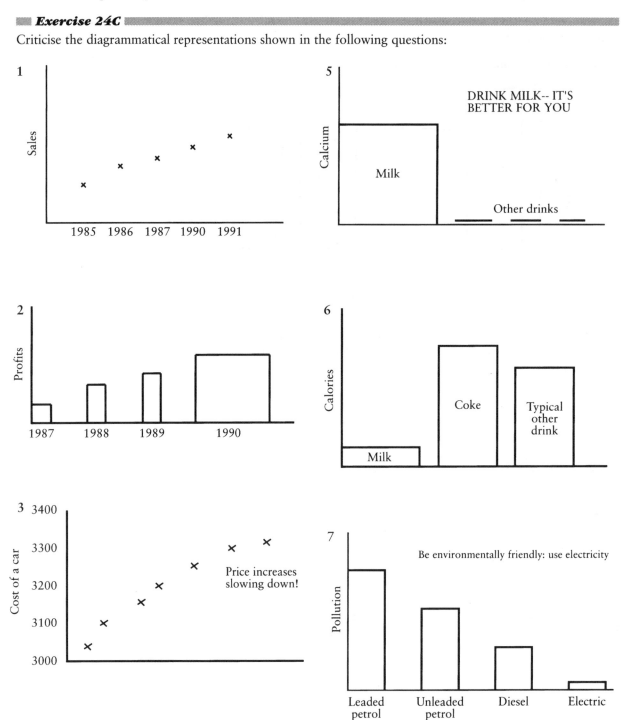

1

Sales

1985 1986 1987 1990 1991

5

Calcium

DRINK MILK-- IT'S
BETTER FOR YOU

Milk

Other drinks

2

Profits

1987 1988 1989 1990

6

Calories

Milk

Coke

Typical
other
drink

3

3400
3300
3200
3100
3000

Cost of a car

x x
x
x x
x

Price increases
slowing down!

7

Pollution

Be environmentally friendly: use electricity

Leaded
petrol

Unleaded
petrol

Diesel

Electric

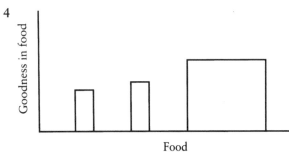

4

Goodness in food

Food

24.4 Interpretation of statistical information

All statistical information needs to be considered carefully before any inference can be made. It is rare for a simple set of statistics to prove anything.

An advertisement states '60% of dog owners, who expressed a preference, liked Dogamix'. It could be that out of 1000 owners, only 5 showed a preference; i.e. 3 out of 1000 liked it and 2 out of 1000 disliked it.

If a packet states average net weight of 250 grams, about half of the packets bought will have weight less than 250 grams, and half will have weight over 250 grams. Hence it is difficult for shoppers to tell whether they are being sold too little.

Professional statisticians are often involved in work of this nature. They are often employed to obtain, and present, the information in as favourable a way as possible for their employer. Detailed work in this area is beyond the scope of this course, but readers should always approach statistical information with care.

25 Averages

Sets of data can be compared, as mentioned previously, by comparing their frequency distributions or frequency polygons. It is also very useful to be able to compare a single, 'typical' statistic from one set of data with a single, 'typical' statistic from another set of data.

This statistic must be representative of the distribution. For this reason it is usually located at or near the centre of the distribution and is called a **measure of central location** or **average**.

The most commonly used averages are the **mean, mode** and **median**.

25.1 The arithmetic mean

The **arithmetic mean**, which is usually just referred to as the **mean**, is the most widely used average.

To calculate the mean, the total of the values is found and this is 'shared out' equally by dividing by the total number of values.

Example

Felicity sat six test papers and her marks, out of 50, were 17, 23, 27, 29, 30, 36.

Find her mean mark.

$$\text{Total marks} = 17 + 23 + 27 + 29 + 30 + 36$$
$$= 162$$
$$\text{Number of marks} = 6$$
$$\therefore \quad \text{Mean mark} = \frac{162}{6} = 27$$

This means that if Felicity's performance had been the same in all the tests, she would have gained 27 marks for each one.

The mean is often given in formula form:

$$\text{Mean} = \frac{\Sigma x}{n}$$

Σ is a capital Greek letter (sigma). Σx means the sum of all the terms and n is the number of terms.

Exercise 25A

1   The weekly wages of ten workers were:

£110, £115, £135, £141, £119, £152, £144, £128, £117, £139.

Find the mean wage.

2   The speeds (in mph) of sixteen cars were recorded as:

83, 75, 61, 72, 64, 51, 41, 89, 92, 71, 68, 66, 67, 64, 69, 72.

Find the mean speed.

3   The wind speed (in mph) at 8 am on a particular day, was recorded at a number of measuring stations as:

88, 74, 61, 92, 48, 59, 71, 80, 70, 51, 48, 45, 75, 80, 82.

Find the mean wind speed.

4   A rugby team scores 37, 21, 64, 0, 18, 7, 35, 49, 28, 51, 82, 71 points in 12 successive matches.
What is its mean score?

5   Eight people were asked their ages, and the replies were 37, 41, 29, 17, 15, 21, 32, 38. John claims that their average age is over 29. Why is he correct?

6   Nine workers have a mean wage of £190 per week.

a   What is their total wage?

A new worker starts at the firm, and is given a wage of £110 per week.

b   What is the total wage for the ten workers?

c   What is the mean wage for the ten workers?

7   Five men have a mean height of 1.95 m.

Four women have a mean height of 1.72 m.

What is the mean height of the nine people?

8   A train of 12 carriages separates for the last part of its journey. There was a mean of 52 people in each carriage before it separated, and the four carriages going to Weymouth contained a mean of 63 people per carriage.

What was the mean number in the carriages *not* going to Weymouth?

25.2   The mode

The **mode** is the number which occurs most frequently. Suppose seven students scored as follows in a test: 2, 3, 6, 7, 7, 8, 9

The mean score here is $\frac{42}{7} = 6$.

The **mode**, the number which occurs most frequently, is 7.

Some distributions can have more than one mode. For example, the numbers of people in twenty minibuses were recorded as:

3, 3, 4, 5, 7, 8, 8, 8, 10, 11, 13, 13, 14, 15, 15, 15, 17, 17, 18, 18.

Both 8 and 15 are modes.

This distribution is said to be **bimodal**.

─ **Example** ─

Find the mode of the numbers:

5, 7, 12, 12, 12, 17, 17, 19, 21.

12 occurs three times, which is more often than any other number.
Hence the mode is 12.

NOTE: for qualitative data, the mode would be qualitative (e.g. a colour) and not numerical.

Modal class

When the values are grouped the mode is replaced by a **modal class**, which is the group of values occurring most frequently.

Example

What is the modal class in the following table?

Height of trees (m)	Frequency
0–2	4
2–4	8
4–6	11
6–8	4
8–10	2

There are 11 trees in the 4–6 category.

This is the greatest frequency.

∴   The class 4–6 is the modal class.

Exercise 25B

1   A die was thrown 12 times, and the scores were

2, 4, 1, 3, 4, 1, 5, 6, 6, 4, 2, 5.

What is the modal score?

2   The price (in pence) of a loaf of bread at 10 shops was found to be 44, 49, 51, 68, 62, 44, 69, 51, 44, 47. What is the mode?

3   The heights of 40 trees were measured, and the data was:

Height (m)	Frequency
2–4	2
4–6	7
6–8	11
8–10	13
10–12	7

What is the modal class?

4   The IQs of 70 students were recorded as:

IQ	Frequency
95–99	5
100–104	7
105–109	18
110–114	21
115–119	7
120–124	4
125–129	5
130–134	1
135–140	2

What is the modal class?

5 Fifteen cars have the following colours: blue, black, red, white, red, white, gold, blue, red, grey, red, grey, black, blue, purple.

What is the modal colour?

6 Thirty two children each timed how long it took a pendulum to make ten swings.

The results of the first twenty four children are shown in the observation sheet below.

Time for ten swings (seconds)	Tally	Frequency
19.70 and less than 19.80	||	
19.80 and less than 19.90	|||	
19.90 and less than 20.00	Ⅲ̶	
20.00 and less than 20.10	Ⅲ̶ ||||	
20.10 and less than 20.20	|||	
20.20 and less than 20.30	||	

The last eight results were

20.14, 19.80, 20.07, 20.19, 19.92, 20.25, 19.86, 19.95.

a Use these results to complete the tally and frequency columns of the observation sheet.

b Draw a frequency diagram to show the results.

c Write down the modal class of these data. (SEG Sp96)

25.3 The median

The **median** is the value of the 'middle' item when the items are placed in numerical order.

To calculate the median, you put all the quantities given in order (usually ascending), and the median is the value of the middle one.

> **Example 1**
>
> Find the median of the numbers 17, 18, 24, 27, 28.
>
> The 'middle' number is 24, so the median is 24.

There is a complication with the definition of a median when there is an even number of quantities, as the following example shows.

Example 2

Find the median of the numbers 13, 24, 35, 18, 32, 15.

Rearrange the numbers in ascending order:

$$13, 15, 18, 24, 32, 35$$

There are 6 numbers, so there is no 'middle' number. The third number is 18, the fourth is 24, and the convention is to average these two 'middle values'.

The average (mean) of 18 and 24 is $\dfrac{18 + 24}{2} = 21$.

∴   The median is 21.

A quick rule for finding the middle quantity

To find the middle quantity, add one to the number of quantities given, then divide by 2, and take the quantity corresponding to this position.

In the first example above, there were 5 items. Add 1 to get 6. A half of 6 is 3. The 3rd number is 24, which is the median.

In the second example above there were 6 items, add 1 to get 7. A half of 7 is $3\frac{1}{2}$, which is half way between 3 and 4. Therefore the median lies half way between the 3rd and 4th number.

25.4   The use of mean, mode, and median

The three averages are useful in different contexts.

- If you were an employer considering the production capacity of your works, it would be helpful to use the *mean* of past production, as this would give you a good idea of the number of goods you can produce.
- If you were a shopkeeper wanting to keep a minimum stock of shirts to sell, the *mode* would be the best to use, as this will tell you which shirts you are most likely to sell.
- If you were a union wage negotiator, the *median* salary would be appropriate to use, because the few high wage earners would not then affect your 'average' of the wages paid.

Exercise 25C

1   The numbers of people on nine buses were recorded as:

17, 31, 11, 3, 51, 49, 52, 47, 34.

Find the median number of people.

2   The numbers of cars per hour on a country road during the hours of daylight were recorded as:

11, 13, 15, 9, 17, 12, 18, 14, 7, 9, 14, 16, 7, 11.

Find the median number of cars.

3   A die was thrown 12 times, and the scores were:

2, 4, 1, 3, 4, 1, 5, 6, 6, 4, 2, 5.

What was the median score?

4   The price of a loaf of bread (in pence) at 10 shops was found to be:

44, 49, 51, 68, 62, 44, 69, 51, 44, 47.

What was the median price?

5   The weights of parcels (in kg) delivered to a library were:

7.4, 8.2, 11.1, 7.8, 2.5, 5.6, 7.1, 8.9, 2.3, 2.7, 2.9, 4.1.

Find the median weight.

6   In each of the following situations, decide which of the **mode**, the **median** and the **mean** would be the most appropriate to use. (You are not required to find any values.)

   a The electricity consumption for a week is known for each house in a terrace of 20 and a typical value is required.

   b Witnesses to a bank robbery are asked to remember how many robbers were present. The answers are as follows:

   4, 5, 5, 5, 5, 5, 6, 6.

   c A student tries to decide if he is fairly paid for his part-time job. He asks some friends how much per hour they are paid for similar work.

   d A group of students are doing mathematics homework together. They compare answers to one question to decide on a correct value. The answers suggested are:

   £5.27, £5.27, £5.29, £6.31.

   e Witnesses to a car accident are asked to estimate the distance between two cars. Their estimates are:

   25 m, 25 m, 30 m, 35 m, 40 m.

   Which average would you use to establish as accurately as possible the true distance?

   f Ten people are asked what percentage of their earnings they save. A typical value is required.

   g Some students are asked how long they spent on a mathematics assignment and a typical value is needed. The answers are (in hours):

   3, 3, 3, 4, 4, 10.

25.5   The mean of a grouped distribution

In this example the data is recorded in a frequency table:

Example 1

Find the arithmetic mean of the following scores:

Score	1	2	3	4	5
Frequency	3	5	11	1	5

The score of 2, for example, occurred five times, but instead of totalling $2 + 2 + 2 + 2 + 2$, it is quicker to multiply 2 by 5.
Similarly, instead of totalling $3 + 3 + 3 + \ldots$ eleven times, it is quicker to calculate 3×11.

The table can be set out like this:

Score x	Frequency f	Score × Frequency xf
1	3	3
2	5	10
3	11	33
4	1	4
5	5	25
Totals:	$\Sigma f = 25$ (Total frequency)	$\Sigma xf = 75$ (Total of 25 scores)

Mean score $= \dfrac{\Sigma xf}{\Sigma f} \left(\text{i.e. } \dfrac{\text{Total of 25 scores}}{\text{Total frequency}} \right) = \dfrac{75}{25} = 3$

The mean is often denoted by $\bar{x}$.

Example 2

Thirty households were surveyed, and the number of children in each household was recorded. Find the mean and the mode.

No. of children in each family	0	1	2	3	4	5
Frequency	4	6	13	4	2	1

No. of children in each family x	Frequency f	xf
0	4	0
1	6	6
2	13	26
3	4	12
4	2	8
5	1	5
Totals:	30	57

$$\text{Mean} = \frac{\Sigma xf}{\Sigma f}$$

$$= \frac{57}{30}$$

$$= 1.9 \text{ children}$$

Mode $= 2$

Note that the mean need not be any of the original figures, nor indeed even a result which is possible. You cannot have 1.9 children in a family! The mean only indicates the average or 'central tendency'.

Calculating the mean of a grouped distribution is more complicated. How to deal with the values when they are in classes is shown in the following example.

Example 3

Thirty bushes were measured. Their heights (in cm) were grouped as shown.
Find the mean height and the modal class.

Height	5–15	15–25	25–35	35–45	45–55
Frequency	6	4	15	3	2

Since there is a spread in each group, it is impossible to determine the exact mean. We can only find an approximation to the mean.

It is assumed that each bush has a height equal to the middle of the range in which it lies; for instance, the 4 bushes with height 15–25 cm are all assumed to have height 20 cm, which is the **mid-interval** of the range 15–25 cm.

Length (cm)	Mid-interval (cm) x	Frequency f	Mid-interval × Frequency xf
5–15	10	6	60
15–25	20	4	80
25–35	30	15	450
35–45	40	3	120
45–55	50	2	100
Totals:		30	810

$$\text{Mean} = \frac{\text{Total of 'mid-interval} \times \text{Frequency'}}{\text{Total frequency}}$$

$$= \frac{\Sigma xf}{\Sigma f}$$

$$= \frac{810}{30}$$

$$= 27 \text{ cm}$$

The modal class is 25–35 cm. (There are 15 bushes in this class, more than in any other class.)

Example 4

In one hour, twenty planes arriving at Heathrow were early. The number of minutes early (to the nearest minute) were grouped as shown. Find the mean number of minutes early.

Time (min)	0–3	4–8	9–13	14–18
Frequency	4	7	8	1

As in Example 3 there is a spread in each group. The lower and upper class boundaries need to be found carefully. The group 0–3 (minutes) will range from 0 minutes (early) to 3.5 minutes (early). Therefore, the four planes in this group are all assumed to be 1.75 minutes early, which is the mid-interval of the range 0–3.5.

The group 4–8 ranges from 3.5 to 8.5 minutes early. Hence, these seven planes are all assumed to be 6 minutes early. (6 is the mid-interval of 3.5 and 8.5.)

No. of minutes early	Mid-interval (in minutes) x	Frequency f	Mid-interval × Frequency xf
0–3	1.75	4	7
4–8	6	7	42
9–13	11	8	88
14–18	16	1	16
Totals:		20	153

$$\therefore \quad \text{Mean} = \frac{\Sigma xf}{\Sigma f}$$

$$= \frac{153}{20}$$

$$= 7.65 \text{ minutes}$$

Note. In Examples 3 and 4 the distributions are continuous.

No problems are caused in Example 3 by including the height 15 cm in the group 5–15 and in the group 15–25 for heights of bushes, since the probability of finding a bush exactly 15.000...m high is nil.

When a distribution is discrete, you should *not* use 5–15 and 15–25, etc., since if an item is exactly 15, it is not clear into which group it should be placed.

If you were considering the number of people on a bus, you could use 5–14, 15–24, 25–34, etc. The range 5–14 would have a mid-interval of 9.5, which would be used to find the mean.

Exercise 25D

1 The numbers of children per family on a housing estate were recorded as follows:

No. of children	0	1	2	3	4
No. of families	12	15	5	2	1

Find the mean number of children per family.

2 An agricultural researcher counted the numbers of peas in a pod in a certain strain as follows:

No. of peas	3	4	5	6	7	8
No. of pods	5	5	20	35	25	10

Find the mean number of peas per pod.

3 The Ace Bus Company went through a bad patch when its buses always left the city centre late. This grouped distribution table shows how late:

Minutes late	0–10	10–20	20–30	30–40	40–60
Frequency	5	8	21	14	5

Find the mean number of minutes late.

4 The number of words per sentence on a page of a book were:

No. of words	1–3	4–6	7–9	10–12	13–15
Frequency	3	38	59	27	4

Find the mean length of a sentence.

5 The weekly wages of a firm's employees were:

Wage in £	50–69.99	70–89.99	90–99.99	100–149.99
Frequency	7	9	15	25
Wage in £	150–199.99	200–249.99	250–299.99	300–350
Frequency	38	41	7	2

Find the mean weekly wage.

6 The pollution levels, measured in parts per million (ppm), of 60
   specimens of tap water are recorded in the table.

Pollution level (ppm)	Mid-point	Number of specimens
0 and less than 10		18
10 and less than 20		22
20 and less than 30		14
30 and less than 40		5
40 and less than 50		1
50 and over		0

a Calculate an estimate of the mean level of pollution for the 60
   specimens.

b Explain why your answer to **a** is only an estimate of the mean level of
   pollution for the 60 specimens.   (SEG Sp96)

26 Cumulative Frequency

The cumulative frequency is the total frequency up to a particular class boundary. It is a 'running total', and the cumulative frequency is found by adding each frequency to the sum of the previous ones.

26.1 The cumulative frequency curve (or ogive)

Virtually all cumulative frequency curves (or ogives) have an 'S' shape. How an ogive is built up will be seen in the following example.

Example

The marks obtained by 100 students in an examination were as shown below.

Marks	No. of students (frequency)
0–10	1
11–20	2
21–30	13
31–40	24
41–50	32
51–60	16
61–70	11
71–80	1

Draw the cumulative frequency curve for this data.

First, we construct a new table. This keeps a running total of the frequencies in the second column.

Marks	Cumulative frequency
0–10	1
0–20	$1 + 2 = 3$
0–30	$3 + 13 = 16$
0–40	$16 + 24 = 40$
0–50	$40 + 32 = 72$
0–60	$72 + 16 = 88$
0–70	$88 + 11 = 99$
0–80	$99 + 1 = 100$

From the cumulative frequency column we can see that one student has 10 marks or less, three students have 20 marks or less, sixteen students have 30 marks or less, and so on.

Next we plot the cumulative frequencies against the **upper class boundaries**. For example, we plot 3 (students) against 20 (marks) and 16 (students) against 30 (marks). The completed graph then looks like this:

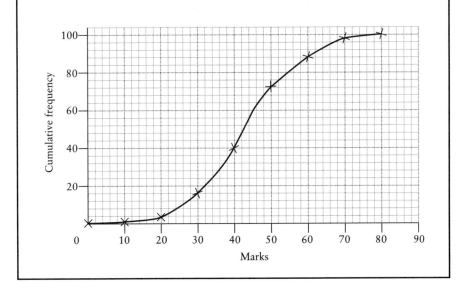

26.2 The median

The median mark is the mark obtained by the middle student. For example, if there are 100 students, the median student is the 50th (half-way point). (The strict definition would of course be the $50\frac{1}{2}$th student, but this accuracy cannot be obtained from a graph, and it is unnecessary at this stage.)

Suppose we wanted to find out the median mark in the example from Section 26.1. We would look across from the 50 on the cumulative frequency axis to the curve, then read down vertically to the number of marks. In this case the median is 43 marks. This is illustrated below:

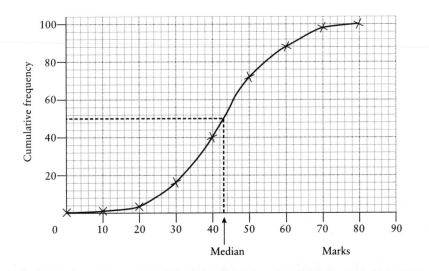

26.3 The interquartile range

(i) The **lower quartile** is the mark obtained by the student $\frac{1}{4}$ of the way along the distribution.

There are 100 students in the example on page 196, so the 25th student is $\frac{1}{4}$ of the way up the cumulative frequency axis. From the graph, the 25th student has 35 marks, so the lower quartile is 35.

(ii) The **upper quartile** is the mark obtained by the student $\frac{3}{4}$ of the way up the cumulative frequency axis. This is the 75th student, whose mark is 52, so the upper quartile is 52.

The upper and lower quartiles are illustrated below:

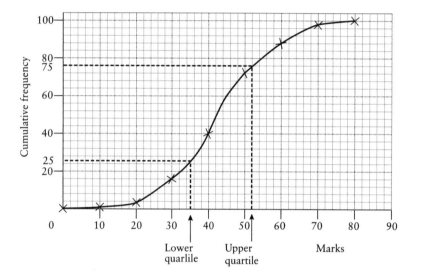

(iii) The **interquartile range** is the difference between the upper quartile and the lower quartile.

In our example, the interquartile range is:

$$52 - 35 = 17 \text{ marks.}$$

The **semi-interquartile range** is half the interquartile range, 8.5 in our example.

Note. Half the students have a mark between the lower and upper quartiles. Hence the interquartile range of 17 shows that half the marks obtained lie in an interval of 17. The semi-interquartile range of 8.5 shows that half the population lie (roughly) within 8.5 marks of the median mark.

*26.4 Percentiles

We can use the example on page 196 to define a **percentile**. As the name suggests, percentiles divide the cumulative frequency distribution into 100 parts, just as the quartiles divide it into quarters. The 70th percentile, for example, is the highest mark obtained by 70% of the entry. From the cumulative frequency curve this mark is 49, so the 70th percentile is 49 marks (as shown below).

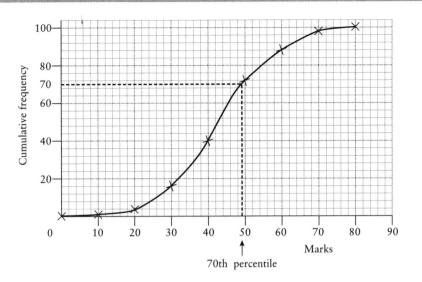

70th percentile

Exercise 26A

In questions 1–4, calculate the cumulative frequencies, and draw the cumulative frequency curve. Hence find:

a the median

b the lower quartile

c the upper quartile

d the interquartile range.

Part **e** is given separately for each question.

1 The times taken for students to complete two questions were recorded as:

Time (in minutes)	0–20	20–25	25–30	30–35	35–40	40–45	45–50
No. of students	3	12	25	49	21	7	3

e Estimate the number of students who took more than 41 minutes.

2 The waist measurements of 100 people were recorded. The data found were:

Waist (cm)	40–50	50–60	60–70	70–80	80–90	90–100
Frequency	2	11	25	41	20	1

e (i) Estimate the number of people with waist measurement less than 53 cm.
   (ii) Estimate the number of people with waist measurement more than 72 cm.

3 The annual gross wages of employees at a factory were:

Wage (£)	0–	8000–	10 000–	12 000–	14 000–	16 000–18 000
Frequency	1	15	34	58	27	3

*e Estimate the 80th percentile.

4   The marks gained by 90 students on a test (out of 100) were:

Marks	0–20	21–40	41–60	61–80	81–100
Frequency	2	15	32	33	8

e   (i) What percentage of students passed the exam if the lowest pass mark was 37?
   (ii) Six students are given a distinction.
     What was the lowest mark to obtain a distinction?
 *(iii) Find the 90th percentile.

5   Potatoes are supplied to a greengrocer's shop in 50 kg bags. Each of the potatoes in one of these bags
was weighed to the nearest gram, and the following table was drawn up:

Mass (g)	50–99	100–149	150–199	200–249	250–299	300–349
No. of potatoes	5	53	87	73	33	3

a   (i) Complete the cumulative frequency for these potatoes.
   (ii) Draw the cumulative frequency graph.

b   From your graph, estimate:
   (i) the median mass
   (ii) the number of potatoes weighing at least 225 g each.

c   For a party, Harry requires 50 baking potatoes, each weighing at least 225 g.
The greengrocer sells the potatoes without special selection.

   (i) Use your answer to **b**(ii) to estimate how many kilograms Harry will have to buy.
   (ii) Taking the mean mass of the 50 baking potatoes to be 260 g, estimate how many kilograms of
potatoes he will have left over.

6   The time for 160 adults to complete an exercise schedule is shown on the frequency diagram.

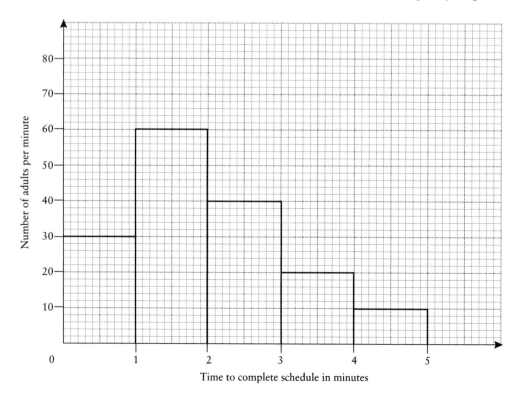

a Construct a cumulative frequency table for the data.

Time in minutes	Cumulative frequency
$\leqslant 1$	
$\leqslant 2$	
$\leqslant 3$	
$\leqslant 4$	
$\leqslant 5$	

b Copy these axes and draw the cumulative frequency curve.

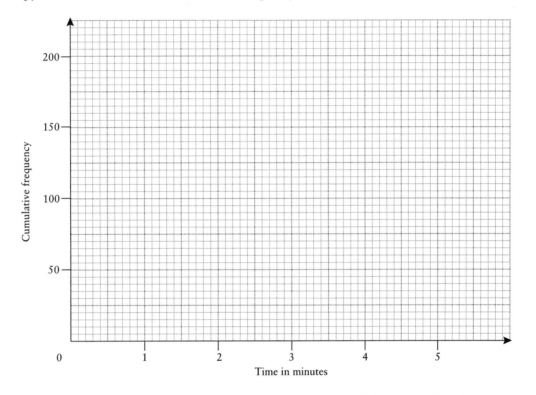

c Use your graph to estimate the time it would take 40% of the adults to complete the exercises.

d Use your graph to find the number of adults who took **more** than 3.5 minutes to complete the exercises.

(SEG W95)

27 Dispersion

Averages are used to represent sets of data or to compare them, but an average on its own does not give sufficient information about the distributions.

Suppose we are comparing the climate of two places in Turkey. Town A has an average yearly temperature of about 12 °C and town Z has an average yearly temperature of about 13 °C. From this, we might suppose that the climates are similar and that town Z possibly lies south of town A.

In fact, town Z is Zonguldak, which is on the Black Sea coast, and town A is Ankara, which is further south in the interior of Turkey.

The monthly temperature distributions for the two towns are:

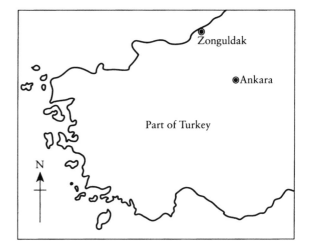

Month	J	F	M	A	M	J	J	A	S	O	N	D
Temperature in Ankara (°C)	−0.2	1.2	4.9	11.0	16.1	20.0	23.3	23.4	18.4	12.9	7.3	2.1
Temperature in Zonguldak (°C)	6.0	6.3	7.0	10.5	15.0	19.2	21.7	21.6	18.4	15.0	11.5	8.5

We can see that the climate in Ankara is more variable than in Zonguldak. There is a larger difference between summer and winter temperatures, whereas in Zonguldak there is a smaller spread of temperatures and a more equable climate.

We need a statistic which will measure this spread of values. The term we use to describe spread is **dispersion**, and there are several ways of measuring it.

27.1 The range

The simplest measure of spread is the **range**, which is the difference between the lowest and the highest values.

The range of temperatures for each town is

Ankara: Range = 23.4 − (−0.2) = 23.6 °C
Zonguldak: Range = 21.7 − 6.0   = 15.7 °C

The problems with using the range are: it only uses two values and so it can be distorted by a very high or low value; it cannot be used in further mathematical calculations.

Exercise 27A

1   The numbers of cars on a hovercraft in one day were:

37, 28, 8, 5, 17, 39, 22, 10.

Find the range.

2   The goals scored by a netball team in a competition were:

10, 7, 5, 1, 12, 8, 6, 7.

Find the range.

3   At the end of a season, the bottom team in a division has 34 points. The range is 52 points. How many points has the top club?

4   The number of cars produced per year per employee by manufacturers in England were:

7.1, 12.1, 8.1, 4.7, 11.6, 9.4, 3.6, 12.5, 3.9.

Find the range.

5   Two trusses of ten tomatoes were taken from two different varieties of plant. The diameters of the tomatoes were measured:

Plant 1 Diameter (cm)	2.1	4.5	3.8	4.7	5.2	6.0	4.9	5.8	4.2	5.9
Plant 2 Diameter (cm)	4.8	4.7	5.2	4.7	4.4	4.6	4.7	4.9	4.8	4.5

a   Find the mean and range of the two sets of data and comment on your results.

b   Which of the two varieties should the grower choose and why?

c   Find the median diameter of each truss.

d   Which of the two averages, mean or median, would give a fair comparison of the two plants?

6   Two book clubs offer 'mystery parcels' of books for £8.50, stating that the minimum value of the contents is £15.

A survey of eight such parcels from each of the two clubs found that the actual value of the contents was:

Club 1 Value (£)	16.00	15.80	16.85	15.80	17.85	15.30	15.75	15.45
Club 2 Value (£)	15.95	16.90	16.85	17.85	15.25	17.50	17.00	17.25

a   Calculate the mean and range of the data and comment on your results.

b   Find the median value of the contents for each club.

c   By comparing the mean and median, describe the distribution of each set of data.

27.2 The interquartile range

The problem of distortion by an extreme value can be overcome by calculating the range of the central part of the distribution. This is called the **interquartile range**, which we met in the previous chapter.

The interquartile and semi-interquartile ranges are always associated with the median.

To find the median and interquartile range, the values must first be written down in numerical order. The list is divided in half and then into quarters.

Ankara: −0.2 1.2 2.1 | 4.9   7.3 11.0 | 12.9 16.1 18.4 | 20.0 23.3 23.4
Zonguldak: 6.0 6.3 7.0 | 8.5 10.5 11.5 | 15.0 15.0 18.4 | 19.2 21.6 21.7

Half the values lie between the lower quartile and the upper quartile.

For Ankara:

$$\text{Median} = \frac{11.0 + 12.9}{2} = 12.0\,°C$$

Similarly:

Lower quartile $= 3.5\,°C$

Upper quartile $= 19.2\,°C$

Interquartile range $= 19.2 - 3.5 = 15.7\,°C$

For Zonguldak:

Median $= 13.3\,°C$

Interquartile range $= 11.1\,°C$

*27.3 The standard deviation

Neither of the previous measures of spread take into account all the data. To do this, first we calculate the amount by which each value differs from the mean.

This is called the **deviation from the mean**.

For Ankara, the mean is $11.7\,°C$.

The deviations from the mean are:

−11.9 −10.5 −6.8 −0.7 4.4 8.3 11.6 11.7 6.7 1.2 −4.4 −9.6

The total of the deviations is zero, which is not surprising because the mean is the average which 'evens out' all the differences.

Of course, we could just ignore the negative signs and find new totals. Dividing these totals would give us the average of the deviations and, in fact, this is a measure of dispersion called the **mean deviation**. However, it is not a very useful measure mathematically, so we solve the problem of the negative signs by *squaring* all the deviations.

The calculation for Ankara is:

Temperature (°C)	Deviations $(x - \bar{x})$	Squared deviations $(x - \bar{x})^2$
−0.2	−11.9	141.61
1.2	−10.5	110.25
4.9	−6.8	46.24
11.0	−0.7	0.49
16.1	4.4	19.36
20.0	8.3	68.89
23.3	11.6	134.56
23.4	11.7	136.89
18.4	6.7	44.89
12.9	1.2	1.44
7.3	−4.4	19.36
2.1	−9.6	92.16
Total:		816.14

The total of the squared deviations = $816.14 = \Sigma(x - \bar{x})^2$

Dividing by 12 gives:

$$\frac{816.14}{12} = 68.012 = \frac{\Sigma(x - \bar{x})^2}{n}$$

which is the average of the squared deviations and is called the **variance**.

However, the variance is in square measure.
To obtain an answer in degrees Celsius, we find the square root of the variance and this is called the **standard deviation**.

For Ankara:

Standard deviation = $\sqrt{68.012} = 8.25\,°C$

$$= \sqrt{\frac{\Sigma(x - \bar{x})^2}{n}}$$

Now calculate the standard deviation of the temperatures for Zonguldak. You should find that:

Mean for Zonguldak = 13.4 °C,
Standard deviation for Zonguldak = 5.64 °C

This shows that, although the mean temperatures are similar, the monthly temperatures for Zonguldak have a smaller spread than those for Ankara, i.e. they are less variable.

The standard deviation has the same advantages as the mean: it uses all the data and can be used in other mathematical calculations. It is the most widely used measure of dispersion and is always associated with the mean.

Alternative form of standard deviation formula

There is an alternative version of the formula for standard deviation which is more widely used than the formula we have just developed. This second version is:

$$\text{Standard deviation} = \sqrt{\frac{\Sigma x^2}{n} - \left(\frac{\Sigma x}{n}\right)^2} \quad \text{i.e.} \quad \sqrt{\frac{\Sigma x^2}{n} - \bar{x}^2}$$

If the value of the mean used in the calculation is exact (i.e. has not been rounded), the two formulae will give the same value for the standard deviation. If, however, the value of the mean has been rounded, the second version will give a more accurate answer because the mean is only used once in the calculation and not n times as in the first version.

Two additional advantages of the second formula are:

(i) it is quicker and easier to use when calculating from values in tables,
(ii) it is easier to check your results using the SD mode on a scientific calculator.

Example

Use the formula: $\quad \text{Standard deviation} = \sqrt{\frac{\Sigma x^2}{n} - \left(\frac{\Sigma x}{n}\right)^2}$

to find the standard deviation of the temperatures for Ankara.

Temperature (°C) x	x^2
−0.2	0.04
1.2	1.44
4.9	24.01
11.0	21.00
16.1	259.21
20.0	400.00
23.3	542.89
23.4	547.56
18.4	338.56
12.9	166.41
7.3	53.29
2.2	4.84
Total 140.5	**2469.25**

$$\bar{x} = \frac{140.5}{12} = 11.7$$

$$\frac{\Sigma x^2}{n} - \bar{x}^2 = \frac{2459.25}{12} - 11.7^2 = 68.0475$$

$$\text{Standard deviation} = \sqrt{\frac{\Sigma x^2}{n} - \bar{x}^2} = \sqrt{68.0475} = 8.25\,°\text{C (to 3 s.f.)}$$

Note. This is the same result as we found in the previous calculation. However, correct to 4 s.f. the results are 8.247 and 8.249 respectively, the second answer being the more accurate.

Remember. You can use a scientific calculator to calculate the mean and standard deviation. You will also find the values of n, Σx and Σx^2 on your calculator. Consult the instruction book for your calculator to assist you in finding these values.

Exercise 27B

1 In 10 games over a season, a rugby team scored the following number of points:

17, 41, 28, 21, 38, 45, 63, 8, 15, 34.

Find the mean number of points scored, and the standard deviation.

2 The ages of people at a youth club were as follows:

14, 15, 14, 16, 14, 14, 15, 17, 15, 18, 14, 15, 15, 16, 15, 16, 15, 14, 13, 15.

Find the mean age and the standard deviation.

3 The weight, in grams, of the marmalade in 10 jars was found to be:

456, 460, 455, 471, 453, 452, 458, 459, 465, 461.

Find the mean and the standard deviation.

There are 453 grams in 1 lb. Should the marmalade be labelled average weight 1 lb, or minimum weight 1 lb?

4 In a college the numbers of students in the various tutor sets were:

15, 13, 15, 13, 18, 14, 15, 16, 17, 21, 11, 18, 16, 17, 19, 15, 16, 19, 17, 15.

Find the mean and the standard deviation.

5 Dr Emmanuel has a choice of two routes when she drives to work. She may drive through the town centre, or take the longer route on the by-pass. She timed her journeys over a 14-day period and the results, in minutes, were:

Town route	16	20	25	26	29	17	28
By-pass route	20	25	22	25	22	26	21

a Calculate the mean and standard deviation for each route.

b Which route would you recommend Dr Emmanuel to use? Give a reason for your choice.

6 The speeds (in mph) of the first 12 cars passing a point on a dual carriageway after 1 pm were recorded on Friday and Saturday. The results were:

Friday: 72, 74, 81, 84, 65, 51, 52, 88, 74,   69, 75, 53

Saturday: 52, 54, 57, 72, 55, 81, 51, 56, 57,   59, 61, 51

Compare the means and standard deviations.

*27.4 Standard deviation of grouped data

Section 25.5 dealt with more difficult calculations of the mean. In the first case, the data occurred with frequencies greater than one, and, in the second case, the data was grouped into classes. You will sometimes need to calculate the standard deviation of such data.

As with the previous calculations of standard deviation, the first stage is to find the mean of the given data.

The formula used for finding the mean when there are frequencies given is

$$\bar{x} = \frac{\Sigma xf}{\Sigma f}.$$

The formula for standard deviation needs to be adapted in a similar way, giving:

$$\textbf{Standard deviation} = \sqrt{\frac{\Sigma x^2 f}{\Sigma f} - \left(\frac{\Sigma x}{n}\right)^2} \quad \text{or} \quad \sqrt{\frac{\Sigma (x - \bar{x})^2 f}{\Sigma f}}$$

A simpler version is **Standard deviation** $= \sqrt{\dfrac{\Sigma x^2 f}{\Sigma f} - \bar{x}^2}$

Example 1

A survey was carried out on a housing estate to find the number of children per family. The results from 50 families were:

Number of children	1	2	3	4	5	6
Number of families	18	22	6	1	2	1

Calculate the standard deviation.

The mean $(\bar{x}) = 2$

Number of children x	x^2	Number of families f	$x^2 \times f$
1	1	18	18
2	4	22	88
3	9	6	54
4	16	1	16
5	25	2	50
6	36	1	36
Totals		50	262

Standard deviation $= \sqrt{\dfrac{\Sigma 262}{50} - 2^2} = \sqrt{5.24 - 4} = \sqrt{1.24} = 1.11$

As in Section 27.3, you may also use a scientific calculator to find the standard deviation (or to check your answer). You will need to read the instructions to find out how to enter the frequencies, as this varies for different calculators.

If you do not possess a scientific calculator, the whole calculation can be done by tabulation, as shown in the next example.

Example 2

The number of computers sold per day in one 30-day period is shown below:

Number sold	1	2	3	4	5
Number of days	1	2	6	20	1

Calculate the standard deviation of the number sold.

Number sold x	x^2	Number of days f	$x \times f$	$x^2 \times f$
1	1	1	1	1
2	4	2	4	8
3	9	6	18	54
4	16	20	80	320
5	25	1	5	25
Totals		30	108	408

$$\text{Mean} = \frac{\Sigma xf}{\Sigma f} = \frac{108}{30} = 3.6$$

$$\text{Standard deviation} = \sqrt{\frac{\Sigma x^2 f}{\Sigma f} - \left(\frac{\Sigma xf}{\Sigma f}\right)^2}$$

$$= \sqrt{\frac{408}{30} - 3.6^2}$$

$$= 0.8$$

When the data is grouped into classes, the mid-point of a class is taken as the x value. The calculation is then the same as in the previous two examples.

─ **Example 3** ─

The weights of 80 salmon sold in a shop in one day are given below:

Weight (in pounds)	0–2	2–4	4–6	6–8	8–10
Number of salmon	1	14	33	28	4

Calculate the standard deviation.

Weight (in pounds)	No. of salmon f	Mid interval x	x^2	$x \times f$	$x^2 \times f$
0–2	1	1	1	1	1
2–4	14	3	9	42	126
4–6	33	5	25	165	825
6–8	28	7	49	196	1372
8–10	4	9	81	36	324
Totals	80			440	2648

$$\text{Mean} = \frac{440}{80} = 5.5 \text{ pounds}$$

$$\text{Standard deviation} = \sqrt{\frac{2648}{80} - 5.5^2}$$

$$= 1.69 \text{ pounds}$$

Exercise 27C

1–3 Find the standard deviation of the data given in questions **1, 2** and **3** in Exercise 25D.

4 The marks of 60 students who took a French test were as follows

Mark	No. of students
1–10	1
11–20	1
21–30	2
31–40	7
41–50	14
51–60	15
61–70	9
71–80	5
81–90	4
91–100	2

The students are to be divided into 3 sets of 20, the top set consisting of the students with the 20 highest marks.

a Calculate an estimate of the mean and the standard deviation for these top 20 marks.

You may wish to use the following formula

$$S^2 = \frac{\Sigma x^2}{n} - m^2 \text{ where } m \text{ is the mean.}$$

The 20 students with the next best marks are to be put in the second set.

b State, with reasons, whether the standard deviation of the second set will be greater or smaller than the standard deviation of the top set.

The remaining 20 students form a third set.

c Compare the standard deviations of the three sets. (SEG S94)

5   The times taken for 100 people to complete their shopping in a food
    supermarket are shown in the table.

Time in minutes (t)	Frequency			
$0 \leqslant t < 10$	12			
$10 \leqslant t < 20$	44			
$20 \leqslant t < 30$	30			
$30 \leqslant t < 40$	10			
$40 \leqslant t < 50$	4			
$50 \leqslant t < 60$	0			

a Calculate the mean and standard deviation of these times.
   You may find it helpful to use the table above and use the following
   formula.

$$S^2 = \frac{\Sigma x^2}{n} - m^2$$

   where m is the mean and S is the standard deviation.

The supermarket decides to expand the range of its products.
It decides to sell a range of household and garden products.
Another survey of shopping times was made after the introduction of
these products.

b State, with a reason, what change you would expect in the value of
   the mean in the second survey.

The **inter-quartile range** remains unaltered.

c Explain how this is possible.   (SEG W94)

28 Correlation

All the graphs and charts used so far to represent data have been concerned with only one variable, e.g. height, holiday destination, number of goals scored. In many statistical surveys, however, the data collected connect two variables, e.g. height *and* weight; number of police cameras on a stretch of motorway *and* number of speeding convictions; age *and* reaction time.

Data of this type is collected because it is believed that there is a link between the two variables.

28.1 Scatter diagrams

The usual method of representing these types of data is by a **scatter diagram**, also called a **scatter graph** or **scattergram**.

Example

A survey was carried out by a group of eight students in which the height and weight of each student was measured. The results were recorded in pairs (e.g. the student with height **164 cm** weighed **58.2 kg**).

Height (cm)	164	152	173	158	177	173	179	168
Weight (kg)	58.2	50.8	60.3	56.0	76.2	64.2	68.8	60.5

Display this data on a scatter diagram.

Two axes are drawn, one for the heights and one for the weights.
(It does not really matter which is which, but, as a general rule, the first set of data is recorded along the horizontal axis and the second set along the vertical axis.)

Each point is plotted using the paired data as the coordinates, i.e. for the student with height 164 cm and weight 58.2 kg, the coordinates are (**164, 58.2**).

The scatter diagram shows the height/weight of eight students.

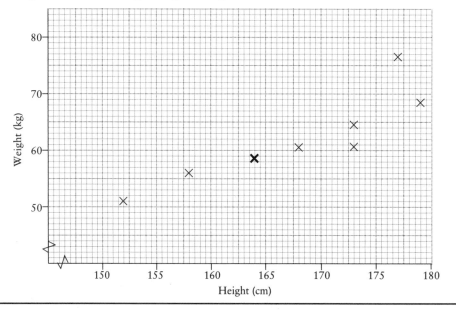

1   Draw scatter diagrams for the following data:

a

Mark on paper 1	15	20	14	5	24	10	19	26	30
Mark on paper 2	22	34	50	20	66	32	50	34	60

b

Time (minutes)	10	15	20	25	30	35	40
Weight of melting ice block (kg)	3.9	3.5	2.7	1.5	1.2	1.0	0.8

c

Weight of parcel (kg)	1.0	4.0	5.5	3.5	5.0	5.5
Length of parcel (cm)	6	15	7	35	10	25
Weight of parcel (kg)	4.4	2.2	2.0	6.0		
Length of parcel (cm)	26	20	30	33		

2   Draw a scatter graph for the following data.

Shoe size	$4\frac{1}{2}$	10	5	9	7	6	$5\frac{1}{2}$
Handspan (cm)	20.2	21.6	17.3	19.6	21.2	21.2	19.4
Shoe size	10	8	9	5	5	11	
Handspan (cm)	22.0	19.5	23.7	19.5	20.2	23.0	

3   The engine capacities, in cubic centimetres (cc), and the corresponding acceleration times, in seconds, for several models of car in the range of one manufacturer are given below:

Engine size (cc)	900	1000	1100	1200	1350	1500	1600
Time (s)	20.0	15.9	13.9	13.3	11.4	12.4	10.7
Engine size (cc)	1750	1800	1900	2000	1100	1350	2000
Time (s)	12.3	7.0	8.1	6.8	14.1	9.4	9.9

4   a   The data given below shows the percentage of breath-tests on motorists during the Christmas/New Year period which proved positive and the number of car accidents which involved injury during the same period. The figures given are for eleven counties in England and Wales.

% of positive tests	9.1	4.1	8.9	9.2	5.5	9.2	7.8
Injury accidents	65	49	46	97	25	96	56
% of positive tests	4.2	5.6	4.9	10.5			
Injury accidents	43	56	42	89			

Draw a scatter graph for these eleven points.

b The same variables for another four counties are given below.

% positive tests	9.0	7.6	1.7	17.5
Injury accidents	5	112	63	43

Draw a scatter diagram for all 15 points and compare the two graphs.

28.2 Correlation

As might be expected, the scatter diagram in the example on page 212 shows that there is some link between height and weight but that it is not a very close one. For example, two people who are the same height may have very different weights. In general, however, the taller you are, the more you will weigh.

There are three basic types of scatter diagram:

Diagram (i)
The points are widely scattered. This shows that there is no link between the variables. For example, having a small head does not mean you are less intelligent than someone with a large head.

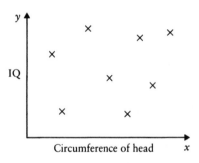

Diagram (ii)
This diagram might represent the number of police cameras on a motorway against the number of convictions for speeding. The points lie quite close together and show an upward trend. We might conclude that there is a weak connection, i.e. that the more cameras there are, the greater the number of motorists caught speeding.

However, the purpose of the cameras is not so much to catch motorists speeding as to deter them from speeding.

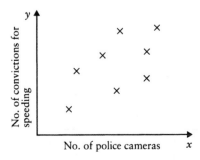

Diagram (iii)
This could represent the same variables after the cameras have been in place for a while and motorists are aware of them. There is still a link, but now the trend is downward, and we might conclude that the more cameras that are known to be operating, the less likely motorists are to drive fast and be caught speeding.

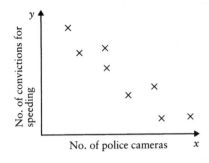

The closer a set of points lies to a straight line, the stronger the relationship between the variables, or, in statistical terms, the higher the **degree of correlation**.

Diagram (i) indicates that there is no correlation
Diagram (ii) indicates quite a weak positive correlation
Diagram (iii) indicates quite a high negative correlation.

Exercise 28B

1–4 For each of the questions in Exercise 28A, state the type of correlation (none, positive, negative) and give some indication of the degree of correlation.

5 For each of the following sets of data, state whether there is no correlation, positive correlation or negative correlation. (It is not necessary to draw the scatter diagrams.)

a

Day	1	2	3	4	5	6
No. of ripe courgettes	3	5	4	7	9	9

b

Height (cm)	166	156	152	160	159	158	163
Age (years)	20	31	25	28	22	32	32

c

% seeds germinating in compost 1	51	60	32	70	20	78
% seeds germinating in compost 2	55	75	70	58	82	50

d

W.p.m. typing test part 1	56	43	49	31	58	60	35	35	44	38
W.p.m. typing test part 2	58	44	47	37	57	57	30	31	44	43

6   The following table shows the amount of cod landed in the years 1981–86 and percentages of babies vaccinated against measles for the same years.

Cod landed (1000 tonnes)	116	114	112	91	90	77
% vaccinated against measles	54	56	59	63	68	71

a   Draw the scatter graph for the above data.

b   Can we deduce from this graph that the greater the number of babies vaccinated, the smaller the cod catch will be?

c   Explain the high degree of correlation.

It is important to realise that correlation between two sets of data does not mean that the changes in one variable **cause** the changes in the other variable. In other words, being a particular weight does not **cause** you to be a particular height, nor does being a particular height decide what weight you will be.

Although height is linked to weight, there are many other factors which determine your weight.

There is a story that a survey carried out in Sweden paired the number of storks nesting on houses and the number of babies born in the area. The results showed a positive correlation.

We cannot conclude from this that a large number of storks in the area causes a large number of babies to be born, or vice versa.

The reason for the seeming connection is likely to be a third factor, the number of new houses. Storks prefer to nest on new chimneys and new housing estates attract young couples starting families.

When analysing scatter graphs, care should be taken not to jump to conclusions.

It is possible to find a high degree of correlation between variables which are very unlikely to be connected (particularly if only a few points are plotted).

28.3   Line of best fit

If there is a perfect (linear) correlation between two variables, the points plotted will lie on a straight line.

This means that one variable is proportional to the other and an equation of the form $y = mx + c$ can be found to fit the line through the points.

When experimental data is collected, however, there is usually some error in the readings and so the points plotted will not lie exactly along a straight line but close to one.

Example 1

An experiment was carried out to measure the height to which a rubber ball bounced after being dropped from various heights.

Height of drop (m)	0.50	0.75	1.00	1.25	1.50	1.75	2.00
Height of bounce (m)	20	25	50	60	75	75	100

Scatter graph:

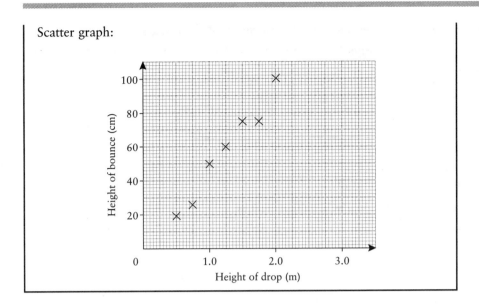

The scatter graph shows that there is a close linear relationship between height of drop and height of bounce. It is not possible to draw a straight line through all the parts so the line which **best fits** all the points is drawn.

All lines of best fit pass through the point $(\bar{x}, \bar{y})$ where $\bar{x}$ and $\bar{y}$ are the means of the first and second sets of data respectively. It is not necessary, at GCSE level, to find these means before drawing the line of best fit unless asked to do so.

Example 2

Draw the line of best fit for the data of Example 1 and use the line to predict the height of bounce when the drop is 1.85 metres.

Draw, by eye, the line which looks to be the 'line of best fit'.

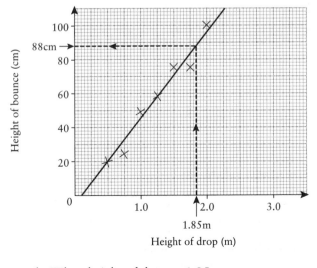

From the graph: When height of drop $= 1.85$ m
  height of bounce $= 88$ cm.

Exercise 28C

1 a Draw the scatter diagram for the following data or use your answer to question **1 b** of Exercise 28A.

Time (minutes)	10	15	20	25	30	35	40
Weight of melting ice block (kg)	3.9	3.5	2.7	1.5	1.2	1.0	0.8

  b Calculate $\bar{x}$ and $\bar{y}$.

  c Draw the line of best fit on your scatter graph.

  d From the graph, find the values of the time when the weight is:
    (i) 2.0 kg   (ii) 3.0 kg

2 a The graph below shows the mean weight for a man of medium build and given height.

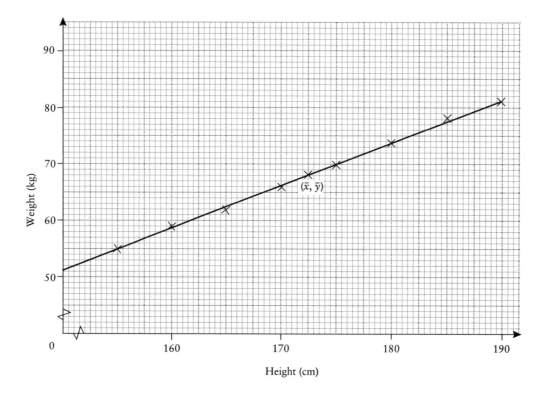

From the graph, find the mean weight for a man whose height is:
(i) 167.5 cm   (ii) 180 cm

  b The corresponding values for a woman's mean weight are:

Height (cm)	142	147	152	157	162	167	172	177
Weight (kg)	46	49	51	54	58	62	65	69

    (i) Plot the scatter graph.
    (ii) Draw the line of best fit.
    (iii) Find the mean weight for a woman of height 165 cm.

3   A mathematics test consisted of two parts: a piece of coursework and a written test. The marks for a class of twelve students are given below:

Student	1	2	3	4	5	6	7	8	9	10	11	12
Coursework (Max 40)	5	7	14	11	20	25	20	29	32	25	32	37
Written test (Max 100)	20	27	32	50	55	62	69	abs	70	90	91	98

a   Draw the scatter graph and line of best fit.

b   One student completed the coursework, and scored 29 marks, but was absent for the written test.

Use your line of best fit to determine a written mark for this student.

4   The amount of energy used by a human body just to exist is given below in terms of the body's surface area.

Age (years)	1	3	5	7	9	11	13	15	19
Energy (kcal/h/m²)	53	51	49	47	45	43	42	42	39

a   Draw the scatter graph and find the line of best fit.

b   Find the energy required by a person of age 17 years.

c   Find the energy required by a person of age 25 years.

d   The energy needed by a 25-year-old is actually 37.5 kcal/h/m². Explain why the graph gives an inaccurate answer for this age.

5   A county in Southern England recorded information, over a seven year period, showing what pupils did after the age of compulsory education.

The scatter diagram represents these data.

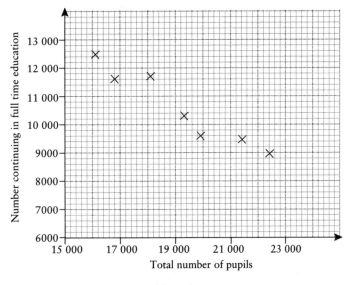

a   On the scatter diagram draw the line of best fit.

b   Use your line to estimate the number of pupils who would **not** continue in full-time education if the total number of pupils in this age group reached 23 000.   (SEG W95)

6   The gestation period and life expectancy for various mammals is represented by this scatter graph.

   a   Copy the diagram and draw the line of best fit on the scatter graph.

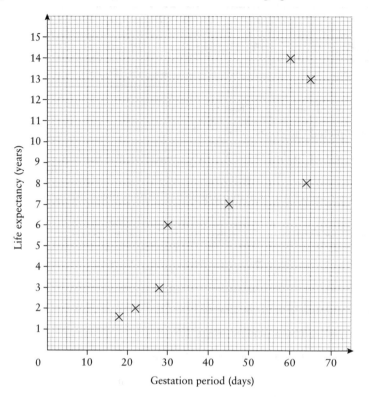

   b   Use your line of best fit to estimate
      (i)   the life expectancy of a mammal that has a gestation period of 35 days.
      (ii) the gestation period of a mammal whose life expectancy is 4.5 years.   (SEG S94)

7   A study of the number of storks and the number of babies born in nine regions of Sweden in 1952 gave
   the following results.

Number of storks ×10 000	Number of babies ×10 000
2	15
3	15
4	18
5	22
6	25
7	30
8	37
9	40
10	40

   a   Plot a scatter graph of these results on a graph.

   b   What does your scatter graph **appear** to suggest about the relationship between the number of storks
       and the number of babies born?   (SEG S95)

8 The scatter diagram shows the population (in millions) and the number of reported cases of an illness, for fourteen regions of Great Britain.

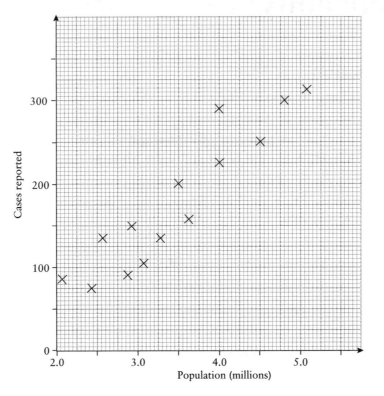

a What relationship does the scatter diagram show?

b For the data above, the mean population is 3.5 million and the mean number of reported cases of the illness is 178.5.

Copy the scatter diagram.
(i) Plot this mean point on the scatter diagram.
(ii) Draw the 'line of best fit' passing through the mean point.

c Use your line to predict the number of reported cases of the illness for
(i) Trent, with a population of 4.7 million;
(ii) the West Midlands, with a population of 5.2 million.

The actual figures for Trent and the West Midlands were

Trent	129 reported cases
West Midlands	145 reported cases.

d Compare these figures with your predictions. (SEG Sp96)

29 Probability

29.1 Introduction

Probability is a measure of how likely something is to occur. What occurs is an **outcome**.

This likelihood or probability can be represented on a sliding scale from the probability when an event is certain to occur to the probability when the event cannot occur.

Suppose twenty children in class 2X are studied. It is found that 19 do not wear glasses, and 1 does wear glasses. 11 of the 20 children are girls and 9 are boys.

Suppose a child is picked at random. The likelihood of the following events occurring fits into the following pattern.

Event	Likelihood
The child studied is in class 2X	Certainty
The child does not wear glasses	High probability
The child is a girl	Just over half
The child is a boy	Just under half
The child wears glasses	Highly unlikely
The child's age is over 70	Impossible

Since 19 out of 20 children do not wear glasses, the likelihood that a child selected at random does not wear glasses is 19 out of 20, or $\frac{19}{20}$. Therefore, the probability that a child selected at random is not wearing glasses is $\frac{19}{20}$.

If all 20 children are wearing shoes, the likelihood that a child selected at random is wearing shoes is 20 out of 20, i.e. $\frac{20}{20}$, which is 1. Therefore, the probability is 1.

None of the children have green hair. The probability that a child selected at random has green hair is 0 out of 20, i.e. $\frac{0}{20}$, which is 0.

If an event is bound to occur, then its probability is 1.

If an event cannot occur, then its probability is 0.

The probability of a child not wearing glasses is $\frac{19}{20}$.

The probability of a child wearing glasses is $\frac{1}{20}$.

The probability of a child wearing glasses or not wearing glasses is 1 (i.e. a certainty).

Note that $\frac{19}{20}$ (not glasses) $+ \frac{1}{20}$ (glasses) $= \frac{20}{20} = 1$ (certainty).

From this we can work out that:

(i) **The total probability for all possible outcomes is 1.**
(ii) **Probability of event happening = 1 − Probability of it not happening**

The events in the initial example occur with the following probabilities:

Event	Likelihood	Probability
The child studied is in class 2X	Certainty	1
The child does not wear glasses	Highly probably	$\frac{19}{20}$
The child is a girl	Just over half	$\frac{11}{20}$
The child is a boy	Just under half	$\frac{9}{20}$
The child wears glasses	Highly unlikely	$\frac{1}{20}$
The child's age is over 70	Impossible	0

29.2   Probability from theory and experiment

Probability can be found either by theory or by experiment.

The theory method relies on logical thought; the experimental method relies on the result of repetition of the event producing results which are taken to be typical.

Example

Find the probability of getting a head when tossing an unbiased coin.

Theory method
'Unbiased' means both events (head, tail) are equally likely to occur.

Heads and tails are the only outcomes,
∴   Probability of head + Probability of tail = 1
(since total of all possible outcomes is 1)

Probability of head = Probability of tail
∴   Probability of tail = $\frac{1}{2}$.

Experimental method
Toss an unbiased coin 100 times, and count the number of tails you obtain. You might obtain 49 tails.
∴   The experimental probability of getting a tail is $\frac{49}{100}$.

If the event is repeated many times, it is usual for the experimental probability to be close to the theoretical probability, but not to be exactly the same.

For the probability of a person being right-handed, there is no theory to use. This would be found by experiment. For example, ask 100 people and see how many are right-handed. The more people you ask, the more likely it is that your probability is accurate, but it must be appreciated that the probabilities found by experiment are not exact. The result may also be different if the experiment is repeated with another group.

Similarly, if the probability of a car being red is $\frac{1}{4}$, the expected number of red cars in a car park containing 80 cars is

$$\frac{1}{4} \times 80 = 20$$

Exercise 29A

1 The probability that a street light is lit is $\frac{3}{8}$.
What is the probability that it is not lit?

2 You toss a coin, marked with a head on both sides.
What is the probability of getting a tail?

3 The probability of a plane being late is $\frac{2}{3}$.
What is the probability of a plane not being late?

4 Jane has a bag containing 6 oranges and 4 apples. She selects a fruit from
the bag at random.
What is the probability that it is a pear?

5 Probabilities can be estimated **either** by making a subjective estimate **or**
by making use of statistical evidence.
State the method which would be used in the following cases.

   a The probability that more women than men will read the news on
   television.

   b The probability that man will live on the moon in the year 2050.

   c The probability that the next vehicle passing a college will be a motor
   cycle.

   d The probability that the next book issued at a library will be a novel.

   e The probability that there will be a cure for deafness within 10 years.
   (SEG S95)

6 In an experiment a drawing pin is thrown. The number of times it lands
point up is recorded.

The drawing pin is thrown 10 times and lands point up 6 times.

   a From these data estimate the probability that it lands point up.

The drawing pin is thrown 100 times and lands point up 57 times.

   b From these data estimate the probability that it lands point up.

   c Give a reason why the second answer is a more reliable estimate.
   (SEG S94)

29.3 Simple probabilities

When n different events are equally likely to occur in an experiment, then the probability of each event occurring is $\frac{1}{n}$.

The general formula when results are equally likely is:

$$\text{Probability of event} = \frac{\text{Number of results giving event}}{\text{Number of possible results}}$$

Example 1

An unbiased cubical die, marked 1 to 6, is thrown. What is the probability of getting a 5?

There are 6 equally likely results.

$\therefore$ The probability of getting a 5 is $\frac{1}{6}$.

Example 2

A bag contains seven balls, identical in shape and size. Three balls are white and four are blue. One ball is selected from the bag at random. What is the probability that the ball is white?

There are seven balls. The probability that any one ball is selected is $\frac{1}{7}$.

There are three white balls.

$\therefore$ The probability of selecting a white ball

$$= \frac{\text{No. of white balls}}{\text{Total no. of balls}}$$

$$= \frac{3}{7}$$

Exercise 29B

1   Jane has a bag containing only 7 oranges and 3 apples. She selects a fruit from the bag at random. What is the probability that it is an apple?

2   A die, numbered 1 to 6, is thrown. What is the probability of getting a 6?

3   A die, numbered 1 to 6, is thrown. What is the probability of getting an even number?

4   From a pack of cards, one card is drawn. What is the probability that it is:

   a red   **b** a Jack   **c** the Queen of Spades?

5   What is the probability of picking an even number from the numbers 10 to 20 (inclusive)?

6   A bag contains 7 red balls, 8 blue balls, 3 yellow balls, and the remainder of the 20 balls are white. What is the probability of drawing:

   a a red ball,

   b a white ball?

7   Out of 20 lamp bulbs, three are faulty. You select a bulb at random. What is the probability that it works?

29.4 Possibility space

When more than one event takes place, you can either write down every possible result, or use the rules outlined on pages 227 and 228.

The possibility space diagram is a means of identifying every possible result. It is used when two or more events occur, and where each of the outcomes is equally likely. Here is an example.

Example

Two unbiased dice, numbered 1 to 6, are thrown, and the score is found by adding the scores on the two dice.
What is the probability of obtaining a total of 10?

Second die	First die					
	1	2	3	4	5	6
1	2	3	4	5	6	7
2	3	4	5	6	7	8
3	4	5	6	7	8	9
4	5	6	7	8	9	10
5	6	7	8	9	10	11
6	7	8	9	10	11	12

The table shows each of the 36 possible results obtained by adding the score on each line. Each of these 36 is equally likely.

The total score of 10 occurs three times:
6 followed by 4; 5 followed by 5; 4 followed by 6.

∴ The probability of a total of 10 is $\frac{3}{36} = \frac{1}{12}$.

Exercise 29C

1 Two dice, numbered from 1 to 6, are thrown, and the score is found by multiplying the scores on the two dice.
What is the probability of a score of:

   a 12   **b** 20   **c** 30?

2 Two coins are thrown. What is the probability of obtaining:

   a two heads,   **b** one head and one tail?

3 Two tetrahedral dice, with faces marked 2, 4, 6 and 8, are thrown together, and the score is found by adding the scores on the two dice.
What is the probability of obtaining a score of:

   a 10   **b** 12?

4 A die, numbered from 1 to 6, and a coin are thrown. If a head is obtained, the score is double that shown on the die. If a tail is obtained, the score is that shown on the die. What is the probability of obtaining a score of:
   a 6   **b** 8?

5   A card is selected from a pack of 52 cards, and a dice, numbered 1 to 6, is thrown. If the card is a heart, the score is twice that shown on the dice. If the card is a diamond, the score is 5. If the card is black, the score is that shown on the dice. What is the probability of obtaining a score of:

a 5   b 6?

6   Two coins are tossed and a dice rolled.

When both coins show heads, the score recorded is twice the value on the dice.

When one coin shows heads and one shows tails, the score recorded is the same as the value on the dice.

When both coins show tails, the score recorded is 0.

a   Complete the table, showing the possible scores.

SCORE		Dice					
		1	2	3	4	5	6
Coins	HH	2					
	HT	1					
	TH	1					
	TT	0					

b   What is the probability that the score is more than 4?

(SEG Sp 96)

29.5   Simple laws of addition and multiplication

Mutually exclusive events

Two events are said to be **mutually exclusive** when both cannot happen at the same time.

Suppose you toss a coin. The event 'obtain a head' and the event 'obtain a tail' are mutually exclusive, as you cannot obtain both a head and a tail at the same time.

Independent events

Two events are said to be **independent** when the result of one does not affect the result of the other.

Suppose you toss a coin *and* throw a die. Whether you obtain a head or a tail from the coin clearly does *not* affect the score on the die. These events are independent of each other.

Addition law for mutually exclusive events

If one event takes place, and you require the probability of A *or* B occurring (and *both* cannot occur), use the **addition law**:

Probability (A *or* B) = Probability (A) + Probability (B).

Example

From a pack of cards, one card is drawn. What is the probability that it is an Ace or a Jack?

$$\text{Probability (Ace)} = \frac{4}{52}$$

$$\text{Probability (Jack)} = \frac{4}{52}$$

$$\therefore \quad \text{Probability (Ace or Jack)} = \frac{4}{52} + \frac{4}{52}$$

$$= \frac{8}{52}$$

$$= \frac{2}{13}$$

Multiplication law for independent events

If two independent events take place, and you require the probability that result A (from the first event), *and* result B (from the second event) both occur, use the **multiplication law**:

Probability (A *and* B) = Probability (A) × Probability (B).

Example

A die is thrown, and a coin is tossed. What is the probability of getting a 5 and a head?

$$\text{Probability (5 on die)} = \frac{1}{6}$$

$$\text{Probability (Head on coin)} = \frac{1}{2}$$

$$\therefore \quad \text{Probability (5 and head)} = \frac{1}{2} \times \frac{1}{6}$$

$$= \frac{1}{12}$$

Exercise 29D

1 A card is selected from a pack of 52 cards. What is the probability that the card selected is a 7 or a 10?

2 Mary drives along a road which has two sets of traffic lights. The probabilities of their being green are $\frac{3}{4}$ and $\frac{2}{5}$ respectively.
What is the probability of Mary finding:

a both sets green,

b both sets not green?

3   The probability of Sway Wanderers winning a match is $\frac{2}{3}$. They play two matches. What is the probability that they win *both* matches?

4   A biased die has a probability of $\frac{1}{5}$ of getting a six. What is the probability that in two throws, you do not get a six?

5   A man throws a die numbered from 1 to 6. What is the probability that he gets a multiple of 5 or a multiple of 3?

6   John selects a pen at random from 6 blue pens and 5 black pens. He then selects a sheet of paper at random from 10 plain sheets and 20 lined sheets. What is the probability that he then writes with a blue pen on lined paper?

7   In a game of bingo there are 20 red, 3 black, 15 blue, and 12 white balls left. What is the probability that the next ball picked is either a red ball or a blue ball?

29.6   Tree diagrams

When two or more events take place, it is often simpler to find probabilities by means of diagrams that show all the possible events and their probabilities. A **tree diagram** is one such diagram.

Suppose a bag contains 5 blue and 7 white balls. A ball is selected at random, its colour noted, and it is then returned to the bag. The bag is shaken, and a second random selection is made.

There are two possible results of the first selection (either a blue ball or a white ball is obtained), which are represented by two branches on a tree diagram:

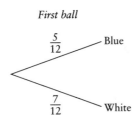

(Notice that the end-points of the branches are labelled with the event, and that the corresponding probabilities are written beside the branches.)
Now consider the second selection:

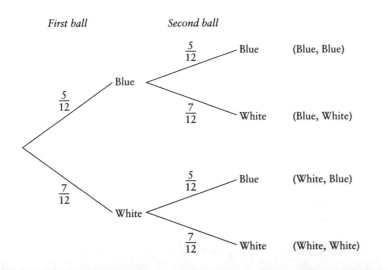

The above tree diagram shows both selections. The very top branches correspond to both selections being a blue ball, shown by (Blue, Blue). The probability of this is:

$$P = \frac{5}{12} \times \frac{5}{12} = \frac{25}{144}$$

This result is obtained by multiplying the probabilities on the two branches.

Similarly:

$$P\text{ (White, Blue)} = \frac{5}{12} \times \frac{7}{12} = \frac{35}{144}$$

$$P\text{ (Blue, White)} = \frac{7}{12} \times \frac{5}{12} = \frac{35}{144}$$

$$P\text{ (White, White)} = \frac{7}{12} \times \frac{7}{12} = \frac{49}{144}$$

The completed tree diagram for this experiment is:

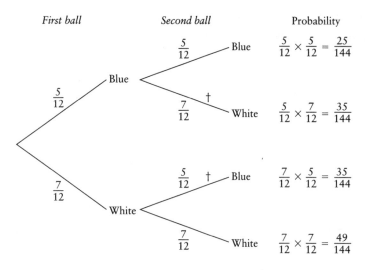

Notes
   (i) The probability of any event represented by consecutive branches of the tree is the product of the probabilities along those branches.
   (ii) The probability of any event which can occur in two or more ways is the sum of the probabilities at the end of the relevant branches.

The two sets of branches marked with a dagger (†) represent picking one white ball and one blue ball.

$$\therefore \quad \text{The probability of one of each colour} = \frac{35}{144} + \frac{35}{144}$$

$$= \frac{70}{144}$$

$$= \frac{35}{72}$$

(For questions on this section, see Exercise 29E, questions 1–7, pages 232–3.)

*Tree diagrams for 'without-replacement' questions

In the example above, the ball was replaced before the second selection was made. This made the probabilities for the colour of the second ball the same as those for the first ball.

When the ball is *not* replaced, it is necessary to amend each probability for the second ball. Here is an example.

— *Example* —

A bag contains 5 blue and 7 white balls. A ball is selected at random, its colour is noted, and it is *not* returned to the bag. What are the probabilities that:

a both are blue,
b both are white,
c one of each colour is chosen?

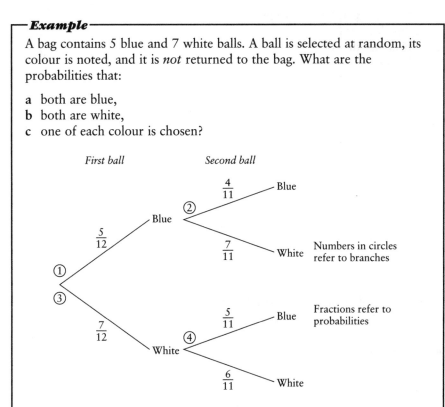

The bag contained 5 blue and 7 white balls. At branch 1, a blue ball has been removed. This leaves only 4 blue and 7 white balls in the bag. Hence the probability of a blue ball on branch 2 is $\frac{4}{11}$. Similarly, in branch 3, a white ball has been removed, leaving 5 blue and 6 white balls, and therefore the probability of a blue ball in branch 4 is $\frac{5}{11}$.

The required probabilities are:

a $\quad P\text{ (Blue, Blue)} = \frac{5}{12} \times \frac{4}{11} = \frac{5}{33}$

b $\quad P\text{ (White, White)} = \frac{7}{12} \times \frac{6}{11} = \frac{7}{22}$

c $\quad P\text{ (One of each)} = \left(\frac{5}{12} \times \frac{7}{11}\right) + \left(\frac{7}{12} \times \frac{5}{11}\right)$

$\qquad\qquad = \frac{35}{132} + \frac{35}{132}$

$\qquad\qquad = \frac{70}{132} = \frac{35}{66}$

1 A bag contains 6 blue discs and 8 red discs. A disc is withdrawn, its colour is noted, and then it is replaced. A second disc is with drawn and its colour is noted. What is the probability that the two discs are:

   a both blue
   b both red
   c one of each colour?

2 The probability of Mark oversleeping is $\frac{1}{3}$. If he oversleeps, the probability that he does not eat breakfast is $\frac{4}{5}$.
   If he does not oversleep, the probability that he has breakfast is $\frac{3}{4}$.
   Find the probability that Mark has breakfast.

3 Victoria travels to college either by bus (probability $\frac{3}{5}$), or by train. If she travels by bus, the probability that she is on time is $\frac{3}{4}$, but if she travels by train, the probability that she is on time is $\frac{1}{5}$.
   What is the probability that Victoria is on time?

4 Two dice numbered 1 to 6 are thrown. What is the probability of obtaining:

   a two sixes
   b exactly one six?

5 A bag contains 5 white, 2 blue and 3 yellow balls. John selects a ball, notes its colour and replaces the ball. He withdraws a second ball, and notes its colour.
   What is the probability that John selects:

   a two white balls
   b two balls of different colours?

6

An aeroplane has two engines. From past experience the probability of failure of an engine on any journey is $\frac{1}{1000}$.
A plane can safely land with just one engine working with a probability of $\frac{9}{10}$.
It will always land safely if both engines are working.
Calculate the probability that a plane will safely finish a journey.

(SEG W95)

7   A manufacturer makes switches for electric circuits. The probability that
    a switch is faulty is 0.1.
    William buys 200 of these switches.

    a   (i) How many of these switches can be expected to be faulty?
        (ii) Explain why your answer is only an approximation.

    b   Draw and label a probability tree diagram to show the possible
        outcomes when two switches are chosen at random.

    c   Calculate the probability that
        (i) both switches are **not** faulty.
        (ii) exactly **one** switch is faulty.   (SEG Sp 94)

*8 A bag contains 8 white and 6 black discs. Two discs are taken out and
    their colour noted.
    What is the probability that the two discs are:

    a   both white
    b   of different colours?

*9 On Mothering Sunday Mrs Myers is given a box of hand-made
    chocolates, half of which are dark chocolate and half of which are milk
    chocolate. The box contains 10 chocolates. Mrs Myers eats two
    immediately, taking them out of the box at random.

    a   Draw a tree diagram to show the possible outcomes of the type of
        chocolate eaten.

    b   What is the probability that Mrs Myers eats two milk chocolates?

    c   What is the probability that Mrs Myers eats one of each kind of
        chocolate?

    d   Mrs Myers eats a third chocolate. What is the probability that she has
        then eaten three dark chocolates?

*10 Kate has a box containing 15 chocolates which look identical. 10 have
    soft centres and the rest have hard centres.
    She picks a chocolate at random, eats it and then she picks a second
    chocolate at random and eats it.

    a   Draw the probability tree diagram to represent the outcomes when
        the **two** chocolates are eaten.

    b   Calculate the probability that she eats **two** chocolates of different
        types.

    Kate picks at random and eats a third chocolate.

    c   Calculate the probability that she eats three soft centred chocolates.
                                                              (SEG Sp 95)

30 Algebra

30.1   The basics of algebra

From words to symbols

Every Saturday the milkman delivers the following to 16 Cromarty St:
3 loaves of bread, 4 pints of milk and a dozen eggs.

He then presents his bill.

A sentence such as 'the cost of three loaves of bread, four pints of milk and a dozen eggs' is difficult to deal with mathematically.

Algebra is a mathematical language which enables us to deal with problems more easily.

Suppose we let:

b stand for 'the cost of a loaf of bread'
m stand for 'the cost of a pint of milk'
and e stand for 'the cost of a dozen eggs'.

The milkman's bill for 3 loaves of bread, 4 pints of milk and 1 dozen eggs becomes

$$3b + 4m + 1e \text{ or } 3b + 4m + e$$

where $3b$ means 3 times b (or $3 \times b$).
We can miss out the 1 in front of the e (e means $1 \times e$).

The foundation of algebra, as a science of equations, was laid down in the year 825 by Muhammad Ibn Musa al-Khwarizmi, an Islamic mathematician.

TERMINAL MODULE

Example 1

A holiday has two prices, one for adults and the other for children. Write down an algebraic expression for the cost of a holiday for two adults and three children.

Let the cost for one adult be £A and the cost for one child be £C:

Cost for 2 adults	$= £2A$
Cost for 3 children	$= £3C$
Cost of the holiday	$= £2A + £3C = £(2A + 3C)$

Example 2

Don thinks of a number, multiplies it by 3, adds 4 and then divides this answer by the original number.
Write his final number in algebraic terms.

Let his first number be	n
Multiply by 3:	$3n$
Add 4:	$3n + 4$
Divide by the original number:	$\dfrac{3n + 4}{n}$

$$\text{Final number} = \frac{3n + 4}{n}$$

Exercise 30A

Simplify the following phrases and sentences by translating them into algebraic expressions:

1   The cost of 3 pounds of apples, 4 pounds of bananas, and 2 pounds of cherries.

2   The entrance fee to the leisure centre for 3 adults and 5 children.

3   The weight of 6 tablespoonsful of flour and 3 of sugar.

4   The sum of seven times x and five times y.

5   The difference between q and twice p.

6   Double the number a and divide by the number b.

7   A number is formed by multiplying a number x by five and then subtracting eight.

8   A number is formed by dividing a number y by four and then adding six.

9   Think of a number, multiply by two, add three and divide by eight.

10 Think of a number, multiply by four, subtract one and then divide by twice the original number.

Collecting like terms

Mrs Goodman empties her son's pockets on washing day and finds two sweets, one dirty hanky, two pencils, two more sweets and another pencil.

The contents of the pocket can be listed in algebraic terms as

$$2s + 1h + 2p + 2s + 1p$$

The list can be simplified if articles of the same kind are added together. This is called **collecting like terms**. If we collect the like terms from Master Goodman's pocket, the list becomes $h + 3p + 4s$ (which cannot be simplified further as each term is different).

Addition and subtraction

Only like terms may be added or subtracted.

Like terms are those which are multiples of the same algebraic variable.

For example, $3a$, $7a$ and $-8a$ are all like terms.

The expression $3a + 7a - 8a$ can be simplified to $2a$.

The terms $7a$, $3b$ and $4c$ are unlike terms and the expression $7a + 3b + 4c$ cannot be simplified any further.

Example 1

Simplify the following expression:

$$2a + 3c + a + 4b + 2c + b$$

$$2a + 3c + a + 4b + 2c + b = (2a + a) + (4b + b) + (3c + 2c)$$
$$= 3a + 5b + 5c$$

Example 2

Simplify the following by collecting like terms:

$$5x + 2y - 4z + 3x - y + z$$

$$5x + 2y - 4z + 3x - y + z = (5x + 3x) + (2y - y) + (z - 4z)$$
$$= 8x + y - 3z$$

Exercise 30B

Simplify the following by collecting terms:

1   $2x + 3x$

2   $8n - n$

3   $3a - a + 6a$

4   $2b - 3b + 4b$

5   $5y - 6y + 3y$

6   $2x + 3x + y + 4y$

7   $3x + 4y - 2x + y$

8   $4a - 2b - a + 3b$

9   $x + 2y + z + 3y + 8z$

10   $3a + 2a + c + b + 2a + 3c + 5b$

11   $5p + 3q + 2r - q + 7p - 3p - r$

12   $8x - 3y - 4x + 5y - 3z$

13   $2x - y + 3x - 2y + z$

14   $a + b - 2a + c - 2b + 3c$

15   $\frac{1}{2}p + \frac{1}{2}q - \frac{1}{4}p + \frac{3}{4}q$

16   $3x - y + \frac{1}{2}y - 2\frac{1}{2}y + 8y$

Substitution

The letters in an algebraic expression stand for numbers or amounts.

For example, in the expression $3b + 4m + e$, b stands for the cost of a loaf of bread, m for the cost of a pint of milk and e for the cost of a dozen eggs.

The advantage of the letters is that the amounts are not fixed, they can change or vary.

The letters are called **variables**. The numbers, which are constant, are called **coefficients**.

If a loaf of bread costs 54p, a pint of milk 29p and a box of eggs 63p (i.e. $b = 54$, $m = 29$, $e = 63$), then

Weekly bill $= 3b + 4m + e$

$$= (3 \times 54) + (4 \times 29) + 63$$

$$= 341 \text{ pence}$$

If, however, prices rise and $b = 56$, $m = 31$, $e = 66$, then

Weekly bill $= 3b + 4m + e$

$$= (3 \times 56) + (4 \times 31) + 66$$

$$= 358 \text{ pence}$$

Example 1

Find the value of $4x + y$ when $x = 3$ and $y = -1$.

$$4x + y = (4 \times 3) + (-1)$$
$$= 12 - 1$$
$$= 11$$

Example 2

Evaluate $\dfrac{3x - y}{x + y}$ when $x = 3$, $y = -1$.

$$\frac{3x - y}{x + y} = \frac{(3 \times 3) - (-1)}{3 + (-1)} = \frac{9 + 1}{3 - 1} = \frac{10}{2} = 5$$

Exercise 30C

1   Evaluate the following expressions when $x = 3$, $y = 5$, $z = 2$.

a $2x + y$ d $\dfrac{12}{z}$ g $2x - 5z$

b $3x - 2y$ e $\dfrac{y}{10}$ h $xy + z$

c $y + 4z$ f $4z - 6$

2   Evaluate each of the expressions in question 1 when $x = -1$, $y = 3$, $z = -2$.

3   Evaluate the following expressions if $x = 4$, $y = 3$, $z = -2$:

a y^2 **d** $\dfrac{x}{z}$ **g** $\dfrac{x+y}{x-y}$

b xyz **e** $xz + y$ **h** $x(x+y)$

c $2z^2$ **f** $2x^2 - yz$ **i** $\dfrac{x + 2z}{y}$

Multiplication and division

Multiplication of algebraic terms is usually easier than multiplication in arithmetic. For example:

$$a \text{ multiplied by } 4 = a \times 4 = 4a$$

$$p \text{ multiplied by } q = p \times q = pq$$

$$x \text{ divided by } y = x \div y = \frac{x}{y}$$

$$a \text{ multiplied by } a = a \times a = a^2$$

Algebraic multiplication is commutative, i.e. $ab = ba$.

However, algebraic terms are usually listed in alphabetical order with the constant first.

The rules for directed numbers are the same as on page 3. For example:

$$(+x) \times (+y) = xy$$

$$(-x) \times (+y) = -xy$$

$$(+x) \div (-y) = -\frac{x}{y}$$

$$(-x) \div (-y) = \frac{x}{y}$$

Example 1

Simplify $2 \times x \times 3 \times y$.

Collect numbers and letters separately:

$$\begin{aligned}
2 \times x \times 3 \times y &= (2 \times 3) \times (x \times y) \\
&= 6 \times xy \\
&= 6xy
\end{aligned}$$

Example 2

Simplify $5 \times p \times 6 \times q \div 3 \div r$

Collect numbers and letters separately:

$$5 \times p \times 6 \times q \div 3 \div r = (5 \times 6 \div 3) \times (p \times q \div r)$$

$$= \frac{10pq}{r}$$

Example 3

Simplify $(3x) \times (-2y) \times (-z)$.

$$(3x) \times (-2y) \times (-z) = (3 \times -2 \times -1) \times (x \times y \times z)$$
$$= 6xyz$$

Exercise 30D

Simplify:

1	$5 \times x$	5	$a \times b \times c$	9	$(-2x) \times (-3y)$
2	$2 \times 3 \times a$	6	$2 \times p \times 3 \times r \times q$	10	$(4p) \div (-2q)$
3	$3 \times m \times n$	7	$2 \times y \div z$	11	$(6x) \times (-3y) \div (-2z)$
4	$4 \times x \times 2 \times y$	8	$c \div 2d \times 8b$	12	$(-2a) \times (-b) \div (-b)$

Indices

$$a \times a \times a = a^3$$

The number 3 is called an **index**. It shows, or indicates, the number of as which have been multiplied together to give the third power of a.

$$a^1 \times a^2 = a \times a \times a = a^3, \text{ i.e. } a^{(1+2)}$$
$$a^3 \times a^2 = (a \times a \times a) \times (a \times a) = a^5 \text{ i.e. } a^{(3+2)}$$

In general:
$$a^x \times a^y = a^{(x+y)}$$

$$a^3 \div a^2 = \frac{a \times a \times a}{a \times a} = a, \text{ i.e. } a^{(3-2)}$$

$$a^5 \div a^3 = \frac{a \times a \times a \times a \times a}{a \times a \times a} = a \times a = a^2, \text{ i.e. } a^{(5-3)}$$

In general:
$$a^x \div a^y = a^{(x-y)}$$

$$(a^2)^3 = a^2 \times a^2 \times a^2 = a^6 \text{ i.e. } (a^2)^3 = a^{2 \times 3}$$

In general:
$$(a^x)^y = a^{xy}$$

$$a^3 \div a^5 = \frac{a \times a \times a}{a \times a \times a \times a \times a} = \frac{1}{a \times a} = \frac{1}{a^2}$$

but
$$a^3 \div a^5 = a^{(3-5)} = a^{-2}, \text{ i.e. } \frac{1}{a^2} = a^{-2}$$

In general:
$$a^{-x} = \frac{1}{a^x}$$

$$a^3 \div a^3 = \frac{a \times a \times a}{a \times a \times a} = 1$$

but
$$a^3 \div a^3 = a^{(3-3)} = a^0$$

In general:
$$a^0 = 1$$

The rules for indices can be summarised as follows:

$$a^x \times a^y = a^{(x+y)}$$
$$a^x \div a^y = a^{(x-y)}$$
$$(a^x)^y = a^{xy}$$

$$a^{-x} = \frac{1}{a^x}$$

$$a^0 = 1$$

Example 1

Simplify $x^5 \div x^3 \times x^2$.

$$x^5 \div x^3 \times x^2 = \frac{x \times x \times x \times x \times x \times x \times x}{x \times x \times x} = x^4$$

or

$$x^5 \div x^3 \times x^2 = x^{(5-3+2)} = x^4$$

Example 2

Simplify $3a \times (2b)^2 \times (-a^2)$.

$$3a \times (2b)^2 \times (-a^2) = (3 \times 2^2 \times -1) \times a \times b^2 \times a^2$$
$$= -12a^3b^2$$

Example 3

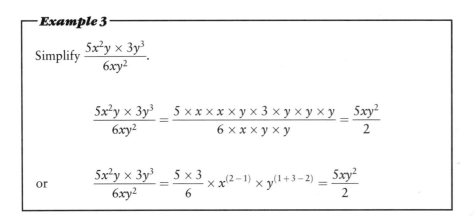

Simplify $\dfrac{5x^2y \times 3y^3}{6xy^2}$.

$$\frac{5x^2y \times 3y^3}{6xy^2} = \frac{5 \times x \times x \times y \times 3 \times y \times y \times y}{6 \times x \times y \times y} = \frac{5xy^2}{2}$$

or

$$\frac{5x^2y \times 3y^3}{6xy^2} = \frac{5 \times 3}{6} \times x^{(2-1)} \times y^{(1+3-2)} = \frac{5xy^2}{2}$$

Exercise 30E

Simplify the following expressions:

1 $x \times x \times x$

2 $x \times x \times y$

3 $x^2 \times x$

4 $x^3 \times x^2$

5 $x^5 \times x^3$

6 $x^3 \times x \times x^2$

7 $x^7 \div x$

8 $x^4 \div x^3$

9 $x^5 \times x^2 \div x^4$

10 $(x^2)^2$

11 $(2x^2)^3$

12 $(-2x^3)^2$

*13 $3x \times y^2 \times x^2$

*14 $2a^2b \times -3bc$

*15 $3ab^2 \times 2ab \times bc^2$

*16 $(-3a)^2 \times b$

*17 $(-2ab) \times (3b)^2$

*18 $(-2x) \times (y) \div (-y)$

*19 $x^2 y \times xy^2$

*20 $12ab \div (-4c^2) \times (3ac)$

30.2 Brackets

The milkman's order for 16 Cromarty Street is three loaves of bread, four pints of milk and one dozen eggs per week. After five weeks the milkman will have delivered five times this amount.

Suppose the cost of a loaf of bread is b, the cost of a pint of milk is m, and that a dozen eggs costs e. Then suppose that we want to work out the total cost after five weeks.

The neatest way to write this cost in algebraic terms is to use a bracket:

Cost after 5 weeks $= 5(3b + 4m + e)$

To 'remove' the bracket from the expression, each term must be multiplied by 5:

$$5(3b + 4m + e) = 15b + 20m + 5e$$

If the number outside the bracket is a negative number, take care: the rules for multiplication of directed numbers must be applied.

Example 1

Remove the bracket from: **a** $4(3x - 2y)$ **b** $-2(x - 3y)$.

a $\quad 4(3x - 2y) = 4 \times 3x - 4 \times 2y$

$\qquad\qquad\qquad = 12x - 8y$

b $\quad -2(x - 3y) = -2 \times x - (-2) \times 3y$

$\qquad\qquad\qquad = -2x + 6y$

Example 2

Remove the brackets and simplify:
a $\quad 3(2x - y) - 2(x - 4y)$ **b** $\quad a(a + b) - b(a + 2b)$.

a $\quad 3(2x - y) - 2(x - 4y) = 6x - 3y - 2x + 8y$

$\qquad\qquad\qquad\qquad\qquad = 4x + 5y$

b $\quad a(a + b) - b(a + 2b) = a^2 + ab - ba - 2b^2$

$\qquad\qquad\qquad\qquad\qquad = a^2 - 2b^2 \text{ (since } ba = ab\text{)}$

Exercise 30F

Remove the brackets and simplify:

1 $\quad 3(x + y)$	8 $\quad -2(3p - 6q)$	15 $\quad 2(x - 2y) + 3(x + 2y)$
2 $\quad 2(3x + y)$	9 $\quad 5(x - 3y + 2z)$	16 $\quad y(y + 4) - 3(y + 4)$
3 $\quad 4(x - y)$	10 $\quad -(a - b - c)$	17 $\quad 5(2a + 3b) - (a - 2b)$
4 $\quad 7(a + 2y)$	11 $\quad x(3x - y + 2z)$	18 $\quad 2(p - 6q) + 3(2p - q)$
5 $\quad 6(p - 2q)$	12 $\quad -a(a + b - c)$	19 $\quad 5x(x - y) - 2x(2x + 3y)$
6 $\quad 4(5x + 2y)$	13 $\quad x(2x^2 + 3x + 2)$	20 $\quad (x + y - z) - 3(x - y + z)$
7 $\quad 3(2a - 4b)$	14 $\quad 4y(y^2 - 3y + 1)$	21 $\quad 4(x + y) - 3(x - y) + 2(x - y)$

30.3   Common factors

The process which is the reverse of multiplying out brackets is called **factorising**.

Factorising is a very important technique in algebra. It enables you to simplify expressions and hence makes the solving of problems easier.

Factors have already been met in Section 1.5 on page 13 and a factor in algebra is the same as a factor in arithmetic. Remember: a factor is a number which will divide exactly into a given number.

Consider the number $2x + 6y$.

2 is a factor of each part (or term) of this number and therefore of the whole number.

$$\therefore \quad 2x + 6y = 2 \times x + 2 \times 3 \times y$$
$$= 2 \times (x + 3 \times y)$$
$$= 2(x + 3y)$$

$(x + 3y)$ and 2 are both factors of $2x + 6y$.

Example 1

Factorise $6p + 3q + 9r$.

The factor which is common to each term is 3.

$\therefore \quad 6p + 3q + 9r = 3(2p + q + 3r)$

Example 2

Factorise $x^2 + xy + 6x$.

The factor which is common to each term is x

$\therefore \quad x^2 + xy + 6x = x(x + y + 6)$

Example 3

Factorise $2x^2 - 4xy$.

This expression has more than one common factor.
Both 2 and x are common factors.

$\therefore \quad 2x^2 - 4xy = 2x(x - 2y)$

$2x^2 - 4xy$ therefore has three factors: 2, x and $(x - 2y)$.

Exercise 30G

Factorise the following expressions:

1 $4x + 12y$	6 $12s + 20t$	11 $6xy + 3x$	16 $7p - 14q + 7r$
2 $3p - 6q$	7 $xy + xz$	12 $5x - 10x^2$	17 $2a^2 + 4ab - 8a$
3 $5a + 10$	8 $xy + x^2$	13 $2a + 8b - 4c$	18 $3p^2 + 6pq - 9p$
4 $10b - 5$	9 $y^2 - 2y$	14 $9x - 3y - 6z$	19 $x^2y + xyz + xy^2$
5 $14m - 21n$	10 $2y^2 - 4y$	15 $15x - 5y + 10$	20 $4pqr - 12p^2q$

*30.4 The addition and subtraction of fractions

The rules for adding and subtracting algebraic fractions are exactly the same as for fractions in arithmetic:

Method	Arithmetic fraction	Algebraic fraction
To add	$\dfrac{2}{3} + \dfrac{2}{5}$	$\dfrac{a}{3} + \dfrac{a}{5}$
(i) Find the LCM of the denominators (see page 26)	LCM = 15	LCM = 15
(ii) Convert each fraction to an equivalent fraction with this denominator	$\dfrac{10}{15} + \dfrac{6}{15}$	$\dfrac{5a}{15} + \dfrac{3a}{15}$
(iii) Collect like terms, i.e. add the numerators	$\dfrac{16}{15}$	$\dfrac{8a}{15}$

Example 1

Express as a single fraction $\dfrac{4}{a} + \dfrac{3}{a}$.

The fractions have a common denominator a:

$$\frac{4}{a} + \frac{3}{a} = \frac{4+3}{a} = \frac{7}{a}$$

Example 2

Express as a single fraction $\dfrac{2}{a} + \dfrac{3}{b} - \dfrac{1}{c}$.

The common denominator is abc and the equivalent fractions are:

$$\frac{2}{a} \times \frac{bc}{bc} = \frac{2bc}{abc} \qquad \frac{3}{b} \times \frac{ac}{ac} = \frac{3ac}{abc} \qquad \frac{1}{c} \times \frac{ab}{ab} = \frac{ab}{abc}$$

In equivalent fractions:

$$\frac{2}{a} + \frac{3}{b} - \frac{1}{c} = \frac{2bc}{abc} + \frac{3ac}{abc} - \frac{ab}{abc} = \frac{2bc + 3ac - ab}{abc} \text{ as a single fraction.}$$

As there are no 'like terms' the answer cannot be simplified further.

Example 3

Express as a single fraction $\dfrac{(x-1)}{2} - \dfrac{(2x+1)}{6}$.

The LCM of 2 and 6 is 6:

$$\dfrac{(x-1)}{2} - \dfrac{(2x+1)}{6} = \dfrac{3(x-1)}{6} - \dfrac{1(2x+1)}{6} \quad \text{(as equivalent fractions)}$$

$$= \dfrac{3(x-1) - (2x+1)}{6} \quad \text{(as a single fraction)}$$

$$= \dfrac{3x - 3 - 2x - 1}{6} \quad \text{(multiplying out brackets)}$$

$$= \dfrac{x-4}{6} \quad \text{(collecting like terms)}$$

Exercise 30H

Express each of the following as a single fraction and simplify where possible:

1. $\dfrac{2x}{5} + \dfrac{x}{5}$

2. $\dfrac{2a}{3} + \dfrac{a}{3}$

3. $\dfrac{x}{2} - \dfrac{x}{4}$

4. $x + \dfrac{3x}{4}$

5. $\dfrac{x}{4} + \dfrac{x}{3}$

6. $\dfrac{2x}{5} - \dfrac{x}{3}$

7. $\dfrac{x}{2} + \dfrac{x}{3} + \dfrac{x}{4}$

8. $\dfrac{3a}{4} + \dfrac{a}{3} - \dfrac{5a}{6}$

9. $\dfrac{y}{6} + \dfrac{y}{2} - \dfrac{y}{3}$

10. $\dfrac{2y}{5} - \dfrac{y}{3}$

11. $\dfrac{1}{x} + \dfrac{2}{x}$

12. $\dfrac{3}{5x} - \dfrac{1}{2x}$

13. $\dfrac{4}{y} - \dfrac{5}{2y}$

14. $\dfrac{2}{3y} + \dfrac{5}{6y} - \dfrac{1}{y}$

15. $\dfrac{1}{4a} + \dfrac{2}{5a} - \dfrac{1}{2a}$

16. $1 + \dfrac{2}{x} + \dfrac{3}{2x}$

17. $\dfrac{(x+2)}{3} - \dfrac{x}{4}$

18. $\dfrac{(2x-1)}{2} + \dfrac{(x+3)}{5}$

19. $\dfrac{(5-2x)}{6} + \dfrac{(4x-1)}{3}$

20. $\dfrac{4x}{5} - \dfrac{(x+3)}{2}$

30.5 Equations

Forming equations

Many problems are solved more quickly if they are first written in algebraic terms.

An equation is, in algebra, the equivalent of a sentence. All equations must contain an equals sign.

The following problem is solved by first translating into algebra.

Example 1

My brother is twice as old as I am and the sum of our ages is 42. How old am I?

English	Algebra
My age	x (years)
My brother's age	$2x$
The sum of our ages	$x + 2x$
The sum of our ages is forty-two	$x + 2x = 42$

In algebra the problem becomes:

$$x + 2x = 42$$
$$3x = 42 \text{ (collecting like terms)}$$
$$x = 14 \text{ (dividing by 3)}$$

This translates back into English as:
My age is 14 years.

Example 2

I think of a number, treble it, add seven and the answer is 19. Form an equation using this information.

Let the number thought of be n.
Then

treble the number add seven is nineteen becomes
$$3n \quad + \quad 7 \quad = \quad 19$$

Exercise 30I

For each of the following, rewrite the problem as an algebraic equation:

1 Five added to a number gives the answer twelve.

2 Seven subtracted from a number leaves thirteen.

3 If 4 is added to twice a number, the answer is equal to 10.

4 Seven times a number is twenty-one.

5 Six subtracted from five times a number gives the answer twenty-nine.

6 Ten added to half of a number gives the answer twenty-two.

7 Think of a number, double it, subtract three and the answer is three.

8 Think of a number, divide by four, subtract three and the answer is three.

9 A rectangle has a length which is double its width. The perimeter of the rectangle is eighteen.

10 The length of a rectangle is 4 cm more than its width. Its perimeter is 28 cm.

Solving equations

Once a problem has been written as an algebraic equation, the problem can be solved by solving the equation.

To solve $3n + 7 = 19$ means finding the value of n which makes the equation true.

To do this, the 7 and 3 must be eliminated from the LHS (left-hand side) of the equation to leave $n =$ the solution.

An equation must always be balanced, i.e. the LHS must always equal the RHS (right-hand side).

In the equation above

$3n + 7$ balances 19

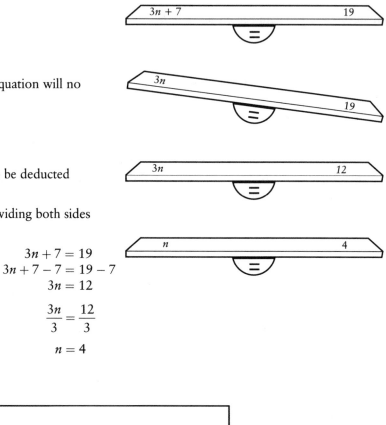

If 7 is deducted from the LHS, the equation will no longer be balanced.

To maintain the balance, 7 must also be deducted from the RHS, giving $3n = 12$.

$3n$ must now be reduced to $1n$ by dividing both sides by 3, giving $n = 4$.

The stages are: $\qquad\qquad\qquad 3n + 7 = 19$
Subtract 7 from both sides: $\qquad 3n + 7 - 7 = 19 - 7$
giving $\qquad\qquad\qquad\qquad\qquad 3n = 12$

Divide both sides by 3: $\qquad\qquad \dfrac{3n}{3} = \dfrac{12}{3}$

giving $\qquad\qquad\qquad\qquad\qquad n = 4$

$\therefore$ The number was 4.

Example 1

Solve the equation $3x - 4 = 5$.

$$3x - 4 = 5$$
Add 4 to each side: $\qquad 3x - 4 + 4 = 5 + 4$
$$3x = 9$$
Divide each side by 3: $\qquad\qquad x = 3$

Example 2

Solve the equation $2(x - 3) = 5$.

$$2(x - 3) = 5$$
Multiply out the bracket: $\qquad 2x - 6 = 5$
Add 6 to each side: $\qquad\qquad 2x = 11$
Halve each side: $\qquad\qquad\qquad x = 5\frac{1}{2}$

Exercise 30J

1–8 Solve the equations formed in questions 1–8 of Exercise 30I to find the numbers.

9 Solve the equation formed in question 9 of Exercise 30I to find the width of the rectangle.

10 Solve the equation formed in question 10 of Exercise 30I to find the length of the rectangle.

In the following questions, solve the equations to find the value of x:

11 $x + 7 = 11$

12 $5x = 15$

13 $4x - 5 = 19$

14 $4x - 3 = 3$

15 $\dfrac{3x}{4} = 6$

16 $12 - x = 9$

17 $17 - 2x = 5$

18 $2(x - 6) = 8$

19 $3(x - 4) = 15$

20 $\dfrac{x}{2} + 3 = 7$

21 $\dfrac{x}{3} - 2 = 1$

22 $\dfrac{2x}{5} + 2 = 10$

23 $5(2x + 1) = 19$

24 $2(3x - 1) = 7$

Harder equations

The method used above can be very time-consuming and, when solving more difficult equations, a quicker method is used.

For the equation

$$3n + 7 = 19$$

when 7 is subtracted from both sides, we know it will disappear from the LHS; therefore it is only necessary to write it on the RHS:

$$3n = 19 - 7$$

i.e. the 7 has *changed sides* and has *changed sign* from + to −.

We now have:

$$3n = 12$$

Similarly, when both sides are divided by 3 we know that the LHS will be reduced to n.

Showing the division on the RHS only we have:

$$n = \frac{12}{3} = 4$$

i.e. the 3 has *changed sides* and *changed sign* from × to ÷.

The rule for eliminating a quantity from one side of an equation is

change to the opposite side and change to the opposite sign.

This process is called **transposing**.

Example 1

Solve the equation $5x - 4 = 3x + 12$.

In this type of equation, the x terms should be collected on one side of the equation and the numerical terms on the other, i.e. $3x$ must be eliminated from the RHS and -4 from the LHS.

$$5x - 4 = 3x + 12$$

Transposing $3x$ and -4 gives:
$$5x - 3x = 12 + 4$$
$$2x = 16$$
$$x = 8$$

Answers to algebraic equations should *always* be checked by substituting back into the LHS and RHS of the original equation, as shown below:

$$\text{LHS} = 5x - 4 = 5 \times 8 - 4 = 36$$
$$\text{RHS} = 3x + 12 = 3 \times 8 + 12 = 36$$
$$\text{LHS} = \text{RHS, so the solution is correct.}$$

Example 2

Solve $2(2x - 5) = 3(x - 4)$

$$2(2x - 5) = 3(x - 4)$$

Multiply out the brackets:
$$4x - 10 = 3x - 12$$

Transpose $3x$ and -10:
$$4x - 3x = 10 - 12$$
$$x = -2$$

Example 3

Solve $4(x + 3) - 2(x - 5) = 46$

$$4(x + 3) - 2(x - 5) = 46$$

Multiply out the brackets:
$$4x + 12 - 2x + 10 = 46$$

Collect terms:
$$2x + 22 = 46$$

Transpose 22:
$$2x = 24$$

Divide by 2:
$$x = 12$$

Exercise 30K

Solve the following equations to find the value of x:

1. $7x + 3 = 5x + 11$

2. $5x - 2 = 2x + 7$

3. $6x - 4 = 10 - x$

4. $3 - 2x = 4 - 5x$

5. $3(x - 5) = 12$

6. $5 + 2(x + 1) = 11$

7. $2(x + 7) + 4 = 18$

8. $2(x - 4) = (x + 2)$

9. $3(2x - 1) = 7(x - 1)$

10. $(x + 5) + 2(3x - 2) = 8$

11. $4(x + 2) + 2(x + 3) = 32$

12. $2(2x + 1) - 3(3x - 4) = 40$

Further problems

---**Example 1**---

I buy a pizza and cut it into three pieces. When I weigh the pieces, I find that one piece is 8 g lighter than the largest piece and 5 g heavier than the smallest piece.
If the whole pizza weighs 360 g, how much does the smallest piece weigh?

Let the weight of the smallest piece be x grams
The weight of the other two pieces are $(x + 5)$ grams
and $(x + 5 + 8)$ grams

$$\text{Total weight of the three pieces} = x + (x + 5) + (x + 13)$$

$$\therefore \quad x + (x + 5) + (x + 13) = 360$$
$$3x + 18 = 360$$
$$3x = 342$$
$$x = 114$$

The weight of the smallest piece is 114 grams.

---**Example 2**---

This year, Dawn is three times as old as her brother Marcus, but in four years' time she will be twice as old. How old are Dawn and Marcus now?

Let Marcus' age now be $\quad x$ years
Then Dawn's age now is $\quad 3x$ years
In four years' time:
    Marcus will be $\qquad (x + 4)$ years
    Dawn will be $\qquad (3x + 4)$ years

Dawn will then be twice as old as Marcus

$$\therefore \quad (3x + 4) = 2(x + 4)$$
$$3x + 4 = 2x + 8$$
$$3x - 2x = 8 - 4$$
$$x = 4$$

Marcus is 4 years old and Dawn is 12 years old now.

Exercise 30L

1 A rubber costs 25p more than a pencil. Twelve pencils and ten rubbers are bought for a bran tub. The cost of a pencil is x pence.

  a Write down, in terms of x:
    (i) the cost of a rubber
    (ii) the cost of 12 pencils
    (iii) the cost of 10 rubbers.

  b The total cost of the 12 pencils and 10 rubbers is £6.90. Using this information:
    (i) write down an equation in terms of x
    (ii) solve the equation to find x
    (iii) find the cost of one rubber.

2 A shop assistant accepts a £5 note from a customer and, in return, hands the customer two boxes of paper hankies and £2.50 change.

  a If x is the cost of one box of hankies, write down an expression for what the customer receives.

  b Write down an equation and solve it to find the cost of one box of hankies.

3   Mr Wilson regularly attends football matches when his team plays at home.

   a   For x number of games he buys a seat in the stands. Each seat costs £9.
       Write down an expression for the cost of these x games.

   b   For the remaining matches he buys a ticket for the terraces. Each ticket costs £5.
       If he attends 20 games in a season, write down an expression in x for the cost of tickets for the terraces.

   c   Using your answers to **a** and **b**, write down and simplify an expression for the total cost for the season.

   d   The cost for the season was £148.
       Write down an equation in terms of x.

   e   Solve the equation to find the number of matches he watched from the stands.

4   Eighteen theatre seats were bought for a school party. Seats in the circle cost £x, seats in the stalls cost £1.80 less. Eight of the party sat in the circle, the remainder in the stalls.

   a   What was the cost of a seat in the stalls in terms of x?

   b   Write down and simplify an expression for the total cost of the seats, in terms of x.

   c   The total cost of the seats was £136.80.
       Write down an equation in terms of x and solve it to find the cost of:
       (i)   a seat in the circle
       (ii) a seat in the stalls.

5   A pound of apples costs 5p more than a pound of pears. The cost of 5 pounds of apples and 3 pounds of pears is £4.65.
    What is the cost of one pound of pears?

6   In a mathematics test, Ann scored 3 more marks than Brian who scored 5 marks more than Carol.

    The total of their three scores was 139.
    How many marks did they each score?

7   Alan weighs 3 kg less than Barry who weighs 4 kg less than Colin.

    Barry weighs x kilograms and the total weight of the three boys is 193 kg.
    How much does Colin weigh?

8   Apples cost 30p per pound more than bananas and 72p per pound less than cherries.

    The cost of 2 pounds of apples and 3 pounds of bananas is the same as the cost of 2 pounds of cherries.
    What is the cost of a pound of cherries?

9   Walking to college takes me twice as long as cycling. Taking a bus takes me 25 minutes longer than cycling. One week I walked five times, and the next week I cycled twice and took the bus three times. The total times for travelling to college were the same for the two weeks.

    How long does it take me to cycle to college?

30.6   Formulae

Substitution in formulae

A second important use of algebra is to enable a relationship between two or more quantities to be expressed in a short but easily understood form. This is called a **formula**.

A formula may be given in words or in symbols.
For example:

The volume of a cylinder is found by multiplying the area of the circular cross-section by its height.

This is more neatly expressed by the formula

$$V = \pi r^2 h$$

In order to calculate the volume, the given values of r (radius) and h (height) are substituted into the formula.

Example 1

The area of a rectangle is found by multiplying the length of the rectangle by its breadth.
Find the area when the length is 12 cm and the breadth is 15 mm:

Breadth = 15 mm = 1.5 cm
Area = Length × Breadth
Area = $12 \times 1.5 \text{ cm}^2$
    = 18 cm^2

Example 2

The distance travelled by a vehicle in a given time can be found by using the formula:

$$s = ut + \tfrac{1}{2}at^2$$

where s is the distance travelled in metres (m), u is the starting velocity in metres per second (m/s), a is the acceleration in metres per second per second (m/s^2), and t is the time in seconds (s).

Find the distance travelled by a sports car:

a in 4 s if $u = 20$ m/s and $a = 2.5$ m/s^2

b in 10 s starting from rest with an acceleration of 3 m/s^2.

a $t = 4, \quad u = 20, \quad a = 2.5$

   Substituting in $s = ut + \tfrac{1}{2}at^2$

   gives $s = (20 \times 4) + (\tfrac{1}{2} \times 2.5 \times 4^2)$
        $= 80 + 20$
        $= 100$

   The distance travelled was 100 m.

b $t = 10, \quad u = 0, \quad a = 3$

   Substituting in $s = ut + \tfrac{1}{2}at^2$

   gives $s = 0 + \tfrac{1}{2} \times 3 \times 10^2$
        $= 150$

   The distance travelled was 150 m.

Exercise 30M

1 a Find the area of a rectangle with length 15 cm and breadth 7 cm.

b The length of a rectangle is the area divided by the width.
Find the length of a rectangle with area 22.5 cm^2 and width 7.5 cm.

c The perimeter of a rectangle is found by adding the length and the breadth and then doubling the answer.
Find the perimeter of a rectangle with length 11 ft and breadth 5 ft.

d The circumference of a circle is approximately three times the diameter. Find an approximate value for the circumference when the diameter is
(i) 6.7 cm (ii) 9.5 ft.

e The speed of a car is found by dividing the distance travelled by the time taken for the journey.
What was the speed of a car which travelled 42 miles in $1\tfrac{1}{2}$ hours?

2   An appropriate rule to change temperature measured in degrees Celsius into degrees Fahrenheit is

double the degrees Celsius and then add 30

a   David says that if the Celsius temperature is 70 degrees then the Fahrenheit temperature is 160 degrees.
Is he right? You must show all your working.

b   Use this rule to change 20 degrees Celsius to degrees Fahrenheit.

c   An approximate rule to change temperature measured in degrees Fahrenheit into degrees Celsius is

subtract 30 from the degrees Fahrenheit and then halve the result

Use this rule to change 100 degrees Fahrenheit to degrees Celsius.   (SEG W94)

3   a   The speed, or velocity, at which the sports car in Example 2 on page 251 travels after a certain time, is given by the formula:

$$v = u + at$$

Calculate the speed, in m/s, when
(i) $t = 4$, $u = 20$, $a = 2.5$
(ii) $t = 10$, $u = 0$, $a = 3.6$.

b   The formula $V = 2.25v$ converts v m/s to V mph.
Calculate the sports car's speed in miles per hour for each of parts (i) and (ii) above.

4   The formula for a straight line graph is:

$$y = mx + c$$

Find the coordinate y if $m = 2$, $c = -3$ and $x = 4$.

5   To convert degrees Fahrenheit F to degrees Celsius C, the following formula is used:

$$C = \frac{5}{9}(F - 32)$$

Convert to degrees Celsius:

a   50°F   b   77°F   c   14°F.

6   The surface area of a cylinder is given by the formula:

$$S = 2\pi r(r + h)$$

Find the surface area of a cylinder if $r = 6$ cm and $h = 14$ cm.

7   When an amount of money, £P, is invested at a compound interest rate of r % per annum, the amount in the account after n years is given by:

$$A = P\left(1 + \frac{r}{100}\right)^n$$

Find the amount in an account after 2 years if £1600 is invested at 9.5% per annum.

8   The formula for the area of a trapezium is:

$$A = \frac{h}{2}(a + b)$$

where a and b are the lengths of the parallel sides of the trapezium and h is its height.

Find the area of a trapezium when:

a $a = 5$ cm $b = 7$ cm $h = 4$ cm

b $a = 4.3$ cm $b = 10.5$ cm $h = 5.6$ cm.

Rearranging formulae

The formula which converts degrees Fahrenheit to degrees Celsius is:

$$C = \frac{5}{9}(F - 32)$$

It may, however, be necessary to convert degrees Celsius to degrees Fahrenheit. In this case, the formula needs to be **rearranged, or transposed,** to give F in terms of C, i.e. F is made the **subject** of the formula.

Example 1

a Rearrange the formula $v = u + at$ to find t.

b Hence find t when $a = 4$, $u = 25$ and $v = 40$.

a The method is the same as for solving equations.

$$v = u + at$$

(i) Subtract u from each side: $\quad v - u = at$

(ii) Divide each side by a: $\quad \dfrac{v - u}{a} = t$

$$\therefore \quad t = \frac{v - u}{a}$$

b $t = \dfrac{40 - 25}{4} = \dfrac{15}{4} = 3.75$

Example 2

a Rearrange the formula $C = \dfrac{5}{9}(F - 32)$ to give F in terms of C.

b Use the transposed formula to convert $15°C$ to $°F$.

a The method is the same as above.

$$C = \frac{5}{9}(F - 32)$$

(i) Multiply through by 9: $\quad 9C = 5(F - 32)$

(ii) Remove the bracket: $\quad 9C = 5F - 160$

(iii) Transpose 160: $\quad 9C + 160 = 5F$

(iv) Dividing through by 5 gives the formula for F: $\quad F = \dfrac{9C + 160}{5}$

b When $C = 15$ $\quad F = \dfrac{9 \times 15 + 160}{5}$

$$= \frac{135 + 160}{5}$$

$$= \frac{295}{5}$$

$$= 59$$

$$\left(\text{Check. } C = \frac{5}{9}(F - 32) = \frac{5}{9}(59 - 32) = \frac{5}{9} \times 27 = 15 \right)$$

Exercise 30N

1 Rearrange:

a $P = 2(L + B)$ to make B the subject

b $S = 2\pi rh$ to make h the subject

c $A = \pi r^2$ to make r the subject

d $S = \left(\dfrac{u + y}{2}\right) t$ to make t the subject

e $v = u + at$ to make a the subject.

2 a Rearrange the straight line equation:

$$y = mx + c$$

to give x in terms of y.

b Find x if:

(i) $m = 2.$ $c = 1,$ $y = 5$

(ii) $m = -1,$ $c = 4,$ $y = -3.$

3 a The mean of three numbers a, b and c is given by:

$$M = \frac{a + b + c}{3}$$

Find the mean of the numbers 12, 8 and 16.

b Rearrange the formula to find a in terms of b, c and M.

c If the mean of three numbers is 7 and two of the numbers are 8 and 9, what is the third number?

4 The area of a trapezium is given by:

$$A = \frac{h(a + b)}{2}$$

a Make h the subject of the formula and find h when $A = 40$, $a = 10$, $b = 6$.

b Make a the subject of the formula and find a when $A = 63$, $h = 9$, $b = 6$.

5 The deposit required when booking a self-catering holiday consists of a fixed amount A, which is non-returnable, plus $\frac{1}{20}$ of the cost of the accommodation.

$$D = A + \frac{NC}{20}$$

where C is the cost of the accommodation per person and N is the number of people booking the holiday.

a Find the deposit payable for six people, if the holiday costs £210 per person and there is a non-returnable amount of £50.

b Rearrange the formula to make C the subject.

c Calculate C if $D = £133$, $A = £65$ and $N = 4$.

Exercise 30P

1 Simplify the following:

a $9x - 2y + 5y - 4x$

b $6p - 7q + 4r + 3p + 5q - 6r$

2 If $x = -2$ and $y = 4$, find the value of:

a $4x + 3y$ b $\dfrac{2x - 4y}{5}$ c $x^2 y + x^3$

3 Simplify:

a $3a \times 4b$ d $2a^2 \times 5a^4$

b $(-2a)^2 \times 2b$ e x^0

c $x^{-4} \times x^6$ f $3pq \times (-4q^2 r) \times (-pr^2)$

4 Simplify:

a $x^7 \div x^4$ d $\dfrac{5}{3a^2} \times \dfrac{9a^5}{10}$

b $x^2 \div x^5$ e $\dfrac{3x^2}{2} \div \dfrac{x^3}{4}$

c $x^3 \div x^5 \times x^2$

5 Remove brackets and simplify:

a $3x + 2(3x - 2)$

b $2(4x + 5) + (x - 7)$

c $4(x - 1) - 2(3x - 2)$

6   Factorise:

a $6ab + 9ac + 3a$

b $xyz - xy^2$

c $2abc^2 - 6a^2bc + 4ab^2c$

*7   Express as a single fraction, simplifying where possible:

a $\dfrac{x}{2} + \dfrac{x}{4} - \dfrac{2x}{3}$

b $\dfrac{4}{5x} - \dfrac{3}{10x}$

c $\dfrac{2(x-1)}{3} - \dfrac{(x-3)}{4}$

8   Solve the following equations:

a $9x - 5 = 67$

b $3(6 - x) = 12$

c $4(x + 1) - 3(2 - x) = 12$

*d $\dfrac{3x}{5} + 2 = 20$

9   $S = \frac{1}{2}n(n + 1)$ is a formula which gives the sum of the positive integers from 1 up to n.

a   Find the sum of the numbers from 1 to 20.

b   Find the sum of the first 50 positive integers.

10   a   Find y when $x = -4$, given that
$$y = \frac{3x + 2}{x}$$

b   Rearrange the above expression to give x in terms of y.

c   Find x when   (i) $y = 2.5$,   (ii) $y = 7$.

31 Graphs

31.1 Graphs and curves

There are many different types of graph, some of which you may have met in the Statistics and Probability Module.

This unit deals with line graphs. The lines may be straight or curved, but they illustrate a relationship between two quantities. This relationship must be clear to anyone looking at the graph.

Graphs are often used to convey information quickly and with impact!

For example, these graphs were used by:

a holiday company   a Government department

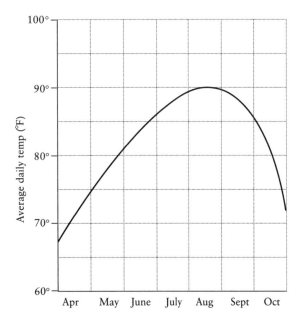

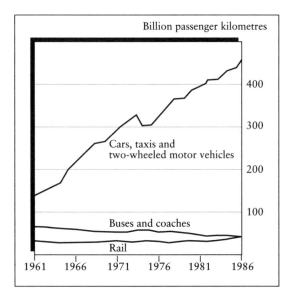

(Source: Department of Transport)

Many graphs involve time, which is always measured along the horizontal axis. The vertical axis then provides a measure of something that changes with time.

Whenever one measurable quantity changes as a result of another quantity changing it will be possible to draw a graph to illustrate these changes.

31.2 The interpretation of graphs

Once a graph has been drawn, information can be found from it very quickly.

Example

The graph shows a cross-section through a river bed. The distance from point A on one bank of the river to point B, directly opposite on the other bank is 20 m.

Soundings were taken of the depth of the water at various points and plotted against the distance from A.

Find, from the graph:

a the depth of the river at a distance of 11.5 m from A

b the distances from A at which the depth of the river is 2.8 m.

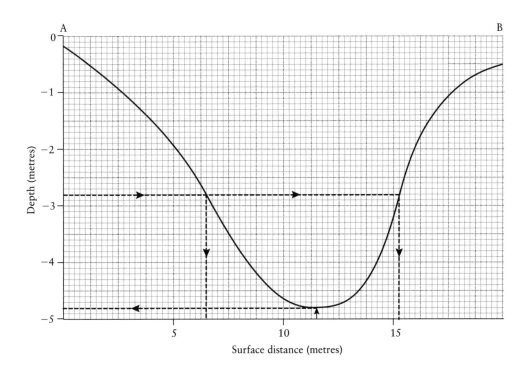

a The distance 11.5 m is located on the horizontal axis and a vertical line is drawn from this point to meet the graph. A horizontal line is then drawn across to the vertical axis and the depth is read off.

The depth is −4.8 m, i.e. the depth of the water is 4.8 m.

b The depth of 2.8 metres is located on the vertical axis at the point −2.8 and a horizontal line drawn from this point meets the curve at two points.
Vertical lines are drawn down to the horizontal axis and the two distances are read off.

The distances from A are 6.50 m and 15.25 m.

In the following exercise the graphs have been drawn for you.
Answer the questions relating to each graph by reading off the values from the graph.

In the first question guide lines have been drawn to make it easier for you.

1 This graph shows the number of births (measured along the vertical axis) in a given year (measured along the horizontal axis).
   (Based on figures from the *Annual Abstract of Statistics*, 1988.)

   a Estimate the number of children born in 1915.

   b In which years were, approximately, 825 000 children born?

   c Which year had the lowest number of births?

   d In which year was there a 'baby boom'?

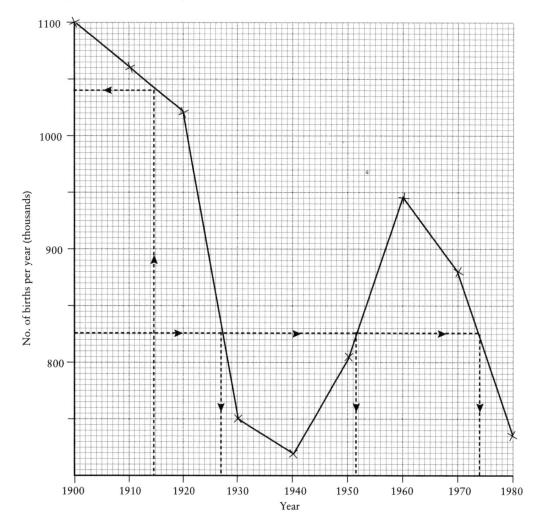

2 The next graph shows the average gross weekly earnings of men from 1970 to 1986. (Based on figures from *Social Trends*, 1988.)

   a Estimate the average weekly earnings in 1974.

   b Estimate the average weekly earnings in 1984.

   c Why is the answer to **b** a better estimate than the answer to **a**?

   d In which year did the average weekly wage reach £100?

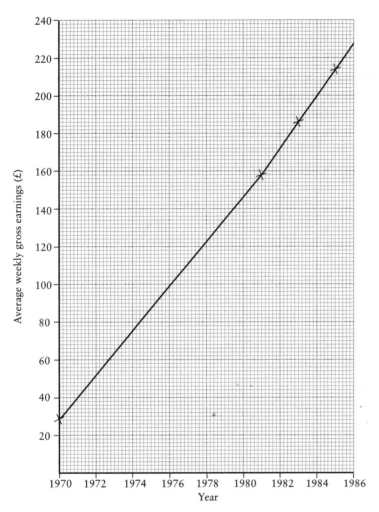

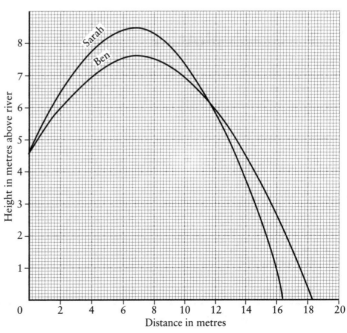

3 Two children, Ben and Sarah, were walking along the river bank when they spotted a bottle floating in the river. They each took turns to throw a stone to see who could hit it first. The graph shows their first attempts.

a The bottle was 17 m from the bank Who came nearest to hitting it?

b How far from the bottle did each stone hit the water?

c On their second attempt they threw their stones at the same time. The stones followed exactly the same paths as before, but this time they collided. How far above the river were they when the collision occurred?

d If the children were both 1.5 m tall, how high above the river was the river bank? (Assume the children's hands were 1.5 m above the ground.)

4 Nessa enjoys writing to her many pen-friends. One pen-friend lives in France and a favourite topic of conversation is the weather. The next graph shows the number of wet days each month in their two home cities.

The solid line shows the number of wet days where Nessa lives and the dotted line is for her pen-friend Louis.

a Which is the wettest month for each city?

b Which is the driest month for each city?

c In which month does the smallest difference in the number of wet days occur?

d Nessa lives in London. Where do you think Louis lives?

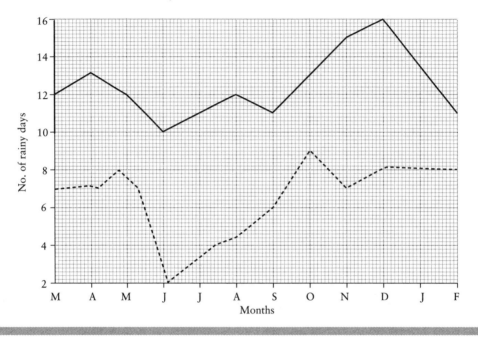

31.3 Graph plotting

When plotting graphs a certain procedure should be followed.

1 Choose the scales.   **4** Plot the points.

2 Draw the axes.   **5** Join the points.

3 Scale the axes.

The information from which the graph is to be drawn is usually given in a table.

Example

£1000 is invested in a savings account for 5 years. The amount in the account at the end of each year is shown on the table:

Number of years (x)	0	1	2	3	4	5
Amount in £ (y)	1000	1100	1210	1331	1464	1610

Plot the graph and find:

a the amount in the account after $2\frac{1}{2}$ years

b the length of time for which the money must be invested to amount to £1500.

1 Choose the scales:

   (i) Find the range of values for each axis.
On the horizontal axis the range is 5.
On the vertical axis the range from 1000 to 1610, i.e. the range is 610.

   (ii) Count the number of large squares in each direction on the graph paper.

The graph paper provided here has 18 large division in each direction and a grid with five small divisions in each large division.

The aim is to draw as large a graph as possible, but the scale must be easy to read.

Taking the 18 large divisions width for the horizontal axis, the range of 5 does not divide exactly into 18.

A convenient scale would be 2 large divisions representing 1 year. On the vertical axis, a range of 610 does not divide into 18 large divisions, but 14 large squares can be divided by 700. There will be little wastage of space if a scale of 2 large divisions representing £100 is chosen. The axis is scaled from £1000 to £1700.

2 Draw the axes.

Once the scales have been chosen, there is no problem in placing the axes. The axes can be placed 2 large divisions from the left and 2 large divisions from the bottom edges of the graph paper.

This allows space for labelling the axes.

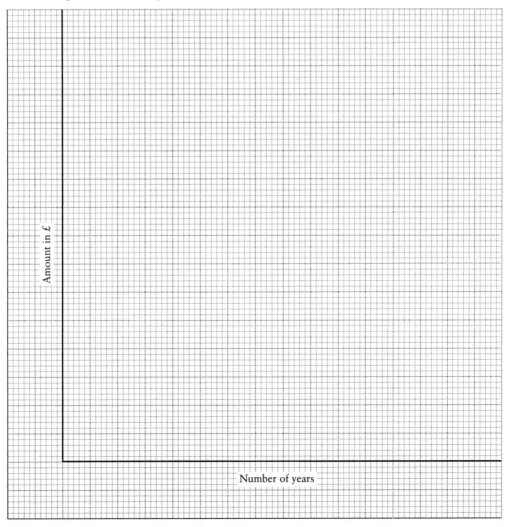

3 Scale the axes:

Mark the position of each layer along the horizontal axis as shown on the following graph.

Mark the position of each £100 along the vertical axis as shown.

4 Plot the points.

Plot the six points given as explained on page 257.

5 Join the points.

The points are joined with a **smooth** curve.

Trace through the points with a pencil, but without touching the paper, to find the shape of the curve. When you are satisfied that the path of the curve is smooth, draw the curve through the points in one movement.

If the values of the coordinates have been rounded, the line may not pass exactly through each point.

6 To find the required information draw appropriate lines on the graph (see pages 257–258) and read off the values:

a £1270   b 4.3 years

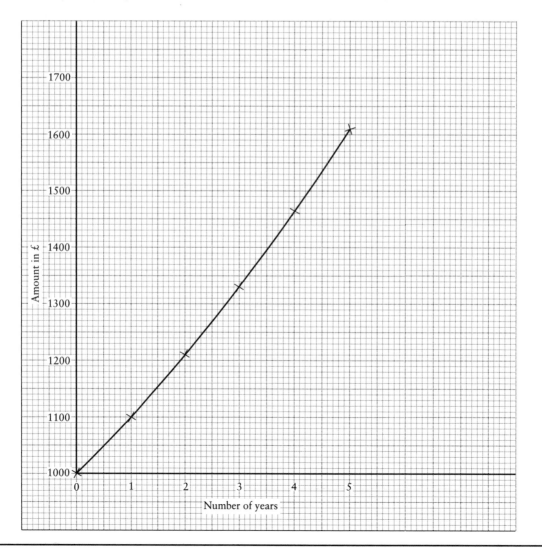

In the following exercises, you are given the information connecting the two variables in a table. In each case plot the points and join them with a smooth curve.

1 Mrs Lister weighed herself now and again through the year. After one Christmas she found her weight was over $10\frac{1}{2}$ stone.

She joined Weightwatchers and was determined to reach her target weight of $8\frac{1}{2}$ stone before the next Christmas.

The table shows her weight in stones and pounds at the beginning of each given month. (1 stone = 14 lb.)

Month/year	Jan/96	May/96	Sep/96	Jan/97	May/97	Sep/97
Weight in st–lb	9–2	9–12$\frac{1}{2}$	10–6	10–8	10–4	9–6

a What weight was Mrs Lister in August 1996?

b What was the heaviest weight she reached?

c During which months did she weigh 9 stone 7 lb?

d Do you think she would have reached her target weight before Christmas 1997?

2 While at a party, William spills a glass of red wine over the table cloth.

The stain spreads rapidly in a circular shape and the table shows how the area increases as the diameter of the stain increases.

Diameter (cm)	0	2	4	6	8	10
Area (cm²)	0	3	12	27	48	75

a What area does the stain cover when the diameter is 9 cm?

b What is the radius of the stain when its area is $50\,\text{cm}^2$?

c The stain stops spreading when its diameter is 10.9 cm. What area does the stain eventually cover?

3 The table shows the average number of hours the sun shone during a day in the middle of each month, recorded at a Spanish resort.

	Jun	Jul	Aug	Sep	Oct	Nov	Dec	Jan	Feb	Mar	Apr	May	Jun
Hours of sunshine	16	15.5	14	12	10	8.5	8	8.5	10	12	14	15.5	16

a During which months might you expect an average of about $10\frac{1}{2}$ hours of sunshine in a day?

b What were the average hours of sunshine at the beginning and end of May?

c Which monthly periods had the smallest increase in hours of sunshine?

4 One day when Heather is having a bath, she notices that the water level is gradually decreasing and realises that the plug is leaking. When she started her bath the depth of water was 300 mm. Plot the points given in the table below to show the depth of water remaining at a given time.

Time (in minutes)	0	1	2	3	4	5
Depth of water (in mm)	300	110	41	15	5.5	2.0

a What was the depth of the water after $2\frac{1}{2}$ min?

b After how long was the bath half empty?

c What was the drop in water level after $1\frac{1}{2}$ min?

5 Two young children have a swing in the garden. On one particular day it is Lucy's turn on the swing and Tony pulls the swing back through an arc of 100 cm and then releases it.

The table shows the position of the swing (distances are measured along the arc of the swing) at different times during its motion.

Time (in seconds)	0	0.2	0.4	0.6	0.8	1.0	1.2	1.4
Distance (in cm)	100	89	57	13	−34	−74	−97	−98
Time (in seconds)	1.6	1.8	2.0	2.2	2.4	2.6	2.8	3.0
Distance (in cm)	−77	−38	9	54	87	100	91	61

a How long does it take for one complete swing?

b In what position is the swing after 0.7 seconds?

c How long does it take to travel through 50 cm?

d How long does it take to travel through 250 cm?

31.4 Conversion graphs

A graph is a very useful means of converting quickly from one quantity to another.

Ranjit is planning a summer holiday in France. He sees in the newspaper that the exchange rate is 10.2 FF = £1, but he wants to be able to compare French prices with prices at home quickly and easily. Before he goes he draws a pocket-size conversion graph which he can keep handy while on holiday.

He knows that:

- A conversion graph is a straight line, so he only needs to plot two points (though a third point is useful as a check).
- Most conversion graphs go through the point (0,0).
- Either variable can be measured along the vertical or horizontal axis.

As Ranjit will mainly be changing from francs to pounds he measures francs along the horizontal axis.

He knows that £1 = 10.2 FF and that the graph goes through the origin (0,0).
He plots these two points. They are very close together so he finds another
point further from the origin.

£10 = 102 FF is an easy one to calculate.

He then draws a straight line through the three points.
(If the line does not pass through all points there is a mistake and the
calculations must be checked.)

This is what his graph looks like:

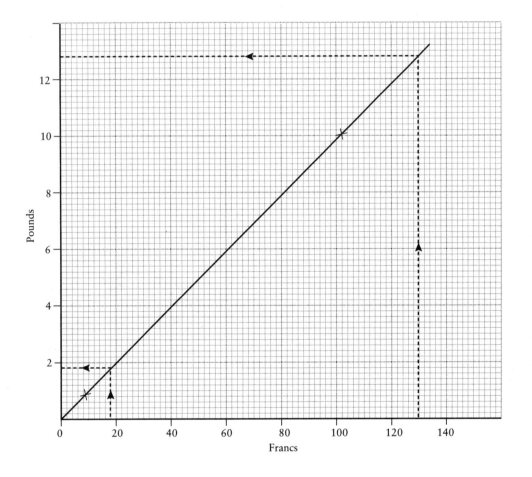

From the graph he can see that an 18-franc bottle of French wine is cheap at
£1.80 and a meal for two costing 130 francs is about £12.80.

Exercise 31C

1   Five miles is approximately equal to eight kilometres.
    Draw a graph to give a rough conversion from kilometres to miles.

    a   How many miles are 35 km?

    b   Convert the distances given on the signboard to miles.

    c   A Frenchman on holiday travels 60 miles. How far has he travelled in
        km?

A10	
ORLEANS	106
TOURS	245
POITIERS	350
BORDEAUX	560

2 Weather forecasts now give all temperatures in degrees Celsius, but some people are more used to degrees Fahrenheit.
Draw a graph to convert between degrees Celsius and degrees Fahrenheit.
$(10°C = 50°F, \quad 20°C = 68°F, \quad 30°C = 86°F.)$

This graph does not go through the origin.

a The freezing point of water is 0°C.
What is this in degrees Fahrenheit?

b Convert a temperature of 25°C to °F.

c Convert a temperature of 83°F to °C.

d Convert a temperature of 14°F to °C.

3 Nessa's Canadian pen-friend sends her a recipe for apple pie.

Unfortunately, some of the ingredients are measured in millilitres (ml).
Nessa has a jug which measures fluid ounces (fl oz) and she finds in her cookery book that 35 fl oz = 1000 ml.
Draw a graph which will change ml to fl oz.

Use your graph to convert the quantities in Nessa's recipe to the nearest fl oz:

175 ml milk   1.5 litres sliced apples
250 ml sugar   30 ml softened butter
125 ml flour   cinnamon and nutmeg

4 Mrs Kurtz works part-time in a fabric shop. The materials are sold by the metre, but people still ask for them by the yard.
Draw a graph to convert yards to metres (1 yard = 0.9 m).
From the graph find (to the nearest metre) how many metres are equivalent to:

a 3 yards   **b** $5\frac{1}{2}$ yards   **c** 12 yards.

32 Coordinate Geometry

32.1 Cartesian coordinates

In the early seventeenth century, the French mathematician René Descartes introduced the idea of a grid for locating and plotting points.

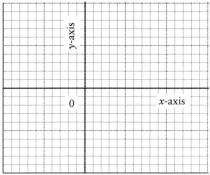

Onto this grid, called a **cartesian** graph, are drawn two straight lines at right angles to one another, called the **rectangular axes**.

Usually the axes cross at the **origin**, O, or starting point. The axes are scaled as number lines with positive numbers to the right and above the origin and negative numbers to the left and below the origin.

The position of any point (A, say) can then be described by two numbers: one referring to the horizontal axis x and one to the vertical axis y.

The two numbers are called the **cartesian coordinates**, after Descartes.

In general, the coordinates are called x- and y-coordinates, written as (x, y), and are plotted by first counting along the horizontal or x-axis and then along the vertical or y-axis.

In the diagram below A is at the point $(3, -2)$.

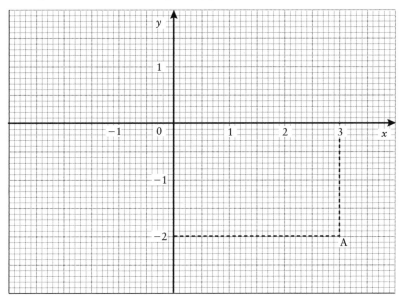

It is usual to plot a point on a graph using a cross (+ or ×) or a dot (·).

32.2 Straight-line graphs

Drawing straight-line graphs

An equation of the form $y = mx + c$ can be represented graphically by a straight line.

To draw a straight line requires **two** points, but it is better to find **three** points as a check.

Example

Draw a graph to represent the line $y = 3x - 2$ for values of x from -3 to 3.

1 Choose the scales.

Choose three values for x between -3 and 3, e.g. $-3, 0, 3$. Calculate the corresponding y-values and write the three pairs of coordinates in a table:

x	-3	0	3
y	-11	-2	7

The range for y is 18 and for x is 6.
Suitable scales are 1 large division to 1 unit on the x-axis
and 1 large division to 2 units on the y-axis.

2, 3 Draw and scale the axes.

As there are the same number of negative values and positive values of x, set the y-axis near the centre of the graph paper.

The x-axis must be drawn above the halfway mark as the negative values of y are numerically greater (at 11) than the positive values of y (at 7).

The axes are scaled in both positive and negative directions.

4, 5 Plot and join the points.

If several graphs are drawn using the same set of axes, each graph should be labelled with its equation.

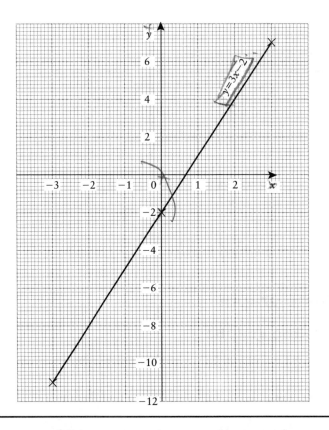

Exercise 32A

Draw on the same axes the graphs of each set of equations in questions 1–5.

For each set of graphs, state what they have in common.

1   Plot the graphs of $y = 3$, $y = -2$ and $y = 0$.

2   Plot the graphs of $x = 3$, $x = -2$ and $x = 0$.

3   Plot the graphs of $y = x$, $y = -2x$ and $y = -3x$.

4   Plot the graphs of $y = x$, $y = x + 5$ and $y = x - 3$.

5   Plot the graphs of $y = 2x$, $y = 2x + 1$ and $y = 2x - 3$.

Draw the graphs of each set of equations using the same axes. (A graphical calculator or computer could be used to save time.)

*6 $y = x$ $y = x + 3$
    $y = x - 1$ $y = x - 2$

*7 $y = 2x$ $y = 2x + 3$
    $y = 2x - 1$ $y = 2x - 2$

*8 $y = -x$ $y = -x + 3$
    $y = -x - 1$ $y = -x - 2$

*9 $y = -3x$ $y = -3x + 2$
    $y = -3x - 1$ $y = -3x - 3$

*10 $y = \frac{1}{2}x$ $y = \frac{1}{2}x + 2$
      $y = \frac{1}{2}x - 1$ $y = \frac{1}{2}x - 3$

*11 $y = -\frac{1}{2}x$ $y = -\frac{1}{2}x + 2$
      $y = -\frac{1}{2}x - 1$ $y = -\frac{1}{2}x - 3$

*12 a What do the equations in each set have in common:
     (i) in their algebraic form,
     (ii) in their graphical form?

    b What conclusion can you draw from this?

*13 a What do the equations $y = x$, $y = 2x$, $y = -x$, $y = -3x$, $y = \frac{1}{2}x$, and $y = \frac{1}{2}x$ have in common?

    b What conclusion can you draw from this?

*14 What does the constant term on the RHS of an equation tell you about the graph of that line?

*15 Compare the graphs in questions 3, 4 and 6. What does a negative coefficient of x tell you about the graph of that line?
(See page 237 if you need a reminder about what a coefficient is.)

Gradients and intercepts

All linear (straight-line) equations are of the general form:

$$y = mx + c$$

where m is the gradient of the line and c is the intercept along the y-axis, providing the origin O is at the intersection of the axes.

The **gradient** of a line is the increase in the vertical value for every unit (i.e. 1) increase in the horizontal value.

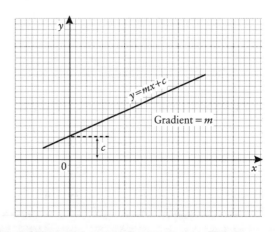

Example 1

Find the gradient of the straight line shown on the following graph:

To calculate the gradient of a line:

1 Find two convenient points on the graph (i.e. the x-values should be integers). Convenient points on the graph given would be $(-2, -4)$ and $(1, 4)$.

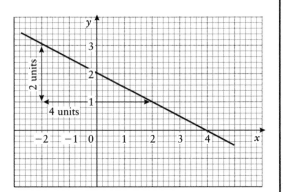

2 Find the horizontal and vertical differences between the two points.

Horizontal difference $= 1 - (-2)$
$$= 3$$

Vertical difference $\quad = 4 - (-4)$
$$= 8$$

3 Calculate the gradient by $m = \dfrac{\text{Increase in } y\text{-value}}{\text{Increase in } x\text{-value}}$

$$= \frac{8}{3}$$

Example 2

Find the equation of the line shown.

1 Gradient of line $= m = \dfrac{\text{Increase in } y\text{-value}}{\text{Increase in } x\text{-value}}$

$$= \frac{1 - 3}{2 - (-2)}$$

$$= -\frac{2}{4} = -\frac{1}{2}$$

Alternatively,

/ is a positive gradient and \ is a negative gradient

From the graph:

Gradient of line $= m = -\dfrac{2}{4} = -\dfrac{1}{2}$ or -0.5

2 The line crosses the y-axis at the point $(0, 2)$, i.e. the intercept along the y-axis is $c = 2$

3 The equation of the line is $y = mx + c$

$$\therefore \quad y = -\frac{1}{2}x + 2$$

$$2y = -x + 4$$

$$x + 2y = 4$$

Exercise 32B

1 What are the gradients of the following lines?

a $y = 3x + 2$ c $y = 4 - 2x$ e $3y = 2x + 6$ g $x + y = -2$ i $x + 3y = 6$

b $y = 2x - 5$ d $2y = x - 2$ f $4y = 2 - 8x$ h $2x + y + 4 = 0$

2 For each of the equations in question 1, state the intercept along the y-axis.

3 The following pairs of points lie on a straight line.
Calculate the gradient of each line.

a $(3, 6); (1, 2)$ c $(4, 6); (2, 5)$ e $(2, -5); (1, -4)$ g $(-1, 6); (1, -2)$

b $(3, 7); (1, 1)$ d $(0, 5); (2, 3)$ f $(4, 7); (-2, -2)$ h $(-4, -1); (0, -3)$

4 The graph of a straight line is given by the equation $y - 2x = 7$.

a Write down the equation of one other straight line which is parallel to $y - 2x = 7$.

b Write down the equation of one other straight line which crosses the y axis at the same point as $y - 2x = 7$.

The graph of a different line is given by the equation

$$3y - 2x = -9$$

c Rearrange this equation into the form $y = mx + c$.   (SEG S94)

5 Find the equation of each of the following straight line graphs:

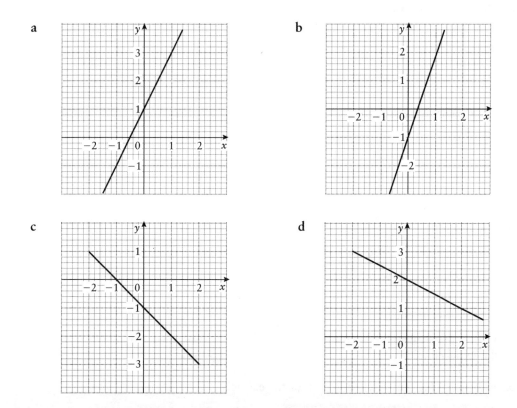

6 Each of the following sketch graphs represents one of the equations:

a $y = 2x + 1$ c $y = 3 - x$ e $x = -1$

b $2y = x$ d $y = 2$ f $x + y = -1$

Which is which?

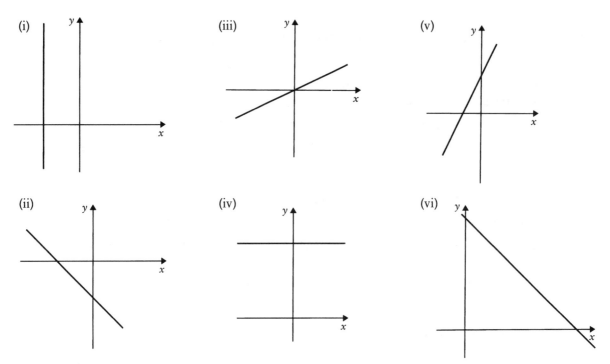

7 Sketch the following graphs marking the point(s) where the graph crosses the x- and y-axes. (Graph or squared paper should not be used.)

a $y = -2$ d $y = 7x$ g $y = 2x + 7$ j $y + 4x = 2$

b $x = 3$ e $y = \frac{1}{2}x$ h $y = 4 - 2x$ k $2y = 4x + 1$

c $y = x$ f $y = x - 4$ i $y - x = 3$ l $2y + x = 2$

8 In an experiment, weights were hung on to a piece of elastic and the stretched length of the elastic was measured.

The results are shown in the table:

Weight (g)	1	2	3	4	5
Length (cm)	28	35	45	50	60

a Plot the points on to a graph and draw a straight line which best fits the points.

(Not all, and possibly only a few, points will lie on the line. Measurements from experiments can be inaccurate.)

b Find the equation of your line.

c Use your equation to calculate the length of the elastic if a weight of 2.8 g is hung on it. Does your graph give the same value?

d What was the original (i.e. unstretched) length of the elastic?

9 The table below shows the pulse rates for healthy people of different ages taking exercise.

Age (years)	20	25	30	35	40	45	50
Pulse rate (beats/minute)	120	117	115	110	108	105	100

a Plot the points and fit a straight line to the points.

b Find the equation of your line.

c What are the appropriate pulse rates for:
   (i) someone 55 years of age?
   (ii) someone 18 years of age?

32.3 Graphs of simple curves

It is sufficient to plot three points if a straight-line graph is to be drawn, but more points are required if a **curve** is to be drawn accurately.

Choose suitable x-coordinates and then calculate the y-coordinates.
The results are displayed on a table of x-values and y-values, as shown in the example below.

After plotting the points, they are joined with a **smooth curve**.

Avoid drawing straight lines between points, and be particularly careful when joining the lowest (or highest) points not to put too sharp a point on the curve or to flatten the curve between these points.

In the diagram below the solid line is correct, the dotted line is incorrect:

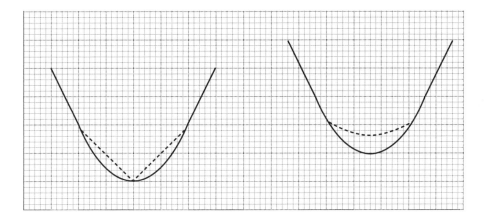

Example

The graph $y = x^2$ can be used to find squares and square roots of numbers.

a Construct a table of values for this graph for $-3 \leqslant x \leqslant 4$.

b Draw the graph of $y = x^2$ for this range of x-values.

c Use the graph to find: (i) the square of 2.5, (ii) the square roots of 8.

a The x values are -3, -2, -1, 0, 1, 2, 3, 4. Calculating $y = x^2$ gives a table of values:

x	-3	-2	-1	0	1	2	3	4
y	9	4	1	0	1	4	9	16

A suitable scale is 1 large division to 1 unit on the x-axis
and 1 large division to 2 units on the y-axis.

As there are no negative values for y, the x-axis is drawn near the bottom of the page. The y-axis is drawn slightly to the left of centre.

b The completed graph is shown below:

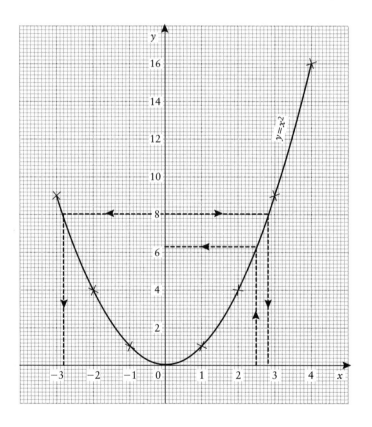

c **(i)** **The square of 2.5 is given by y when $x = 2.5$.**
Vertical and horizontal lines are drawn, as shown on the graph, and the value of y is found to be 6.3.

That is, the square of $2.5 = 6.3$ (to 1 d.p.).

(ii) The square roots of 8 are given by x when $y = 8$.
From the graph, $x = 2.8$ and $x = -2.8$.

That is, the square roots of 8 are 2.8 and -2.8 (to 1 d.p.).

Exercise 32C

1 a Complete the following table of values for the graph $y = x^2 + 2$.

x	-3	-2	-1	0	1	2	3
x^2		4	1				
$x^2 + 2 = y$			3				11

b Draw the graph.

c Use your graph to find y when $x = -1.5$.

2 a Draw the graph of $y = x^2 + 3$ from $x = -3$ to $x = +3$.

b In what way is this graph different from the graph in question 1?

c Use your graph to find:
   (i) the value of y when $x = 2.4$,
   (ii) the square roots of 5 (i.e. the values of x when $y = 8$).

3 a Draw the graph of $y = 4 - x^2$.

b Use your graph to find:
   (i) the value of y when $x = -1.6$,
   (ii) the square roots of 6.

4 Use a graphical calculator, or a computer, to investigate graphs of the type $y = x^2 + c$ and $y = c - x^2$, where c is any rational number.

5 A car starts from rest and accelerates at a steady rate. The distance d metres travelled in t seconds is given by:

$$d = \frac{t^2}{5}$$

a Draw a graph to show the distance travelled during the first 30 seconds.

b From the graph, find:
   (i) the distance travelled after 24 seconds
   (ii) the time taken to travel 50 m
   (iii) the distance travelled between the 8th and 18th seconds
   (iv) the average speed over this interval of time.

6 The reciprocal of a number is given by the relationship:

$$y = \frac{1}{x}$$

a Copy and complete the table of values given below for:

$$y = \frac{1}{x}$$

x	-5	-4	-3	-2	-1	1	2	3	4	5
y	-0.20		-0.33				0.50		0.25	

(There is no value for y when $x = 0$.)

b Plot the points and join with two smooth curves.

c From the graph, find the reciprocal of:
   (i) 1.5   (ii) -3.5 (iii) -0.4 (iv) 0.84

*7 The time taken to complete a job depends on the number of workers employed. The time can be calculated from the formula:

$$T = \frac{k}{N}$$

where T is the time (in hours) and N is the number of employees.

a If three workers take 12 hours to complete the job, find the value of k.

b Using this value of k, draw up a table of values of T against $N = 4, 6, 8, 10$ and 12.

c Draw the graph, joining the points with a smooth curve.

d Use your graph to find how long the job will take if seven workers are employed.

*8 Draw the graph of $y = x^3$ and hence solve $x^3 = 14$ to find $\sqrt[3]{14}$ correct to 1 d.p.

*9 Draw the graph of $y = x^3 - 2$ for $x = -2$ to $x = 2$ and hence find the cube root of 6.

*10 Solve the equation $x^3 + x^2 = 1$ by drawing the graph of $y = x^3 + x^2$.

*11 Sketch the following graphs, identifying the points where the graph cuts the x- and y-axes.

a $y = x^2 - 4$ d $y = x^3 + 8$

b $y = 2x^2 + 3$ e $y = 9 - x^2$

c $y = x^3$

*12 Match each of the following equations to the correct sketch graph:

a $y = x^2 - 1$ e $y = x^3 + x^2$

b $y = 2x^2 + 1$ f $y = x^3 - x^2$

c $y = x^3 - 1$ g $y = x^2 - x^3$

d $y = 1 - x^3$

(i)

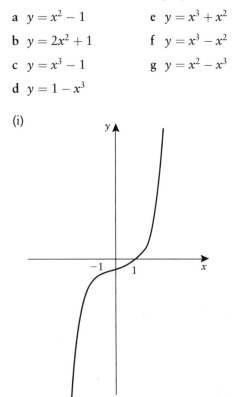

(ii)

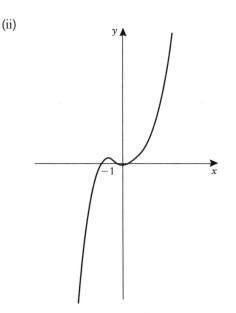

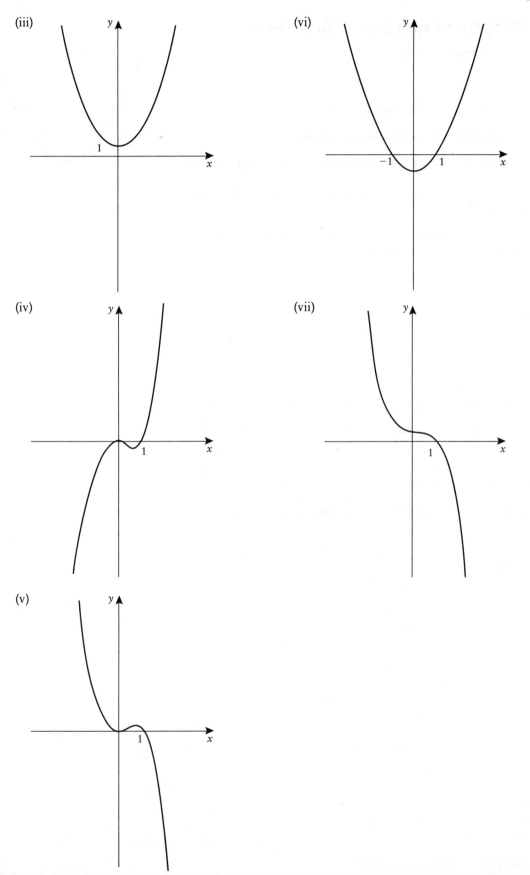

32.4 Graphs of quadratic functions

All quadratic functions of the form

$$y = ax^2 + bx + c$$

have the same general shape when represented by a graph.

This shape is a **smooth** curve and is called a **parabola**.

It is symmetrical about a vertical line through the lowest or highest point.

When drawing up a table of values for the quadratic function, it is easier if the values of y are calculated in stages.

Alternatively, the values for x may be substituted into the equation and the values for y obtained by using a calculator.

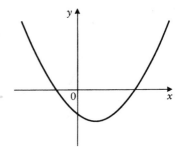

Example

Draw the graph of $y = 2x^2 - 3x - 6$ for values of x between -3 and 4.

The table of values is:

x	-3	-2	-1	0	1	2	3	4
$2x^2$	18	8	2	0	2	8	18	32
$-3x$	9	6	3	0	-3	-6	-9	-12
-6	-6	-6	-6	-6	-6	-6	-6	-6
y	21	8	-1	-6	-7	-4	3	14

Alternatively, when $x = -3$
$$y = 2 \times (-3)^2 - 3 \times (-3) - 6$$
From the calculator $y = 21$.
The other values for y are calculated
in the same way.

The range for x is from -3 to 4, i.e. a range of 7.
The range for y is from -7 to 21, i.e. a range of 28.

The graph paper used in the examination is usually
18 cm wide and 24 cm high.
Therefore use a scale of 2 cm to 1 unit for the x-axis
and 1 cm to 2 units for the y-axis.

The y-axis is drawn to the left of centre, e.g. 8 cm
from the LHS.
The x-axis is drawn 4 cm from the bottom.

The points plotted are joined with a smooth curve,
taking care not to flatten the curve between $x = 0$
and $x = 1$.

The completed graph is shown opposite:

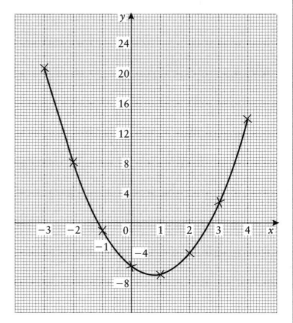

Exercise 32D

1 a Copy and complete the table given below
and hence draw the graph of $y = 4x - 3x^2$.

x	-2	-1	0	1	2	3
$4x - 3x^2$	-12	-4			8	12 -27
y	-20			1	-4	-15

b In what way is this parabola different from
the one in the example on page 278?
What is the reason for this difference?

c Find the value of y when $x = 1.5$.

d Find the value of x when $y = -10$.

2 a Draw up a table of values for x and y, taking
values of x from -1 to 7, for the function:
$$y = x^2 - 6x + 8$$

b Draw the graph of y against x.

c From the graph find:
   (i) the value of y when $x = 1.25$
   (ii) the values of x when $y = 9$
   (iii) the lowest value of y
   (iv) the equation of the line of symmetry.

3 Romeo throws a pebble at Juliet's window. The
path followed by the pebble is given by the
equation:
$$y = 2x - \frac{x^2}{100}$$

where x is the horizontal distance and y is the
vertical distance from Romeo.

a Draw up a table of values, taking x from
0 to 5.

b Draw the graph which shows the path the
pebble takes.

c Juliet's window is 1 metre high and the
window sill is 4 metres above the ground.
Will the pebble hit the window if it is
thrown from a distance of: (i) 3 m (ii) 2.5 m
from the house?

d If Romeo is to hit the window, between
what distances from the house should he be
when throwing the pebble?

4 Stopping distances, in metres, on wet roads can
be calculated using the formula:
$$d = \frac{v}{5} + \frac{v^2}{80}$$

where v is the speed of the car in km/h.

a Draw up a table of values of d against v for
speeds from 0 to 100 km/h.

b Plot the graph of d against v.

c Use the graph to estimate the minimum
distance the car should be from a junction
when the brakes are applied if the car is
travelling on wet roads at:
   (i) 24 km/h   (ii) 64 km/h.

d The driver sees a hazard 100 m ahead and
applies the brakes immediately.
If she is travelling at 80 km/h on wet roads,
can she stop in time?

5 The total surface area of a cylinder of height
9.5 cm is estimated using the formula:
$$A = 6r^2 + 60r$$

a Complete the table below which gives the
approximate areas for radii from 0 to 10 cm.

r	0	2	4	6	8	10
$6r^2$ $60r$						
A						

b Plot the graph A against r.

c Use the graph to estimate:
   (i) the surface area when the radius is
        4.8 cm
   (ii) the radius of the cylinder if the surface
        area is 800 cm².

d The formula for calculating the surface area
of a cylinder of height 9.5 cm is:
$$A = 2\pi(r^2 + 9.5r)$$

Calculate the surface area of the cylinder
(correct to 3 s.f.) when $r = 6$ cm.

e Calculate the percentage error in the
estimated value for the surface area given in
the table, when $r = 6$ from the correct value
calculated in **d**. Give your answer correct to
2 s.f.

*32.5   Intersecting graphs

If two graphs $y = f(x)$ and $y = g(x)$ are drawn on the same pair of axes, the values of x where the two graphs intersect is a solution of $f(x) = g(x)$.

Example 1

Draw $y = x^2 - 2x$ for values of x from -1 to $+3$, and $y = x - 1$.
Hence solve $x^2 - 2x = x - 1$
or $\qquad x^2 - 3x + 1 = 0$.

The table for the two graphs is:

x	-1	0	1	2	3
$y = x^2 - 2x$	3	0	-1	0	3
$y = x - 1$	-2	-1	0	1	2

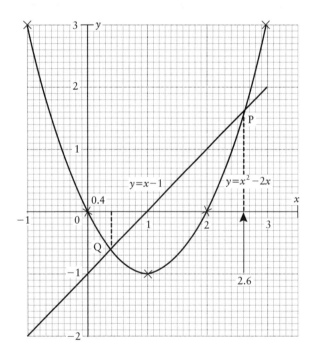

The two graphs intersect at P and Q.
At P, $y = x^2 - 2x$ since P is on the curve.
Also, $y = x - 1$ since P is on the straight line.

Hence $x^2 - 2x = x - 1$ (both equalling y).
∴   The x-coordinate of point P is a solution of
$\qquad x^2 - 2x = x - 1$

or $x^2 - 3x + 1 = 0$.
∴ $x = 2.6$ is a solution of this equation.

Similarly for point Q.
∴   The solution of $x^2 - 3x + 1 = 0$ are $x = 0.4$ and $x = 2.6$

---**Example 2**---

Use the graphs of $y = x^3$ and $y = 3x^2 + 2x - 5$ between the values of $x = -2$ and 4 to solve $x^3 - 3x^2 - 2x + 5 = 0$.

The tables are:

x	-2	-1	0	1	2	3	4
$y = x^3$	-8	-1	0	1	8	27	64
$y = 3x^2 + 2x - 5$	3	-4	-5	0	11	28	51

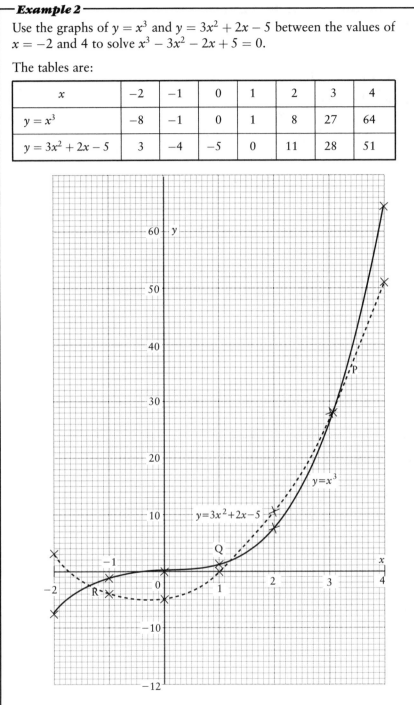

The two graphs intersect at P, Q and R. The x-values of these points are 3.2, 1.1, -1.4.

The two graphs intersect where $x^3 = 3x^2 + 2x - 5$.

This can be rearranged to give $x^3 - 3x^2 - 2x + 5 = 0$.

Hence the solutions of $x^3 - 3x^2 - 2x + 5 = 0$ are:

$$x = 3.1, \ 1.2 \text{ and } -1.3$$

Note

These values are not exact. There is no algebraic method which will solve all equations containing an x^3 term.

1 Draw $y = 2x^2$ for values of x between -4 and $+4$.

    a Hence solve (i) $2x^2 = 9$ (ii) $x^2 = 11$.

    b Draw $y = x + 4$ and hence solve $2x^2 = x + 4$.

    c Draw $y = 3 - 2x$ and hence solve $2x^2 + 2x - 3 = 0$.

2 Draw $y = x^3$ and $y = x - 1$ for values of x between -2 and 2. Hence solve $x^3 = x - 1$.

3 Draw $y = 2x^2 + x - 1$ and $y = x^2 - 4x$ for values of x between -6 and $+1$. Hence solve $2x^2 + x - 1 = x^2 - 4x$.

4 Draw $y = x^3 - 2x$ and $y = x^2 + 5$ for values of x between -1 and 4. Hence solve $x^3 - x^2 - 2x - 5 = 0$.

5 Draw $y = x^2 + x - 2$ and $y = \dfrac{1}{x}$ for values of x between -2 and $+2$.
(What happens to the value of y as x gets nearer 0?)
Hence solve $x^3 + x^2 - 2x - 1 = 0$.

6 **a** Complete the table of values for the function $y = x^3 + 5$ and use it to draw the graph of the function.

x	-3	-2	-1	0	1	2	3
y	-22	-3					

    b (i) On the same axes draw a suitable line to help you to solve the equation

$$x^3 + 5 = 5x + 3$$

    (ii) Write down all the value of x that satisfy this equation.

    c On the graph draw another line, parallel to your line, such that the equation

$$x^3 + 5 = 5x + C$$

only has one solution. Write down the value of C.

(SEG S94)

33 Algebra with Graphs

33.1 Number patterns or sequences of numbers

Sequences

Number patterns can arise from many mathematical situations, particularly investigations.

A number pattern or **sequence** of numbers is an ordered list of numbers which are connected by a rule.

Each number in the sequence is called a **term** and, if the rule connecting two consecutive terms is known, the next term in the sequence can be found, and the one after that, and so on . . .

For example, you will recognise the sequence of numbers

$$2, 4, 6, 8, \ldots$$

as the *even numbers*.

The rule which connects them is: *to find the next term, add two to the previous term.*

The next sequence of numbers is called a *Fibonacci sequence*. The interesting properties of these numbers was first noticed in medieval times by Leonardo Pisano.

The sequence is 1, 1, 2, 3, 5, 8, . . .

Can you see how each new term is formed?

Exercise 33A

For each of the following sequences:

a find the missing terms,

b describe, in words, the rule for finding the next term.

1 1, 3, 5, 7, ?, ?, . . .

2 7, 9, 11, ?, ?, . . .

3 1, 3, 9, 27, ?, ?, . . .

4 ?, ?, 1, ?, −3, −5, . . .

5 0.1, 0.01, ?, 0.0001, ?, . . .

6 1, 1, 2, 3, 5, 8, ?, ?, ?, . . .

7 $\frac{1}{2}, \frac{1}{4}, \frac{1}{8}, ?, \frac{1}{32}, ?, \ldots$

8 $\frac{1}{2}, \frac{3}{4}, ?, ?, \frac{9}{32}, \ldots$

9 1, 4, 9, 16, ?, ?, ?, . . .

10 1, 3, 6, ?, 15, ?, 28, . . .

Using algebra

In the previous exercise, you described the rule for the next term in a sequence in words. It is more mathematical, and neater, to describe the rule using algebra.

The **term** is usually represented by a capital letter with a suffix which tells you the number of the term. For example, T_1 would mean the first term in a sequence while U_5 could represent the fifth term in a different sequence. T_n represents the nth term in a sequence. This is often taken to be the **general term**, i.e. it represents any term in the sequence.

For the sequence of even numbers:

n	1	2	3	4	...
T_n	$T_1 = 2$	$T_2 = 4$	$T_3 = 6$	$T_4 = 8$	...

The rule for obtaining the next term was *add two to the previous term*. This can be written algebraically as

$$T_n = T_{n-1} + 2$$

where T_{n-1} is the term *before* T_n.

Exercise 33B

1–7 Write down algebraic rules for questions 1 to 7 in Exercise 33A.

For each of the following sequences, write down the first four terms. You are given the first term and the rule connecting the terms.

8 $U_1 = 4, U_n = U_{n-1} + 3$ **11** $U_1 = 4, U_n = \frac{1}{2} U_{n-1}$

9 $U_1 = 2, U_n = 4U_{n-1}$ **12** $U_1 = -2, U_n = \frac{1}{2} U_{n-1} + 3$

10 $U_1 = 2, U_n = U_{n-1} - 4$ **13** $U_1 = \frac{1}{2}, U_n = U_{n-1} - 1$

Linear sequences

You will have noticed that sometimes the same rule (i.e. $U_n = U_{n-1} + 2$) may apply to different sequences such as those in questions 1 and 2 of Exercise 33A. Also, it would not be easy to write down a similar rule for the sequences in questions 9 and 10 of Exercise 33A. So this method has its limitations.

If we require a *unique* rule for a sequence, we need to relate *the value of each term* to n, the *number of the term*.

For the even numbers, $T_n = 2n$

n	1	2	3	4	...
$T_n = 2n$	2×1	2×2	2×3	2×4	...
T_n	2	4	6	8	...

The sequences 1, 3, 5, 7, ... and 7, 9, 11, ... were both formed in the same way as the even numbers, i.e. by adding 2 to the previous term, but the formula $T_n = 2n$ will not produce either of these sequences. $T_n = 2n$ is unique to the sequence of even numbers which starts at 2.

How then do we find formulae for these two sequences of odd numbers? The first clue is that the **difference** between the terms is always 2. This tells us that, as for the even numbers, the formula must contain the expression $2n$.

For the sequence 1, 3, 5, 7, 9 we have:

n	1	2	3	4	5
$2n$	2	4	6	8	10
T_n	1	3	5	7	9

We can see that each term T_n is one less than $2n$ so:

n	1	2	3	4	5
$2n$	2	4	6	8	10
$2n - 1$	$2 - 1$	$4 - 1$	$6 - 1$	$8 - 1$	$10 - 1$
T_n	1	3	5	7	9

and the formula is

$$T_n = 2n - 1$$

For the sequence 7, 9, 11, 13 we have:

n	1	2	3	4
$2n$	2	4	6	8
$2n + 5$	7	9	11	13
T_n	7	9	11	13

So the formula is $T_n = 2n + 5$

Example 1

Find a formula, in terms of n (where n is the number of the term), for the sequence
$$15, 11, 7, 3, -1, \dots$$

The sequence is $\qquad\qquad$ 15 $\quad$ 11 $\quad$ 7 $\quad$ 3 $\quad$ -1

The difference between the terms is $\quad$ -4 $\quad$ -4 $\quad$ -4 $\quad$ -4

$\therefore$ the expression $-4n$ is part of the formula.

n	1	2	3	4	5
$-4n$	-4	-8	-12	-16	-20
$-4n + 19$	15	11	7	3	-1
T_n	15	11	7	3	-1

$\therefore$ the formula is $T_n = 19 - 4n$

Check. When $n = 1$, $T_n = 19 - 4 = 15$ and when $n = 2$, $T_n = 19 - 8 = 11$, etc.

An alternative method for finding the formula for this type of sequence is to draw a graph and find the equation of the line.

Example 2

a Plot the first five numbers in the sequence 4, 7, 10, 13, 16, ... and hence draw the straight line which represents the sequence.

b From the graph, find (i) the gradient (ii) the intercept on the y-axis.

c Write down the equation of the graph and hence the formula for the sequence.

a

n	1	2	3	4	5
T_n	4	7	10	13	16

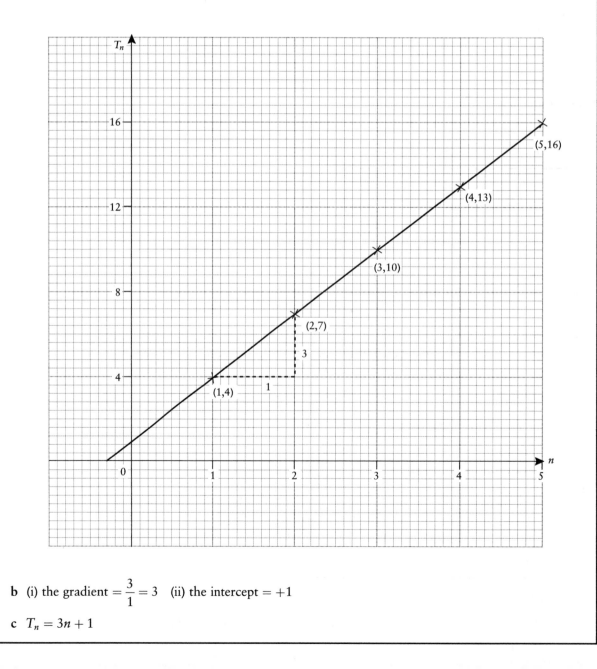

b (i) the gradient $= \dfrac{3}{1} = 3$ (ii) the intercept $= +1$

c $T_n = 3n + 1$

Exercise 33C

1 Write down the first five terms of the following sequences where:

   a $T_n = 2n + 3$ **c** $T_n = \frac{1}{2}n - 2$

   b $T_n = 4 - n$ **d** $T_n = 5 - 3n$

2 Find a formula for the nth term of each of the following sequences:

   a $3, 6, 9, 12, \ldots$ **d** $50, 40, 30, 20, \ldots$

   b $1, 5, 9, 13, \ldots$ **e** $-6, -4, -2, 0, \ldots$

   c $9, 16, 23, 30, \ldots$

3 Use the graphical method to find formulae for the following sequences:

   a $4, 7, 10, 13, \ldots$ **d** $-1\frac{1}{2}, -1, -\frac{1}{2}, 0, \ldots$

   b $3, -1, -5, -9, \ldots$ **e** $4, -2, -8, -14, \ldots$

   c $1, 1\frac{1}{2}, 2, 2\frac{1}{2}, \ldots$

4 In the following investigation, the size of a Tee is the sum of the numbers inside the Tee.

   Tee eleven (T_{11}) is shown on the grid below:

1	2	3	4	5	6	7
8	9	10	11	12	13	14
15	16	17	18	19	20	21
22	23	24	25	26	27	28
29	30	31	32	33	34	35
36	37	38	39	40	41	42
43	44	45	46	47	48	49

   The size of $T_{11} = 10 + 11 + 12 + 18 + 25 = 76$

   a Complete the table below for Tee sizes from 2 to 6:

n	2	3	4	5	6
T_n			41		

   b Find a formula for T_n.

   c (i) Use your formula to find T_{34}.
   (ii) Check your formula by finding T_{34} from the grid.

*5 **a** Continue the above investigation by finding formulae for T_n when the width of the grid, a, is 4, 5 and 6 and completing the table below:

a	4	5	6	7
$T_{n,a}$				$5n + 21$

   b What is the difference between each term?

   c Write down a formula for $T_{n,a}$

6 Change the Tee shape in question 3 to one of your own choice and investigate to find new formulae.

7 Investigate the following patterns to find a sequence and hence a formula.

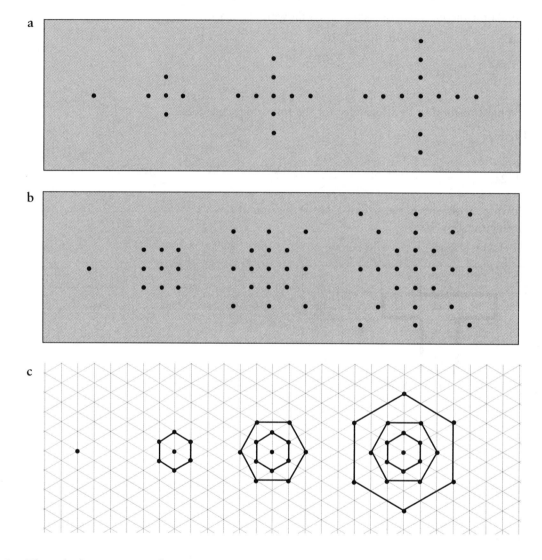

a

b

c

8 The rule for any term of a sequence is

$$\frac{n}{2n-1}, \text{ where } n \text{ is the number of the term.}$$

a Use this rule to write down the first five terms of the sequence.

b The first five terms of a different sequence are

Term	1	2	3	4	5
Sequence	$\frac{1}{2}$	$\frac{2}{5}$	$\frac{3}{8}$	$\frac{4}{11}$	$\frac{5}{14}$

Write down the rule for the nth term of this sequence.

(SEG S95)

*Quadratic sequences

A linear sequence is a sequence whose points all lie on a straight line (see Example 2 on page 286).

The sequence 1, 4, 9, 16, 25, ... is not a linear sequence.

It is the sequence of **square numbers** and $T_n = n^2$.

First we find the differences between terms, as we did for the linear sequences:

Sequence   1   4   9   16   25

Differences (1)   3   5   7   9

Differences (2)   2   2   2

We see that two sets of differences need to be found before the differences between terms are all equal.

If we need to subtract twice to obtain equal differences, then the formula contains the expression n^2. In fact, if the differences are 4, the formula will contain $2n^2$, if the differences are 6, the formula will contain $3n^2$, and so on.

Example 1

Find the formula, in terms of n, for the sequence 3, 6, 11, 18, 27, ...

Sequence: 3   6   11   18   27

Differences (1)   3   5   7   9

Differences (2)   2   2   2

n	1	2	3	4	5
T_n	3	6	11	18	27
n^2	1	4	9	16	25
$T_n - n^2$	2	2	2	2	2

$\therefore \quad T_n = n^2 + 2$

Exercise 33D

1   **a**   (i)   Write down the next two terms in the following sequence:

         1, 4, 9, 16, __, __

      (ii)   Write down an expression for the nth term.

   b   (i)   How is the following sequence derived from the sequence in part **a**?

         6, 9, 14, 21, __, __

      (ii)   Write down an expression for the nth term.

2 **a** The rule for any number in a sequence is $\dfrac{n+1}{n^2+1}$

where n is the number of the term.
Use this rule to write down the first five numbers in the sequence.

b The first five terms of a different sequence are

$$\tfrac{1}{4}, \tfrac{2}{7}, \tfrac{3}{12}, \tfrac{4}{19}, \tfrac{5}{28}$$

Write down the rule for the nth term of this sequence.

3 The numbers in the sequence 2, 8, 18, 32 can be written as
$(1 \times 2), (2 \times 4), (3 \times 6), (4 \times 8)$

a Find the 9th term in the sequence.

b Find an expression for the nth term.

4 **a** The sequence of triangular numbers is 1, 3, 6, 10
Write down the next two terms.

b Another sequence is $(1 \times 2) (2 \times 3) (3 \times 4) (4 \times 5)$
(i) Write down the 10th term of the sequence.
(ii) Write down the nth term of the sequence.

c (i) Write down a rule for the nth term of the sequence of triangular numbers.
(ii) Use your rule to find the 20th triangular number.

5 The nth term of a sequence is given by $U_n = n^3 + 1$.

a Write down the third and fourth terms.

b Use the rule for the nth term to find the number of the term which has value 217.

6 The rule for generating a sequence is given by the function $U(x) = 2^x$
where $U(x)$ is the xth term.
Write down the first five terms of the sequence.
Hence write down an expression, as a function of x, for the xth term of the sequence:

$$7, 9, 13, 21, 37$$

33.2 Solving polynomials by trial and improvement

Polynomials of order 2 (i.e. quadratic equations) and above (cubic equations, etc.) are solved by trial and improvement in the same way as the simpler examples you met in Unit 18 of the Money Management and Number module.

---**Example**---

Use a trial and improvement method to find **one** solution of the equation
$x^3 - x^2 = 5$.
Give your answer (i) to 1 decimal place (ii) to 2 decimal places.

First choose a suitable starting value for x. (We can see that when $x = 1$
the left-hand side $= 0$ so we choose a larger value for x.)

When $x = 2$ $x^3 - x^2 = 4$
      $x = 2.1$ $x^3 - x^2 = 4.851$
      $x = 2.2$ $x^3 - x^2 = 5.808$

The solution lies between $x = 2.1$ and $x = 2.2$. As $x = 2.1$ gives a value
which is nearer to 5,
the solution to (i) is $x = 2.1$.

To find a solution which is correct to 2 decimal places, we need to
search between $x = 2.1$ and $x = 2.2$.

When $x = 2.11$ $x^3 - x^2 = 4.941$
      $x = 2.12$ $x^3 - x^2 = 5.033$

The solution lies between $x = 2.11$ and $x = 2.12$. As $x = 2.12$ gives a
value which is nearer to 5,
the solution to (ii) is $x = 2.12$.

Exercise 33E

1   Use a trial and improvement method to solve the following equations to
the requested number of decimal places:

a $x^2 + 2x = 4$ (1 d.p.)

b $x^2 - 3x = 7$ (2 d.p.)

c $x^3 + 3x = 1$ (2 d.p.)

d $x^3 - 2x^2 = 15$ (2 d.p.)

2   Plot the graph of $y = x^2 - 8x + 2$.

a   From the graph, find the values of x when $y = 0$.

b   Use a trial and improvement method to find **two** solutions to the equation
$x^2 - 8x + 2 = 0$, giving your answers correct to 1 decimal place.
Compare your answers to parts **a** and **b**.

3   a   For the equation $y = x^3 - 4x^2 - 3$, find the value of y when
(i) $x = 4$ (ii) $x = 5$.

b   Explain why there must be a solution to the equation
$x^3 - 4x^2 - 3 = 0$ between $x = 4$ and $x = 5$.

c   Use a trial and improvement method to find a solution to the
equation $x^3 - 4x^2 = 3$, correct to 2 decimal places.

4   a By first trying $x = 3$, use a trial and improvement method to find **one** of the solutions to the equation $x^2 - x = 7$.

Give your answer correct to **one** decimal place.

You must show **all** your working.

The graph of $y = x^2 - x - 7$ is drawn below.

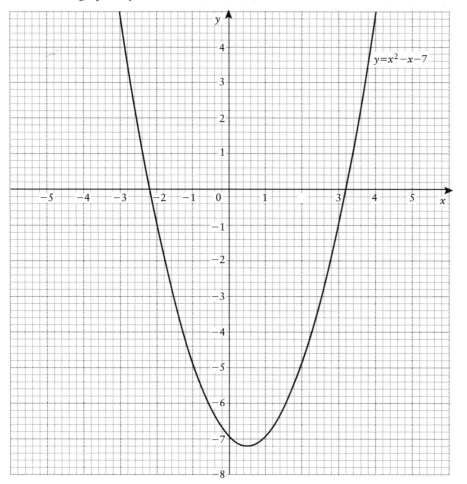

b Use the graph to
   (i) write down the co-ordinates of the point for which y has its minimum value;
   (ii) find the **two** values of x where the curve crosses the x axis.

(SEG W95)

33.3   Mappings and functions

As we have seen, a graph is a 'picture' of a mathematical relationship such as $y = 2x + 1$ (see page 268).

This relationship can also be thought of as a 'rule' for transforming one set of numbers into another. In this case it is called a **mapping** and is written as

$$m\!:\!x \rightarrow 2x + 1$$

We say, 'the mapping m is such that x maps to $2x + 1$' or 'x is changed into $2x + 1$'.

For every number x transformed by the rule 'multiply x by 2 and add 1', a new number y is produced. For example:

$$2 \rightarrow 5 \qquad 6\tfrac{1}{2} \rightarrow 14 \qquad -3 \rightarrow -5.$$

The mapping 'm: $x \rightarrow$ the square root of x' transforms numbers greater than zero to **two** output numbers, while for numbers less than zero there is no output. For example:

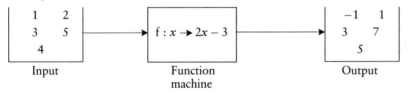

It is not convenient to have two (or more) possible outputs from which to choose (your calculator produces only one answer at a time). A new type of mapping is defined which does not include these awkward mappings. It is called a **function**.

A **function** is defined as a mapping which outputs only **one** new number for every input.

f:$x \rightarrow 2x + 1$ means 'we have a function f such that x, the number input, becomes $(2x + 1)$, i.e. 2 times the number plus 1.'

Alternatively, this can be written as $f(x) = 2x + 1$, where x is the input number and $f(x)$ is the output number.

It is possible for two different input numbers to give the same output:

For f:$x \rightarrow x^2$ $\qquad 4 \rightarrow 16$ and $-4 \rightarrow 16$.

A function can be thought of as a 'machine' which processes numbers. (A calculator acts in this way when the function keys are used.)

The input numbers x are fed into the 'function machine', processed, and the output is the y, or $f(x)$, numbers.

For example:

1 2				−1 1

Input — f : $x \rightarrow 2x - 3$ — Output

```
   1   2              ┌─────────────────┐              −1   1
   3   5   ──────────▶│ f : x → 2x − 3   │──────────▶    3   7
   4                  └─────────────────┘                  5
  Input              Function machine              Output
```

A mapping or function may also be represented by a **mapping diagram**. This links the input on one number line with the output on a second number line.

Example

Represent the function f: $x \rightarrow 2x - 3$ by a mapping diagram.

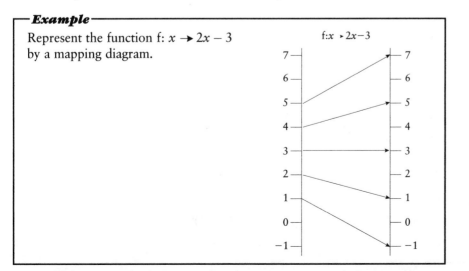

Exercise 33F

1 Find the output for the following:

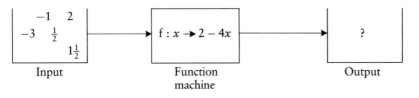

Input	Function machine	Output

2 For the following functions, find the outputs for the x values given.

   a f:$x \to x - 7$ when $x = -2, -1, 0, 1, 2, 3$

   b f:$x \to 3x + 4$ when $x = 8, -5$

   c f:$x \to 2x^2$ when $x = -1, -\frac{1}{2}, 0, \frac{1}{2}, 1, 2$

   d f:$x \to 2x^2 + 3$ when $x = -1, -\frac{1}{2}, 0, \frac{1}{2}, 1, 2$

   e f:$x \to \dfrac{1}{x}$ when $x = 2, -5, \frac{1}{4}, -0.2$

3 Write down the equation represented by this mapping diagram.

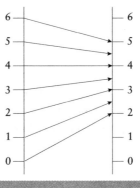

Solving and forming functions

In order to find the outputs for a function more easily, we can separate the function into stages.

For example, to find the outputs when $y = 3x + 11$, the input, x, is multiplied by 3 and then 11 is added.

This rule can be represented by a diagram:

$x \longrightarrow$ | Multiply by 3 | $\longrightarrow$ | Add 11 | $\longrightarrow$ Result y

We can also form a function from a diagram, as shown in the example below.

Example

Find the relationship between x and y shown in the diagram below:

$x \longrightarrow$ | Multiply by 4 | $\longrightarrow$ | Subtract 5 | $\longrightarrow$ | Square | $\longrightarrow$ | Divide by 2 | $\longrightarrow y$

The results after each step will be:

x $4x$ $4x - 5$ $(4x - 5)^2$ $\frac{1}{2}(4x - 5)^2$

$y = \frac{1}{2}(4x - 5)^2$

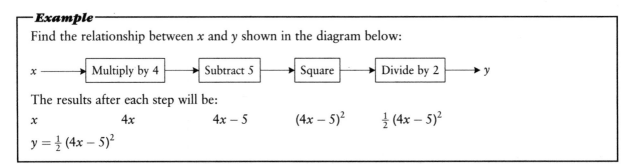

Exercise 33G

1   Draw a diagram to represent each of these equations:

   a $y = 7x + 5$ **b** $y = 3x - 11$ **c** $y = 3x^2 + 5$ **d** $y = (2x + 1)^2$

2   Find the relationship between x and y shown in the diagram.

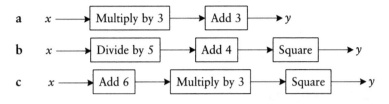

a x ⟶ Multiply by 3 ⟶ Add 3 ⟶ y

b x ⟶ Divide by 5 ⟶ Add 4 ⟶ Square ⟶ y

c x ⟶ Add 6 ⟶ Multiply by 3 ⟶ Square ⟶ y

Inverse functions

Consider the function $y = 3x + 7$, which may also be written as $f : x \rightarrow 3x + 7$ or $f(x) = 3x + 7$.

This function transforms, or maps, 4 (the input x) to 19 (the output y or $f(x)$).

The **inverse function** will map an input $x = 19$ to an output $f^{-1}(x) = 4$.

The diagram for $f(x) = 3x + 7$ is:

x ⟶ Multiply by 3 ⟶ Add 7 ⟶ $f(x)$

To find the diagram for the inverse function $f^{-1}(x)$ you must replace each box by its inverse and start from the right-hand side, i.e.

$f^{-1}(x)$ ⟵ Divide by 3 ⟵ Subtract 7 ⟵ x

This is usually written from left to right, i.e.

x ⟶ Subtract 7 ⟶ Divide by 3 ⟶ $f(x)$

This produces the algebraic inverse: $f^{-1}(x) = \dfrac{(x - 7)}{3}$

Some functions do not have an inverse which is a function.
$g(x) = x^2$ transforms both 3 and -3 into 9. Hence the inverse mapping could go back to 3 or -3.

x ⟶ Square root ⟶ $g^{-1}(x)$ is not a function.

Exercise 33H

1   Draw the diagram showing the inverse of each of the diagrams shown:

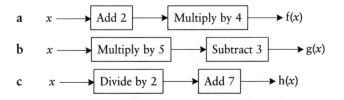

a x ⟶ Add 2 ⟶ Multiply by 4 ⟶ $f(x)$

b x ⟶ Multiply by 5 ⟶ Subtract 3 ⟶ $g(x)$

c x ⟶ Divide by 2 ⟶ Add 7 ⟶ $h(x)$

2   Use diagrams to find the inverse of:

  a $f(x) = 3x - 5$

  b $g(x) = 2x + 7$

  c $h(x) = 7x - 11$

*33.4   Transformations of graphs

What happens if we begin with one function and then transform it into
another function? Is it possible to predict how the graph will change?

It is much easier to investigate this question with the aid of a graphics
calculator or a computer graph package.

Begin with one of the simpler functions, e.g. $x \rightarrow 2x$.
If we transform this function by adding a constant, 7, how is the graph
changed?

This is like processing x through **two** function machines. The first multiplies x
by 2, the second adds a constant to the new value of x.

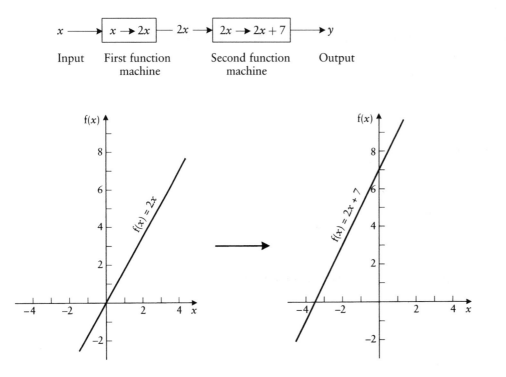

For Exercise 32A, question 7, you have already drawn graphs of the functions
$f(x) = 2x$, $f(x) = 2x + 3$, $f(x) = 2x - 1$, and $f(x) = 2x - 2$.
Describe how the graph changed each time and sketch the four functions.

Does this change apply to all functions?

Exercise 33I

1 Draw the graph of $f: x \rightarrow x^2$.
On the same axes, draw the graphs of
$$f: x \rightarrow x^2 + 3$$
$$f: x \rightarrow x^2 - 1$$
$$f: x \rightarrow x^2 - 2$$

Describe the transformations.

2 Write down a general rule for transforming any function of x by adding a constant, i.e.

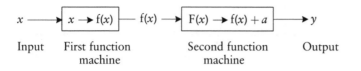

Input   First function   Second function   Output
         machine   machine

3 Sketch the following graphs of functions:

a $f(x) = x$ c $f(x) = x^3$ e $f(x) = \dfrac{1}{x}$

b $f(x) = x^2$ d $f(x) = x^4$ f $f(x) = \dfrac{1}{x^2}$

You may check your answers using a graphics calculator or computer.

4 Use the functions you have sketched in question 3 to investigate what happens when the constant a is added to x, before it is mapped by a given function, i.e.

$$x \longrightarrow \boxed{x \rightarrow x + a} - (x + a) \longrightarrow \boxed{(x + a) \rightarrow f(x + a)} \longrightarrow y$$

Input   First function   Second function   Output
         machine   machine

For example, when $f(x) = x^2$ and $a = 3$, $x \rightarrow (x + 3) \rightarrow (x + 3)^2$.

5 Sketch the graph of $y = x^2$.
From your results for question 4, deduce the graph of $y = (x - 2)^2$ and sketch it on the same axes as $y = x^2$.

6 Investigate the transformations $f(x) \rightarrow af(x)$ and $f(x) \rightarrow f(ax)$.

7 Investigate the transformation $f(x) \rightarrow -f(x)$ and $f(x) \rightarrow f(-x)$.

8 List general results for all the above investigations.

9 Sketch, or draw, the following functions and their reciprocal functions, using the same axes for each pair:

a $f(x) = x$, $f(x) = \dfrac{1}{x}$ c $f(x) = x^3$, $f(x) = \dfrac{1}{x^3}$

b $f(x) = x^2$, $f(x) = \dfrac{1}{x^2}$ d $f(x) = x - 2$, $f(x) = \dfrac{1}{x - 2}$

e Compare each pair of functions and comment.

10 Figure (i) shows a function $y = f(x)$. The remaining figures are
transformations of $f(x)$. Match each of the graphs (ii) to (viii) with one of
the following transformations:

a $2f(x)$ c $f(x-1)$ e $f(x)+1$ g $f(-x)$

b $f(2x)$ d $f(x+1)$ f $-f(x)$

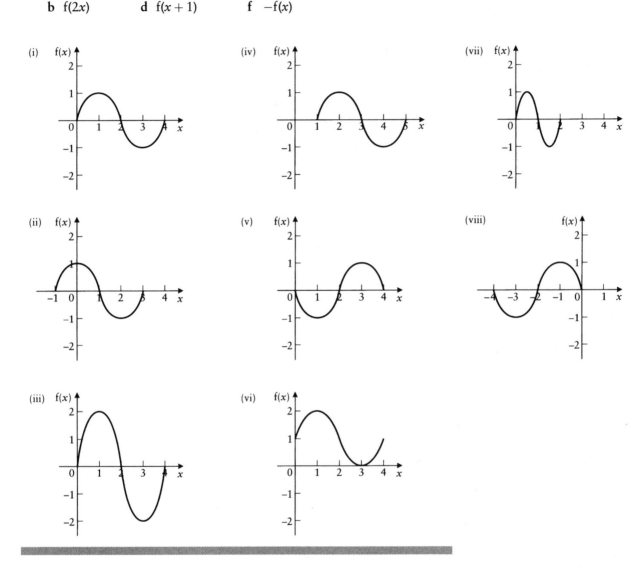

Exercise 33J An investigation

Use the methods of Exercise 32A and 33I to investigate the coefficients a, b,
and c in the general quadratic function $y = ax^2 + bx + c$.
You have already done most of the work for a and c in these exercises.
b is more difficult to investigate.

Look first at functions such as $y = x^2 + bx$ then $y = 2x^2 + bx$ for different
values of b.

Also consider functions like $y = (x-1)^2$ which is the same function as
$y = x^2 - 2x + 1$.

33.5   Inequalities

An **equation** is a mathematical relationship which is balanced, i.e. what is on the left-hand side of the equation is always equal to what is on the right-hand side.

In many situations, the relationship is not one of equality. In an **inequality** one side is larger than the other.

Examples of inequalities are:

1   The average woman's height is 'less than' the average man's height:

$$\text{Height of average woman} < \text{Height of average man}$$

2   On a cumulative frequency graph, the median is 'greater than' the lower quartile:

$$\text{Median} > \text{Lower quartile}$$

3   In order to win a game a player must throw at least a score of four, i.e. the score on the die must be 'equal to or greater than' four:

$$\text{Score} \geqslant 4$$

4   The number of people standing on the bus should be no more than six, i.e. the number standing should be 'less than or equal to' six:

$$\text{Number standing} \leqslant 6$$

Statements such as those above may be represented on a number line by a range of values.

Example 1

Represent the inequality $x \leqslant 3$ on a number line.

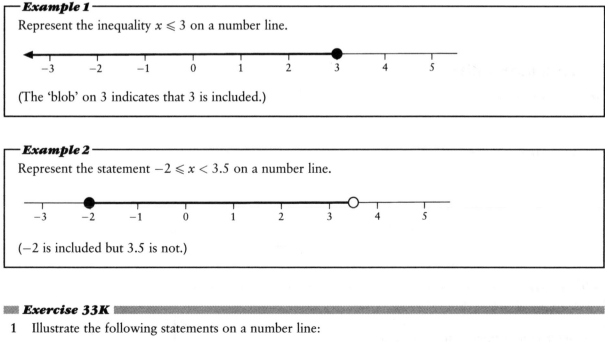

(The 'blob' on 3 indicates that 3 is included.)

Example 2

Represent the statement $-2 \leqslant x < 3.5$ on a number line.

(−2 is included but 3.5 is not.)

Exercise 33K

1   Illustrate the following statements on a number line:

a $x < 5$ c $-3 < x < 3$ e $-1 < x \leqslant 4$ g $x < 5, x > 6$

b $x \geqslant 0$ d $-3 \leqslant x \leqslant 3$ f $-7 \leqslant x < -2$ h $x < -3, x \geqslant 3$

2 Write down the inequalities which are illustrated on the following number lines.

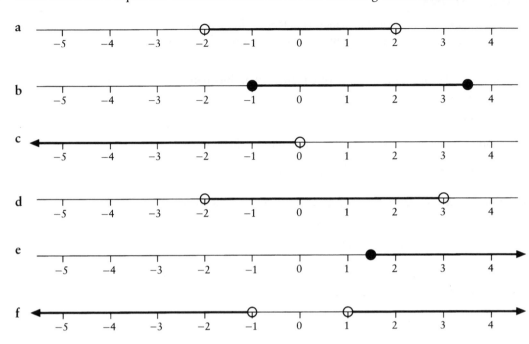

3 If x is an integer, list the values of x which satisfy the following inequalities.

a $7 < x < 13$ **d** $-1 \leqslant x \leqslant 1$ **g** $-24 \leqslant 3x < 6$

b $-3 \leqslant x < 0$ **e** $-1 < x < 1$

c $-6 < x \leqslant -1$ **f** $8 \leqslant 2x \leqslant 12$

Solving inequalities

Some of the rules for solving equations can also be used for solving inequalities:

Example 1

Solve **a** $x - 5 = 9$ **b** $x - 5 \geqslant 9$

a Adding 5 to each side gives $x = 14$.

b Similarly, adding 5 to each side gives $x \geqslant 14$.

Example 2

Solve **a** $x + 5 = 3$ **b** $x + 5 < 3$

a Subtracting 5 from each side gives $x = -2$.

b Similarly, subtracting 5 from each side gives $x < -2$.

Example 3

Solve **a** $4x = 15$ **b** $4x > 15$

a Dividing each side by 4 gives $x = 3\frac{3}{4}$.

b Similarly, dividing each side by 4 gives $x > 3\frac{3}{4}$.

Example 4

Solve **a** $\dfrac{x}{4} = 20$ **b** $\dfrac{x}{4} \leqslant 20$

a Multiplying each side by 4 gives $x = 80$.

b Similarly, multiplying each side by 4 gives $x \leqslant 80$.

But there are some rules for solving equations which, when applied to inequalities, **change the sign**:

1 Interchanging the two sides of the equation:

    e.g.   if $x = y$ then $y = x$

    but   if $x < y$ then $y > x$ (e.g. $2 < 3$ but $3 > 2$)

2 Multiplying (or dividing) by a negative number:

    e.g.   if $-x = -4$ then $x = 4$

    but   if $-x > -4$ then $x < 4$ (e.g. $-5 < -4$ but $5 > 4$)

Remember, when solving inequalities you may:

- add any number to both sides
- subtract any number from both sides
- multiply or divide both sides by any positive number
- multiply or divide both sides by any negative number and change the inequality sign.

You must also take care when finding square roots.

Example 5

a $x^2 = 25$ **b** $x^2 \leqslant 25$ **c** $x^2 > 25$

a Square-rooting each side gives $x = +5$ or -5.

b Square-rooting each side gives $x \leqslant +5$, and $x \geqslant -5$,
so the solution is $-5 \leqslant x \leqslant +5$.

c Square-rooting each side gives $x < -5$ and $x > 5$.

Exercise 33L

Solve the following inequalities:

1 **a** $4x < 12$ **c** $3x + 2 \leqslant 11$ **e** $2x - 7 \geqslant 10$ **g** $\dfrac{x}{4} - \dfrac{1}{2} > \dfrac{3}{2}$ **i** $x^2 \leqslant 2.25$

     b $x + 2 > 9$ **d** $x - 7 \geqslant 10$ **f** $\dfrac{x}{6} > 2.5$ **h** $x^2 > 25$ **j** $7 - 2x < 3$

2   Solve the inequality $x^2 - 1 < 8$ and represent the solution on a number line.

3   a Norris can't afford to pay more than 95p for his lunch. He buys a bag of chips and an apple.
       Write down an inequality.

    b Apples cost 20p and a bag of chips costs 45p, 60p or 90p.
       What is the most he can pay for chips?

4   I think of a number, x, multiply it by 4, and subtract seven.

    a Write down an expression, in terms of x, for this statement.

    b If the answer must be less than 20, write down an inequality.

    c Solve the inequality, if the answer is the largest possible whole number.

Graphs of inequalities

When a linear equation is plotted on a graph, the result is a straight line.

When a linear inequality is plotted on a graph, the result is a **region**.

Example 1

Draw the graph of $x < 3$.

Method

   (i)   Draw and scale x- and y-axes.

   (ii) Draw the graph of $x = 3$ (i.e. replace the inequality sign with an equal sign).

   (iii) Shade the region which is required, i.e. all values of x which are less than 3.

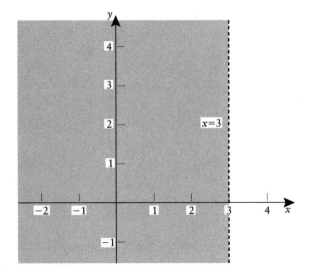

The solution is the shaded area.

(The line $x = 3$ is drawn with dotted lines to show that it is not included in the solution.)

Example 2

Illustrate, with a graph, the inequality $y \geqslant 4 - x$

Draw axes and plot the line $y = 4 - x$, which is the boundary line of the region required.

x	0	2	4
y	4	2	0

Shade the region which is required.
If you are not sure on which side of the line the region you want lies, choose a test point, and proceed as follows:

- Choose the point (4, 3)
- The values $x = 4$ and $y = 3$, substituted into the inequality, give LHS $= 3$, RHS $= 4 - 4 = 0$.
  As $3 > 0$, this fits the inequality and the point (4, 3) is in the required region.
- Shade the region to the right of the line which contains the point (4, 3).
  (This represents the points where $y \geqslant 4 - x$.)

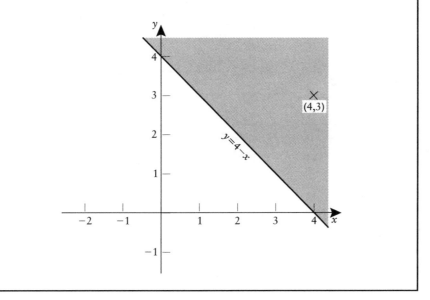

Exercise 33M

1   Illustrate, with a graph, the following inequalities. Shade the regions not required.

    a $x \geqslant 5$ **d** $x \leqslant 0$ **g** $x \leqslant 3$ and $x \geqslant 6$

    b $y < 4$ **e** $-1 < x < 4$ **h** $y < -4$ and $y > -1$

    c $y > -2$ **f** $4 \leqslant y \leqslant 7$

2 Write down the inequalities illustrated by the shaded areas in each of the following graphs:

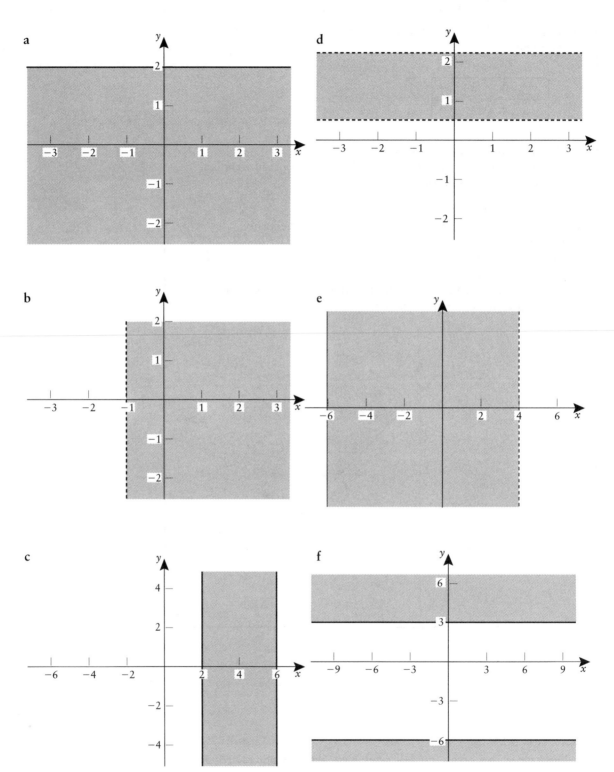

3 **a** Show that the point (1, 8) does not satisfy the inequality $y < 2x + 4$.

b Show that the point (1, 8) does satisfy the inequality $y > 7 - 2x$.

4 Draw graphs for each of the following inequalities. Shade the regions not required.

    a $y \leqslant x + 3$ **c** $y < 2 - x$ **e** $2x + y < 4$

    b $y \geqslant 2x - 1$ **d** $x + y > 5$ **f** $2y \geqslant x + 5$

5 **a** Solve the following inequalities for x.
      (i) $4x + 10 < 19 + 2x$ (ii) $x^2 \geqslant 19$

    b Copy the graphs of $y = 10$ and $x = 3$ drawn on the axes below.
      (i) Draw the line $y = 2x + 10$ on the same axes.
      (ii) Shade the region satisfied by the inequalities $y \leqslant 2x + 10$, $y \geqslant 10$ and $x \leqslant 3$.

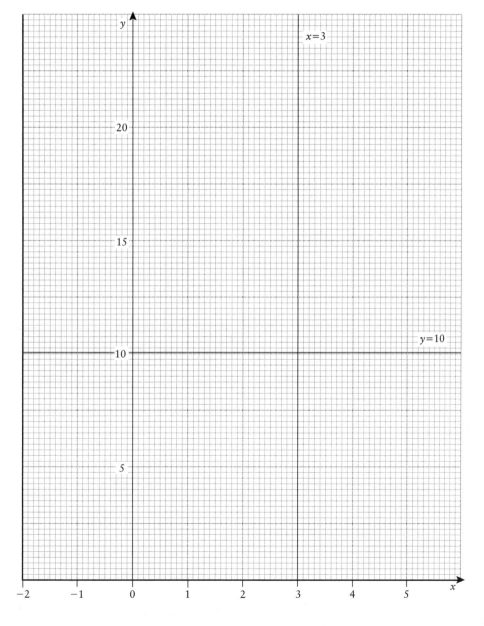

(SEG S95)

*Graphical solution of inequalities

One method of solving linear equations is by drawing graphs
(see Section 35.1 page 330.)

Two or more inequalities can be solved by a very similar method, except that
the solution is not a **point** of intersection but a **region** of intersection.

Example

Solve the inequalities $x \geqslant -2$, $y \geqslant 1$ and $x + y \leqslant 3$.

Method
Draw and scale x- and y-axes,

Plot each line **a** $x = -2$ **b** $y = 1$ **c** $x + y = 3$, in turn, shading the required region.

a Plot $x = -2$ and shade out the *unwanted*
region on the left of the line:

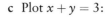

b Plot $y = 1$ and shade the region below the
line:

c Plot $x + y = 3$:

x	0	1	3
y	3	2	0

Shade the region on the right of the line.

The solution is all the points in the unshaded
region, including points on the boundary
lines.

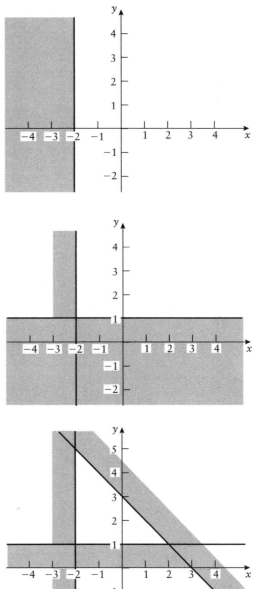

Exercise 33N

1   For each question, show on a graph the solution region for the given set
    of inequalities:

    a $x \leqslant 4, y \geqslant 2, y \leqslant x$ **d** $y < x - 2, x + y > 2, x \leqslant 4$

    b $x \geqslant -1, y \geqslant 0, 2x + y \leqslant 5$ **e** $y \geqslant 2x, 3 \leqslant y \leqslant 6$

    c $x + y \geqslant 3, 2y \leqslant x - 4, y < 3$

2   Write down the set of three inequalities, in terms of x and y, whose
    solution is given by the shaded area in the graph below:

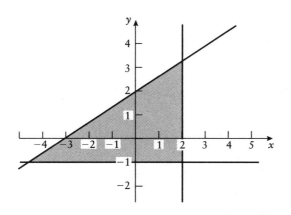

3   Write down the set of four inequalities, in terms of x and y, whose
    solution is given by the shaded area in the graph below:

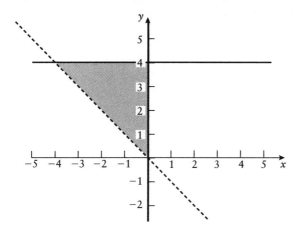

4   Plot the points A(0, 0), B(0, 3), C(2, 3). ABC is a triangle. Write down
    the three inequalities which describe the set of points inside the triangle.

5   **a** Draw a graph to illustrate the following inequalities, shading the
        required region:

    $$2x + y < 7, x > 0, y \geqslant 1$$

    b Given that x and y are both integers, list the coordinates of all
        possible solutions to the above inequalities.

6   The graph below illustrates the inequalities,

$y - 2x \leqslant 0$, $2y + x \geqslant 5$, $x \leqslant 5$.

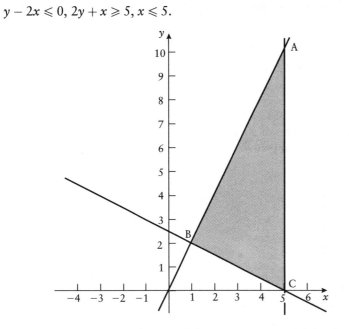

a Write down the coordinates of the vertices A, B and C.

b What is the value of $2y - x$ at (i) A (ii) B (iii) C?

c   (i) At which point does $2x + y$ have the largest value?
     (ii) At which point does $2x + y$ have the smallest value?

7   Michelle has won a number of trophies, some for sprinting, s, and some
     for long jumping, j. On the graph paper the vertical axis is j and the
     horizontal axis is s. The lines $j = 1$, $s = 2$, and $s + j = 7$ are drawn on
     the graph.

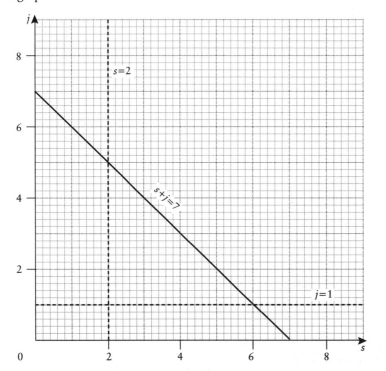

a On the graph shade the regions **not required** for the following
conditions.
  (i) Michelle has more than one trophy for jumping.
  (ii) Michelle has more than two trophies for sprinting.
  (iii) Michelle has no more than seven trophies altogether.

b Write down the **three** inequalities represented by the **unshaded** area
on the graph and solve them to find one possible combination of
Michelle's trophies.

(SEG W94)

33.6 Linear programming

The need for linear programming

Linear programming is a method used in business and industry for solving
problems which involve a large number of conditions that all need to be met
at the same time.

Here are two examples:

- finding how to make the largest possible profit taking into account costs
  such as labour, transport, materials, etc., and the price which can be
  charged for each product;
- finding how to keep costs as low as possible when providing food for
  soldiers, but still meeting all the requirements of nutrition, energy, variety,
  etc., which are needed in a diet.

The second example was one of the earliest linear programming problems to
be attempted. It applied to American soldiers in the Second World War. So
you can see how recent this type of mathematical problem is.

In practice, because there are large numbers of unknown quantities involved,
these problems are solved by computer.

In a simpler problem, where there are only two variables, a problem of this
type can be solved graphically.

Applications

Example

Two friends are planning an exercise programme in order to keep fit. They calculate that they can
spend up to $1\frac{1}{2}$ hours per day on exercise. Some of the time will be spent on aerobics and the
remainder on flexibility exercises. They would prefer to do more aerobics than flexibility, but could not
manage more than 60 minutes of aerobics.

a Letting x be the time spent on aerobics and y be the time spent on flexibility, write down three
  inequalities, other than $x \geqslant 0$ and $y \geqslant 0$, using the information above.

b Plot your three inequalities on a graph, showing clearly the **feasible region** (the region in which all
  possible solutions lie).

c They discover that aerobics use 8 calories per minute and flexibility exercises use 3 calories per
  minute.
  Write down an equation, in terms of x and y, for the calorie loss.

d They wish to lose the maximum number of calories.
At which point on the graph will they find this solution?

e What is the maximum calorie loss?

a Total time spent is less than or equal to 90 minutes:
$$x + y \leqslant 90$$
x must be more than y: $x > y$
x must be less than or equal to 60 minutes:
$$x \leqslant 60$$

b
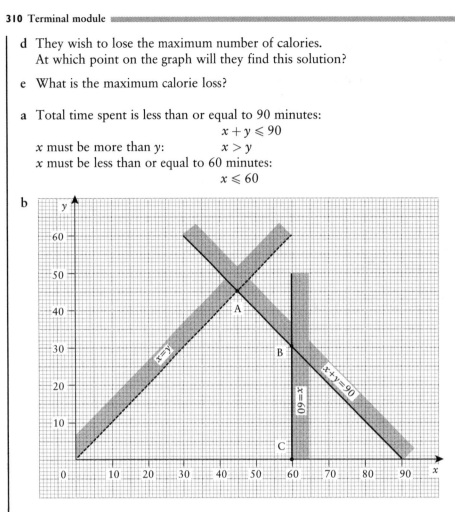

c Calories used $= 8x + 3y$.

d Maximum and minimum values will be found at the points of intersection of the lines, i.e. at 0, A, B or C.
For the largest calorie loss they need to spend the maximum time on exercise; therefore the solution lies on the line AB at point B.

e B is the point (60, 30) which gives a calorie loss $= 570$

(Point A gives a calorie loss of 495, and points between A and B give calorie losses between 495 and 570.)

Exercise 33P

1 Darren has won a prize in a music competition of £150 to spend on CDs and cassettes. He expects to pay £10 for the CDs and £6 for the cassettes. He has storage space for up to 12 discs and up to 9 cassettes.

a Write down three inequalities which fit the information above.

b Draw a graph of your inequalities, shading out the unwanted regions.

c Darren wants to buy as large a total as possible. Use your graph to find two possible solutions to the problem.

d What would you advise Darren to buy?
Give a reason.

2 An American firm owns a fleet of 12 minibuses and 10 taxi cabs. The minibuses seat 8 and the cabs seat up to 4 passengers.
Minibuses and cabs are sent to the airport when a plane is due to land. The firm is expecting to pick up 110 passengers and knows that, for this flight, they will need at least 3 taxis for every 5 minibuses. They do not want to send any more vehicles than necessary.

How many minibuses and how many cabs should be sent?

a If x is the number of minibuses sent and y is the number of cabs, show that two of the inequalities are:

$$4x + 2y \geqslant 55 \text{ and } 5y \geqslant 3x$$

b Write down two more inequalities which fit the information above.

c Draw a graph of your inequalities, shading out the unwanted regions.

d From the graph, find a solution to the problem which uses as few vehicles as possible.

3 A firm intends to sell packets of biscuits which contain a selection of plain and chocolate biscuits.
A plain biscuit weighs 25 g and costs 4p to produce.
A chocolate biscuit weighs 50 g and costs 6p to produce.
The packet contents will weigh at least 500 g but the firm wishes to keep the production costs to under £1.20.

a Write down the equations of the lines AD and HE in the graph below:

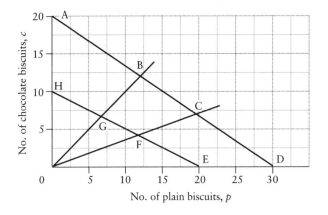

At least a quarter, but no more than half, of the biscuits in each tin must be chocolate, i.e.

$$3c \geqslant p \text{ and } c \leqslant p.$$

The equations $3c = p$ and $c = p$ are represented by the lines OGB and OFC on the graph.

b In which region do the possible solutions lie?

    c Near to which point on the graph will the contents of the packet cost the least to produce?

    d How many plain and how many chocolate biscuits should there be in a packet to keep the cost of production to a minimum?

    e What is the production cost in this case?

4 A window cleaner works on an estate which is a mixture of bungalows and houses and he wishes to know how many of each he should clean in a day. It takes him 40 minutes to clean the windows of a bungalow and 60 minutes for a house.
He charges £4 for a bungalow and £8 for a house.
He does not work more than 6 hours in a day, but needs to earn a minimum of £32 per day.
He prefers to clean more bungalows than houses.

    a If b is the number of bungalows and h is the number of houses whose windows are cleaned, write down three inequalities, in terms of b and h, which fit the information above.

    b Draw a graph of your inequalities and mark on it all possible solutions. How many are there?

    c State the solution, or solutions, which will earn him the most money.

5 Tammy wishes to treat her friends when they meet for a picnic.
She is not sure how many will be there, but there will not be more than fourteen, including herself.
Some of her friends like chocolate bars, which cost 25p and some prefer fruit, which costs 15p.
She decides to buy at least three pieces of fruit and at least five bars of chocolate, but she does not want to spend more than £3.00.

    a Write down four inequalities which fit the information above.

    b Draw a graph of your inequalities, shading out the unwanted regions.

    c Write down all the possible solutions which provide a treat for the most people.

    d Which of these solutions would you choose?
      Give a reason.

6 A baker makes x cakes and y packets of scones each day.

  Restrictions due to availability of ingredients can be expressed as follows.

$$y + 3x \leqslant 180$$
$$2y + 3x \leqslant 240$$
$$x \leqslant 50 \quad x > 0 \quad y > 0$$

    a On a copy of the diagram shade the region which would represent all the possible combinations of the number of cakes and scones that could be made in a day.

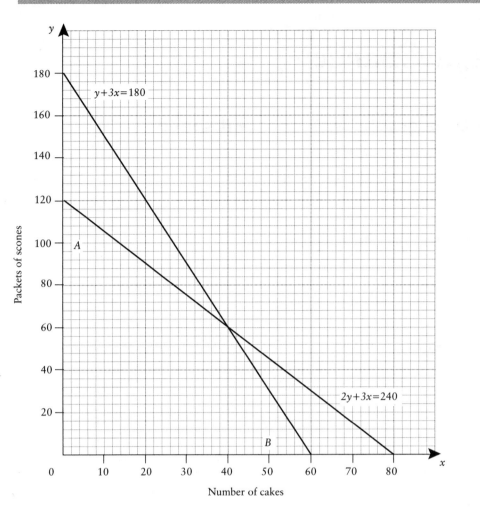

The line *AB* joining *A* (0, 100) to *B* (50, 0) represents all the possible combinations of cakes and scones which when sold would give £100 profit.

b How many packets of scones would be produced if the baker made 35 cakes and a profit of £100? (SEG W95)

34 Rates of Change

34.1   Distance–time graphs with constant speed

Graphs which connect distance and time are called distance–time graphs.

Graphs which connect velocity and time are called velocity–time graphs.

Example

One Sunday at 12 o'clock, a family set out for a picnic at a well-known beauty spot, 25 miles from their home.

They travelled at a steady speed and arrived at their destination at 12.45 pm.

After $3\frac{1}{4}$ hours at the picnic spot they set off for home.

Unfortunately the traffic was very heavy and they only managed to average 20 mph.

The family's day can be represented by a 'travel graph' which measures distance from home along the vertical axis and time along the horizontal axis.

The vertical axis is scaled from 0 to 25 miles and the horizontal axis from 12 noon to 6.00 pm.

At 12 noon they were at 0 miles (home) and at 12.45 pm they were at 25 miles (the picnic spot). These two points are plotted and the line joining them represents the family's outward journey.

For $3\frac{1}{4}$ hours they are not travelling and this is represented by a horizontal line from 12.45 pm to 4.00 pm, 25 miles from home.

If their average speed on the return journey was 20 mph then after 1 hour they were 5 miles from home.

The point (5.00, 5) is plotted and a line is drawn through it to meet the horizontal axis.

The graph is shown in the column opposite.

They arrived home in time for their favourite programme.

Find:

a   the time at which they arrived home

b   at what times they were 10 miles from home

c   their average speed on the outward journey.

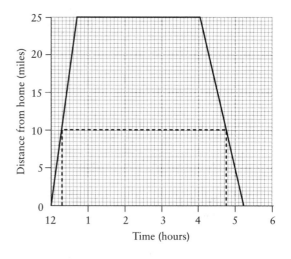

From the graph we can see that:

a   they arrived home at 5.15 pm.

b   on the outward journey they were 10 miles from home after 0.3 hours, i.e. at 12.18 pm; on the homeward journey they were 10 miles from home after 4.75 hours, i.e. at 4.45 pm.

c   Their average speed on the outward journey is given by:

$$\text{Average speed} = \frac{\text{Total distance}}{\text{Total time}} = \frac{25 \text{ miles}}{45 \text{ min}}$$

$$= \frac{25 \text{ miles}}{0.75 \text{ hours}}$$

$$= 33.3 \text{ mph}$$

Exercise 34A

1 The graph below represents Mr Phillip's journey to work one morning. The first 10 miles of his journey is through a built-up area and then he joins the motorway. Unfortunately, there has been an accident and he is held up for some time. He leaves the motorway at the next exit and completes his journey on the ordinary roads.

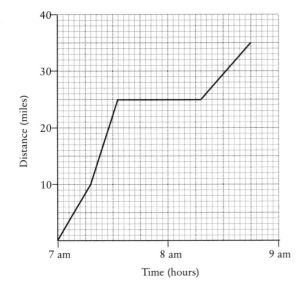

a How long did the journey through the built-up area take him?

b What was his average speed on the motorway section of the journey, until he had to stop?

c For how long was he held up?

d At what time did he arrive at work?

e How far did he travel from home to work?

f What was his average speed for the whole journey?

2 Draw a graph to represent the following journey:

Susan left home at 10.00 am and drove to her friend Valerie's house, 15 minutes away, for coffee. Her average speed was 32 mph. The two friends chatted for 45 minutes and then Susan drove her friend to the local supermarket, 2 miles away, arriving at 11.12 am. They spent 24 minutes shopping. Then Susan drove home, dropping Valerie off on the way. Her average speed for the return journey was 25 mph.

a How far did she drive to her friend's house?

b At what time did they leave Valerie's house?

c At what time did Susan arrive home?

3 Barry has a friend who lives 5 miles away. They decide to spend a day together cycling and agree to meet at a barn 2 miles from Barry's house. His friend, Ali, leaves home at 9.00 am and cycles at a steady speed of 12 mph. Barry does not leave until 10 minutes later. He arrives at the barn 10 minutes after Ali.

Draw a graph to represent their journeys.

a When did each boy arrive at the barn?

b At what speed did Barry cycle?

4 On another occasion they decide that they will keep cycling towards each other until they meet. Ali leaves home at 9.30 am and Barry leaves at 9.35 am, both cycling at 12 mph. Draw a graph to find:

a at what time they meet,

b how far each boy will have travelled.

5 Nessa is invited to stay with her French pen-friend in Nice. Below is Nessa's itinerary for travelling from Calais to Nice along the A6 and a map showing the distances involved.

Arrive Calais	1400	Continental time
Arrive Paris	1700	Dinner stop
Depart Paris	1800	
Arrive Valance	0400	Refreshment stop
Depart Valance	0430	
Arrive Nice	0800	

Draw a travel graph of Nessa's journey and use it to calculate the average speeds, to the nearest km/h, for the three sections of the journey:

a Calais to Paris,

b Paris to Valance,

c Valance to Nice.

Distance in kilometres

*34.2 Distance–time graphs with variable speed

A straight-line graph has a constant gradient, but if the graph is a curve, the gradient is continually changing.

For the function

$$s = 25t + 3t^2$$

which connects distance s with time t,

the distance s is changing at different rates at different times.

The gradient, which is $\dfrac{\text{Distance travelled}}{\text{Time taken}}$, gives the **average** rate of change of distance over the given time interval, i.e. the average speed or average velocity.

At a particular instant, the rate of change of distance, i.e. the **speed or velocity**, is given by the gradient of the curve at that point in time.

The **gradient of a curve** at a given point is the **gradient of the tangent** drawn to the curve at that point.

Example

A car is braking for traffic lights. The lights change to green when the car's speed is 4 m/s. The driver accelerates and, during the next 6 seconds, the distance travelled, s metres, after a time t seconds, is given by the formula:

$$s = 4t + t^2$$

a Draw up a table of values to show the distance travelled by the car in the first 6 seconds.

b Draw the graph of s against t.

c Find the average speed of the car during the third and fourth seconds, i.e. from $t = 2$ to $t = 4$.

d By drawing a tangent to the graph, find the speed of the car after 2 seconds.

a

t(s)	0	1	2	3	4	5	6
s(m)	0	5	12	21	32	45	60

b See opposite.

c Average speed $= \dfrac{\text{Distance travelled}}{\text{Time taken}}$

$$= \frac{32 - 12}{2}$$

$$= 10 \text{ m/s}$$

d Speed after 2 s $=$ Gradient of the tangent at $t = 2$

$$= \frac{36 - 12}{5 - 2} = \frac{24}{3}$$

$$= 8 \text{ m/s}$$

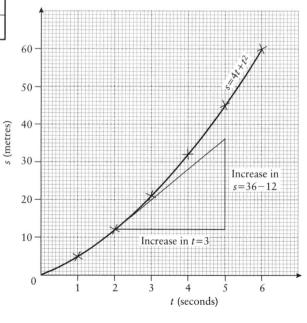

Exercise 34B

1 A cyclist rounds a corner and sees a tree across the road. He brakes and the distance travelled, 5 metres, in the next t seconds is given by
$$s = 6t - t^2$$

  a Draw the distance–time graph for values of t from 0 to 3 seconds.

  b What is the gradient of the curve at $t = 3$?

  c How far does the cyclist travel after he starts braking?

  d By drawing a tangent to the curve, find the speed of the cyclist immediately before he started to brake.

2 a Draw the graph of $y = 2x^2 + 3x - 1$ for $-3 \leqslant x \leqslant 2$.

  b By drawing appropriate lines to the graph, find the gradient of the curve at:

   (i) $x = -2.5$,   (ii) $x = 1.25$.

3 A train entering a station reduces its speed so that the distance, s metres, covered in time t seconds, is given by
$$s = 36t - \tfrac{1}{4}t^2$$

  a Draw the graph of s against t for $0 \leqslant t \leqslant 40$.

  b By drawing an appropriate line on the graph, find the speed after 30 seconds.

4 a Draw the graph of y against x for the function
$$y = 3x^2 + 5x + 2$$
taking values of x from -4 to $+4$.

  b Find the gradient of the curve at the points where:

   (i) $x = -3$,   (ii) $x = 2\tfrac{1}{2}$.

5 A stain is spreading in such a way that the area, $A\,\mathrm{cm}^2$, of the stain after t seconds is given by
$$A = \tfrac{1}{2}t^2 + 3t + 12$$

  a Draw up a table of values for A against t when $t = 0, 2, 4, 6, 8, 10$.

  b Draw the graph of A against t.

  c Find the amount by which the area increased during the third second.

  d By drawing a suitable line on the graph, find the rate of increase of A when $t = 5$.

6 A stone is catapulted vertically upwards and its height, h metres, after some time t second, is given by:
$$h = 20t - 5t^2$$

  a Plot the graph of h against t for values of t from 0 to 4 seconds.

  b Use the graph to find:

   (i) the maximum height to which the stone rises,

   (ii) the times when the stone is at a height of 12 m.

  c By drawing an appropriate line on the graph, find the speed of the stone after:

   (i) 1 second   (ii) 2.5 seconds.

*34.3   Velocity–time graphs

For the function $v = 25 + 6t$, the velocity (v metres per second) changes for various times (t seconds).
The gradient, which is

$$\frac{\text{Change in velocity}}{\text{Time taken}}$$

gives the average rate of change of velocity over a given time interval, i.e. the acceleration.

Note. Velocity, v metres per second, may be written as $v\,\mathrm{m/s}$ (or $v\,\mathrm{m\,s^{-1}}$).
At a particular instant, the rate of change of velocity, i.e. the acceleration, is given by the gradient of the curve at that point in time.

For any graph involving time, the rate of change is given by the **gradient** of the curve at that point in time.

The **area** under the curve shows the product of the velocity and time. Since

Velocity × Time = Distance

the distance travelled is the area under the graph.

For any graph involving time, the area under the graph represents the amount of growth, or change, between the given times.

Example 1

The velocity time graph shows a cyclist travelling at 6 m/s for 10 seconds. The distance travelled is found by use of the formula:

Distance = Velocity × Time
∴ Distance = 6 × 10 m
= 60 m

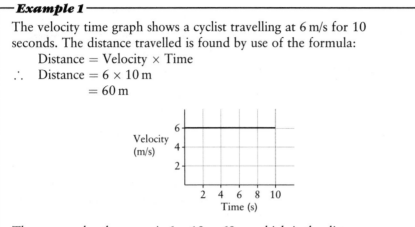

The **area** under the curve is 6 × 10 = 60 m, which is the distance travelled.

Example 2

The velocity of a lorry (v m/s) is given by:
$$v = 10 + 0.9t$$
between times $t = 0$ and $t = 10$ seconds.
Find:

a the acceleration after 3 seconds

b the distance travelled in the first 5 seconds

c the distance travelled in the fourth second.

It is often helpful to draw a velocity–time graph before you start on the question. This graph is shown below:

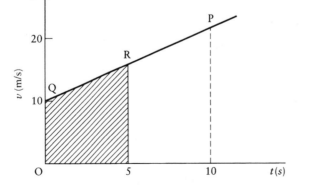

a The graph shows that the velocity is changing during the ten seconds. There is no time when the velocity is constant. Since the line is straight the gradient of the graph is a constant, i.e. the acceleration is constant.

Hence the acceleration is the same at all points between P and Q.

$$\text{Acceleration} = \frac{\text{Change in velocity (between P and Q)}}{\text{Change in time (between P and Q)}}$$

$$= \frac{19 - 10}{10}$$

$$= \frac{9}{10}$$

$$= 0.9 \, \text{m/s}^2$$

b Distance travelled in the first 5 seconds is area OQRS
$$= \tfrac{1}{2} \times 5 \times (10 + 14.5) \text{ (i.e. the area of a trapezium,}$$
$$\text{see page 440)}$$

$$= 61.25 \, \text{m}$$

c Distance travelled in the fourth second
$$= \text{Distance travelled between } t = 3 \text{ and } t = 4$$
$$= \tfrac{1}{2} \times 1 \times (12.7 + 13.6)$$
$$= 13.15 \, \text{m}$$

Example 3

The number of bacteria in a piece of cheese doubles every hour.
Assume there are 100 bacteria present at the beginning, draw a graph to show the number present over the next 6 hours and find:

a the number of bacteria after $2\tfrac{1}{4}$ hours

b the rate of growth after 4 hours.

The table of values for time t and number of bacteria n is:

t(h)	0	1	2	3	4	5
n	100	200	400	800	1600	3200

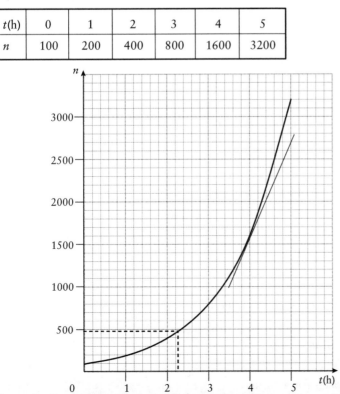

a Read along the dotted line from $t = 2.25$.
   After $2\frac{1}{4}$ hours the number of bacteria is 475.

b The rate of growth is the gradient of the tangent at $t = 4$.

$$\text{Rate of growth} = \frac{2700 - 1000}{5 - 3.5} = \frac{1700}{1.5} = 1133 \text{ bacteria per hour}$$

It is not possible to draw curves, or tangents to a curve, with a high degree of accuracy and, therefore, values taken from graphs are only correct to a limited number of significant figures.
By calculation, the rate of growth is 1109 bacteria per hour. The graphical answer is therefore correct, if given to 2 s.f. (i.e. 1100 bacteria per hour).

Exercise 34C

1 The velocity v m/s of a car is given by $v = 15 + 1.3t$ for values of t from 0 to 20, where t is the time in seconds.
Find:

   a the acceleration when $t = 3$,

   b the distance travelled in the first 5 seconds,

   c the distance travelled in the third second i.e. from $t = 2$ to $t = 3$.

2 The velocity of a parachutist before she pulls the ripcord is v m/s, where v is given by $v = 0.1 + 7t$. She pulls the ripcord when her velocity is 28 m/s.
Find:

   a the time t when she pulls the ripcord,

   b the distance she falls before she pulls the ripcord,

   c the acceleration before she pulls the ripcord.

3 **a** The velocity of a waterskier, v m/s, is given by $v = 8 + 0.6t$ for time $t = 0$ to $t = 15$ seconds.
      Find:
      (i) his acceleration,
      (ii) the distance travelled between $t = 5$ and $t = 15$.

   b He lets go of the rope after 15 seconds, and his velocity changes to $v = 17 - 4T$ until he stops (T is the time, in seconds, from when he lets go of the rope).
      Find the distance until he stops.

4 The velocity of a Tour de France cyclist in kilometres per hour descending an Alpine pass is given by:

$$v = 30 + 180t$$

where t is the time in hours.
The cyclist reaches 48 kph before he brakes for a corner.

Find:

a the time taken until he starts braking

b the distance travelled before he starts braking.

5 The spread of moss in a lawn is represented on the graph below.

a What area of lawn was covered by moss at the beginning?

b After how long had the area of moss doubled?

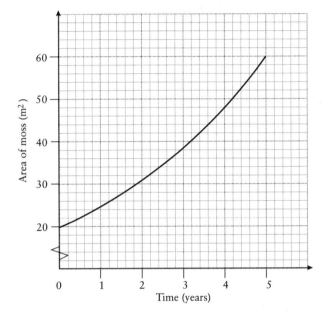

6 A radioactive substance has a *half-life* of one day. This means that at the end of each day the mass of the substance remaining radioactive is half the mass at the beginning of that day.

a If the mass at the beginning of day 1 was 200 g, draw up a table to show the mass over the following 6 days.

b Draw the graph of mass against time and find after how long the mass is 70 g.

7 The value of a car depreciates every year.
The table gives the car's value at the end of each year of ownership.

Time (years)	0	1	2	3	4	5
Value (£)	5000	4000	3200	2560	2048	1638

a Draw the graph of value V against time t.

b What is the value of the car (to the nearest £10) after $2\frac{1}{2}$ years?

c What is the rate of depreciation after: (i) 1 year, (ii) 5 years?

Example

The velocity of a car is given by $v = 7 + 0.2t^2$ between the time $t = 0$ and $t = 10$.

Draw the velocity–time graph and use it to find:

a the acceleration after 5 seconds.

b the distance travelled between $t = 2$ seconds and $t = 6$ seconds.

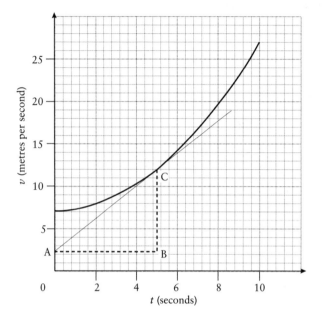

Because the graph is a curve, it is not possible, without using A-level Mathematics, to find the exact values of the acceleration or the distance. To find the acceleration, a tangent must be drawn where $t = 5$, to obtain the gradient of the curve at that point.

Using triangle ABC, gradient is $\dfrac{10}{5} = 2$.

To find the distance travelled, it is necessary to estimate the area under the curve between $t = 2$ and $t = 6$.
Carefully counting the squares, noting how part of two small squares can give a whole square:
this area is 3 large squares and about 30 small squares.
i.e.   4.2 large squares.

One large square is 2 units horizontally and 5 units vertically, and hence is $2 \times 5 = 10$ metres.

$\therefore$ The distance travelled in this time interval is 42 metres.

Exercise 34D

1   The velocity of a motor bike v m/s is given by $v = 12 + 0.6t^2$.
Draw a graph of v against t for values of t from 0 to 6 where t is the time in seconds. Find:

a the acceleration when $t = 4$

b the distance travelled in the first 2 seconds

c the distance travelled in the fourth second (i.e. from $t = 3$ to $t = 4$).

2   A car driver sees a sign 'Warning: Stop Sign Ahead'. He starts braking and his speed in m/s is given by
    $v = 39 - 6t$ until he stops. Draw a graph of v against t for $0 \leqslant t \leqslant 3$.
    Find:

    a   his retardation

    b   the distance travelled until he stops

    c   the speed when he is halfway to the stop sign.

    (*Note.* When the acceleration is negative it is called 'retardation'.)

3   The velocity of a car v m/s is given by $v = 15 + 0.4t^2$. Draw a graph for values of t from 0 to 10.
    Find:

    a   the acceleration when $t = 6$

    b   the distance travelled in the first 3 seconds

    c   the distance travelled in the fifth second (i.e. from $t = 4$ to $t = 5$).

4   The velocity of a lorry v m/s starting from rest is given by $v = 0.7t + 0.6t^2$.
    Draw a graph for values of t from 0 to 10.
    Find:

    a   the acceleration when $t = 5$

    b   the distance travelled in the first 10 seconds.

5   This is a velocity–time graph for a car which took 10 seconds to reduce its speed from 11 m/s to 0 m/s.

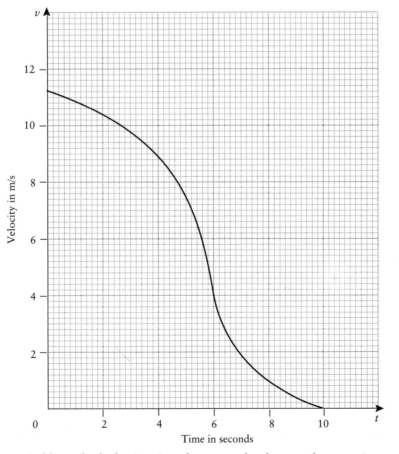

    a   Calculate, by a suitable method of estimation, the area under the curve between 0 seconds and 6 seconds.

    b   What does your answer to part **a** represent?   (SEG S94)

*34.4   The trapezium rule

N.B. The area of a trapezium is:

$\frac{1}{2}$ × Sum of parallel sides × Perpendicular height

(See Section 44.2 for further details of areas of shapes.)

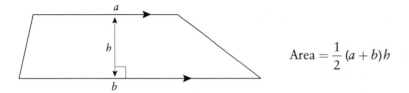

$$\text{Area} = \frac{1}{2}(a+b)h$$

Counting the squares under a velocity–time graph is a very inaccurate method of finding the distance travelled. The accuracy can be improved by dividing the area to be found into strips.

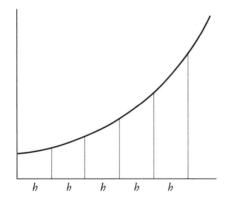

For convenience, each strip should be of the same width, h.

The shape of each strip is approximately a trapezium. (The smaller the value chosen for h, the closer each strip is to a trapezium.)

The area of a trapezium $= \dfrac{h(a+b)}{2}$ where a, b are the lengths of the parallel sides and h is the height.

For the trapezia on the graph, the lengths of the parallel sides are the y values when $x = 0, h, 2h$, etc. We call them y_0, y_1, y_2, etc.

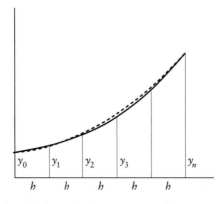

The area of the first trapezium $\qquad = \dfrac{h(y_0 + y_1)}{2}$

The area of the second trapezium $\qquad = \dfrac{h(y_1 + y_2)}{2}$

The area of the last, or nth, trapezium $\quad = \dfrac{h(y_{n-1} + y_n)}{2}$

The total area is given by the sum of the areas of all the trapezia.

$$\text{Total area} = \frac{h}{2}(y_0 + y_1) + \frac{h}{2}(y_1 + y_2) + \ldots + \frac{h}{2}(y_{n-2} + y_{n-1}) + \frac{h}{2}(y_{n-1} + y_n)$$

$$= \frac{h}{2}\{y_0 + y_1 + y_1 + y_2 + y_2 + \ldots + y_{n-2} + y_{n-1} + y_{n-1} + y_n\}$$

$$= h\{\tfrac{1}{2}(y_0 + y_n) + y_1 + y_2 + \ldots + y_{n-1}\}$$

or

Total area = Strip width × {Average of 1st and last y values + other y values}

This formula is known as the **trapezium rule**.

Example

The velocity, in metres, of a car is given by $v = 7 + 0.2t^2$.
Find the distance travelled between $t = 2$ seconds and $t = 6$ seconds.

Take $h = 0.5$. Then the number of trapezia will be $(6 - 2) \div 0.5 = 8$
The values of v from v_0 to v_8 are given in the table below.

n	0	1	2	3	4	5	6	7	8
t_n	2.0	2.5	3.0	3.5	4.0	4.5	5.0	5.5	6.0
v_n	7.8	8.25	8.8	9.45	10.2	11.05	12.0	13.05	14.2

The sum of the first and last velocities $= (7.8 + 14.2)$
$$= 22$$

The sum of the remaining velocities $\quad = 8.25 + 8.8 + 9.45 + 10.2$
$$+ 11.05 + 12.0 + 13.05$$
$$= 72.8$$

The distance travelled $= 0.5\{\tfrac{1}{2} \times 22 + 72.8\}$
$$= 0.5 \times 83.8$$
$$= 41.9 \text{ metres}$$

Note: The exact answer is 41.86 m, so the trapezium rule has given a very good approximation.

Exercise 34E

1 The velocity of a motor bike v m/s is given by $v = 12 + 0.6t^2$.

   a Find the distance travelled in the first 2 seconds taking h equal to
     (i) 0.5 (ii) 0.25

   b Find the distance travelled in the fourth second taking $h = 0.1$.

2 A car driver's speed, in m/s, is given by $v = 39 - 6t$.

   a If she starts to brake at $t = 0$, how many seconds does it take for her
     to stop?

   b Find the distance travelled in this time, taking $h = 0.5$.

3 The velocity of a lorry, in m/s, starting from rest is given by
   $v = 0.7t + 0.6t^2$.
   Find the distance travelled in the first 2 seconds, taking
   (i) $h = 0.5$ (ii) $h = 0.2$.

35 Further Algebra

35.1 Simultaneous equations

Eddie has just bought a new pen and his friend Pete would also like one, but Eddie can't remember how much it cost. He does remember that he bought the pen and a ruler for £2.20.

Writing this information in algebra and working in pence gives:

$$p + r = 220$$

This does not help Pete to work out the cost of the pen as there are too many values of p and r which would fit the equation.

Fiona then remembers that she bought two of the same type of pen and a ruler for £3.50.

Pete can now work out the cost of the pen:

$$p + r = 220$$
and
$$2p + r = 350$$

The extra pen which Fiona bought must have cost (350-220) pence, i.e. 130 pence.

Therefore the pen cost £1.30 and the ruler 90 pence.

If an equation has two unknown values (variables), it cannot be solved on its own.

Two variables require **two** equations which are solved at the same time, i.e. **simultaneously**.

Three variables would require three simultaneous equations, etc.

To solve a pair of simultaneous equations the method of **elimination** is used.

Solving simultaneous equations

Example 1

Solve the equations: $4x + 3y = 24$ (1)
$\qquad\qquad\qquad\quad 2x + 3y = 18$ (2)

The only difference between the left-hand sides of the two equations is that equation (1) has $2x$ more, which must be balanced by the extra 6 on the right-hand side.

Subtracting equation (2) from equation (1) gives:

$$
\begin{aligned}
4x + 3y &= 24 \\
- 2x + 3y &= 18 \\
\hline
2x\qquad &= 6 \\
\therefore\quad x &= 3
\end{aligned}
$$

The value of y can be found by substituting the value of x into one of the equations.

Substitute $x = 3$ into (2):
$$2 \times 3 + 3y = 18$$
$$6 + 3y = 18$$
$$3y = 12$$
$$y = 4$$

The solution is $x = 3$, $y = 4$.

(This solution can be checked using the other equation:

Substitute $x = 3$, $y = 4$ into equation (1)
$$\text{LHS} = 4 \times 3 + 3 \times 4 = 12 + 12 = 24 = \text{RHS}$$
So the solution is correct.)

Example 2

Solve the equations: $3x - y = 11$ (1)
$\qquad\qquad\qquad 2x + y = 9$ (2)

The number of ys in each equation is the same, except for the sign, so, in this case y is eliminated by *adding* the two equations.

Adding equation (1) to equation (2) gives:

$$
\begin{array}{r}
3x - y = 11 \\
+\;2x + y = 9 \\
\hline
5x = 20 \\
\therefore \quad x = 4
\end{array}
$$

Substitute $x = 4$ into (2):
$$2 \times 4 + y = 9$$
$$8 + y = 9$$
$$y = 1$$

The solution is $x = 4$, $y = 1$.

(To check, substitute $x = 4$, $y = 1$ into (1).)

Example 3

Solve the equations: $3x + 4y = 18$ (1)
$\qquad\qquad\qquad 4x - 3y = -1$ (2)

Neither x nor y has the same coefficient in each equation, so this needs to be remedied first.

Method	Working
(i) Decide which variable is to be eliminated.	Eliminate y.
(ii) Multiply one or both equations so that this variable has the same coefficient in each equation (not counting the signs).	Multiply equation (1) by 3 and equation (2) by 4 $$9x + 12y = 54$$ $$16x - 12y = -4$$

(iii) Add or subtract the equations, depending on the signs of the variable to be eliminated	As the coefficients of y have opposite signs, add the equations: $+\ \begin{array}{r} 9x + 12y = 54 \\ 16x - 12y = -4 \\ \hline 25x = 50 \end{array}$
(iv) Solve for the remaining variable.	$x = 2$
(v) Substitute into one of the original equations to find the eliminated variable and hence the complete solution.	Substitute $x = 2$ into (1) $3 \times 2 + 4y = 18$ $4y = 12$ $y = 3$ Solution is $x = 2,\ y = 3$
(vi) Substitute into the other original equation to check the solution.	Substitute into (2) LHS $= 4 \times 2 - 3 \times 3$ $= 8 - 9$ $= -1 =$ RHS

Exercise 35A

Solve the following pairs of simultaneous equations and check your solutions:

1 $2x + y = 8$
   $x + y = 6$

2 $x + 3y = 12$
   $x + y = 10$

3 $x + y = 12$
   $x - y = 2$

4 $3x - 2y = 9$
   $x + 2y = 15$

5 $x + 3y = 6$
   $2x + y = 7$

6 $4x + 2y = 10$
   $3x + 5y = 11$

7 $3x + 2y = 18$
   $4x - y = 2$

8 $2x - 3y = 10$
   $3x + 4y = 15$

9 $x - y = -1$
   $2x + 3y = 28$

10 $7x + y = 9$
    $3x + y = -3$

11 $2x + 5y = 15\frac{1}{2}$
    $3x - 4y = -5\frac{1}{2}$

12 $6x + 5y = -17$
    $3x + 2y = -8$

Solving problems

Example

An ice cream seller charges a customer £3.20 for three ice cream cornets and two choc-ices.
The next customer is charged £4.50 for four ice cream cornets and three choc-ices.
What are the prices of the ice cream cornet and the choc-ice?

Let x pence be the cost of an ice cream cornet and y pence be the cost of a choc-ice.

Then	$3x + 2y = 320$	(1)
and	$4x + 3y = 450$	(2)

Eliminate y:

Multiply (1) by 3:	$9x + 6y = 960$	(3)
Multiply (2) by 2:	$8x + 6y = 900$	(4)

Subtract (4) from (3): $x = 60$

Substituting $x = 60$ in (1): $3 \times 60 + 2y = 320$
$$2y = 140$$
$$y = 70$$

An ice cream cornet costs 60p and a choc-ice costs 70p.

(Check by substituting $x = 60$ and $y = 70$ in (2).)

Exercise 35B

1 Five pounds of apples and three pounds of bananas cost £4.52.

Three pounds of apples and two pounds of bananas cost £2.80.
What is the price per pound of the apples and the bananas?

2 Two numbers are chosen such that twice the first plus the second is 17 and four times the first minus the second is 25.
What are the two numbers?

3 The perimeter of a rectangle is 38 metres. The difference between the length and the breadth is 3 metres.
What is the area of the rectangle?

4 A holiday for 2 adults and 3 children costs £748. The same holiday for 3 adults and 4 children costs £1056.
What would be the cost of the holiday for 2 adults and 2 children?

5 A potter makes two types of ware. Type A takes $\frac{1}{2}$ hour to make and sells for £3 each. Type B takes 1 hour to make and sells for £5 each.

One day he worked for 8 hours and sold the pottery he made for £45.
How many of each type did he make?

6 The ages of a man and his daughter total 27 years.

In 4 years time, the father will be four times as old as his daughter.
How old are they both now?

7 At the beginning of a mathematics lesson a teacher says to her class:
'If you subtract twice the number of sweets in my left hand from the number of sweets in my right hand the answer is 3 sweets, but if you add four times the number in my left hand to the number in my right hand the answer is seven times as many.
The first person to tell me, correctly, how many sweets I have in my hands wins the sweets.'

How many sweets will the winner receive?

Graphical solution

Simultaneous equations can also be solved by a graphical method.

The graphs of the two equations are drawn and the solution to the equations is the coordinates of the point of intersection of the graphs, since these values of x and y will satisfy both equations.

(For the method of plotting straight-line graphs see page 267.)

Example

Solve, graphically, the equations $x + 2y = 8$
$$3x + 4y = 18$$

To draw the lines, use the points where $x = 0$ and $y = 0$.

For $x + 2y = 8$ the points are $(0, 4)$ and $(8, 0)$
For $3x + 4y = 18$ the points are $(0, 4\frac{1}{2})$ and $(6, 0)$

The axes are now drawn and scaled and the points plotted and joined.

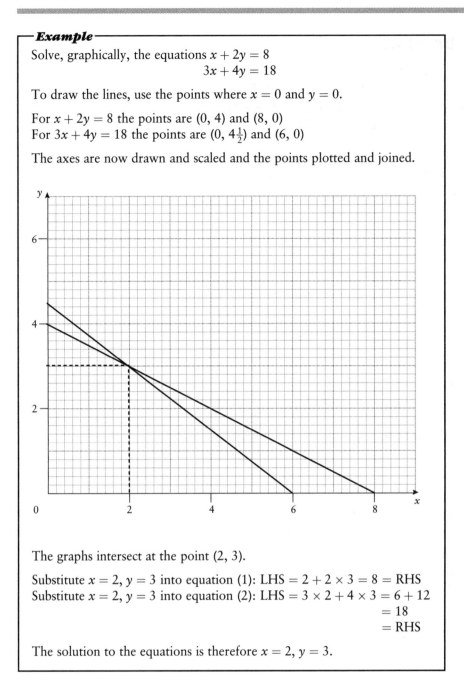

The graphs intersect at the point $(2, 3)$.

Substitute $x = 2$, $y = 3$ into equation (1): LHS $= 2 + 2 \times 3 = 8 =$ RHS
Substitute $x = 2$, $y = 3$ into equation (2): LHS $= 3 \times 2 + 4 \times 3 = 6 + 12$
$$= 18$$
$$= \text{RHS}$$

The solution to the equations is therefore $x = 2$, $y = 3$.

Exercise 35C

Solve the following equations by the graphical method:

1 $x + y = 5$
   $3x + 2y = 12$

2 $x - 2y = 1$
   $x + y = 4$

3 $4x + 5y = 40$
   $x - y = 1$

4 $3x - 2y = 0$
   $3x + 4y = 18$

5 $2x + 5y = 15$
   $2x - y = 3$

6 $x + 3y = 11$
   $2x + y = 7$

7 $y = 3x + 5$
   $2y = 5x + 9$

8 $y = 7x$
   $2y = 5x - 9$

9 $y = 6 - 2x$
   $3y + 4x = 10$

35.2 Removing brackets

To multiply out a bracket in an expression of the form $2x(x + 3)$
each term in the bracket must be multiplied by $2x$, i.e. $2x(x + 3) = 2x^2 + 6x$

To multiply out a bracket in an expression of the form $(2x + 4)(x + 3)$
each term in the second bracket is multiplied by $2x$ and by 4, i.e.

$$(2x + 4)(x + 3) = 2x(x + 3) + 4(x + 3)$$
$$= 2x^2 + 6x + 4x + 12$$

Collecting like terms gives $(2x + 4)(x + 3) = 2x^2 + 10x + 12$

Example 1

Multiply $(3x - 2)$ by $(2x + 1)$

$$(3x - 2)(2x + 1) = 3x(2x + 1) - 2(2x + 1)$$
$$= 6x^2 + 3x - 4x - 2$$
$$= 6x^2 - x - 2$$

Brackets of this type can be removed without the intermediate stages being
shown if the following stages of working are carried out mentally and only
the answer is written down.

To expand $(2x + 4)(x + 3)$:

The final answer will have three terms, i.e. an x^2, an x and a constant term.

The x^2 term is calculated by multiplying the x term in the first bracket by the
x term in the second bracket:

$$(2x + 4) \quad (x + 3) \rightarrow 2x^2$$

The x term is calculated by multiplying the x term in one bracket by the
constant term in the other bracket and adding the two resultant terms.

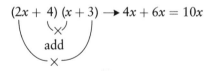

$$(2x + 4)(x + 3) \rightarrow 4x + 6x = 10x$$

The constant term is calculated by multiplying the constant term in the first
bracket by the constant term in the second bracket.

$$(2x + 4)(x + 3) \rightarrow 12$$

$$\therefore \quad (2x + 4)(x + 3) = 2x^2 + 10x + 12$$

Example 2

Expand $(3x - 4)(2x - 3)$

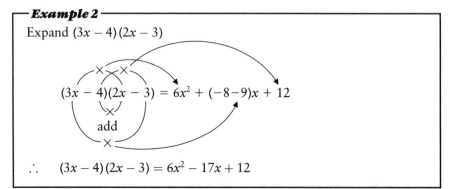

$$(3x - 4)(2x - 3) = 6x^2 + (-8 - 9)x + 12$$

$$\therefore \quad (3x - 4)(2x - 3) = 6x^2 - 17x + 12$$

Example 3

Expand $(2x + y)(x - 3y)$

$$(2x + y)(x - 3y) = 2x^2 - 6xy + xy - 3y^2$$
$$= 2x^2 - 5xy - 3y^2$$

Exercise 35D

Remove the brackets from the following:

1	$(x + 3)(x + 4)$	6	$(x - 3)(x - 7)$	11	$(2x + 5)(2x - 5)$	16	$(2x - 3y)(2x - 3y)$
2	$(x + 2)(x - 1)$	7	$(x + 1)(x - 2)$	12	$(3x - 4)(3x + 4)$	17	$(2x + 1)(x + 4)$
3	$(x - 5)(x - 3)$	8	$(x - 6)(x - 2)$	13	$(x + 2)(x + 2)$	18	$(3x + 2)(x - 2)$
4	$(x + y)(x + 4y)$	9	$(x + 3)(x - 3)$	14	$(x - 1)(x - 1)$	19	$(2x + y)(3x + 2y)$
5	$(x + y)(x - 2y)$	10	$(x + y)(x - y)$	15	$(3x + 1)(3x + 1)$	20	$(3x - 2y)(4x + 3y)$

35.3   Quadratic factors

All the expressions in Exercise 35D were **quadratic expressions**.

In each case, the brackets multiplied out to give **four** terms (which then simplified to three terms) and the final expression always contained an x^2 term, but no higher power of x.

$$\left. \begin{matrix} 2x^2 + 7x - 3 \\ x^2 - 6x \\ 4y^2 - 9 \end{matrix} \right\} \begin{matrix} \text{are all examples} \\ \text{of quadratic} \\ \text{expressions.} \end{matrix}$$

We saw that $(2x + 4)(x + 3) = 2x^2 + 10x + 12$.
Therefore, $(2x + 4)$ and $(x + 3)$ are **factors** of the quadratic $2x^2 + 10x + 12$.

Quadratic expressions and quadratic equations occur in many branches of mathematics and it is necessary to be able to factorise them.

When the coefficient of x^2 is unity (i.e. 1)

Example 1

Factorise $x^2 + 5x + 6$.

Reversing the process of multiplying out brackets, we can see that
$$x^2 + 5x + 6 = (x + ?)(x + ?)$$
and 6 must factorise as either 6×1 or 3×2.

The key to factorising the quadratic expression is the x term.

$5x$ is the sum of two x terms and the coefficients of these two terms must be the factors of 6.

$$3 \times 2 = 6 \text{ and } 3x + 2x = 5x$$
$$\therefore \quad x^2 + 5x + 6 = (x + 3)(x + 2)$$

Example 2

Factorise $x^2 - x - 6$.

The factors of -6, when added together, must equal the coefficient of x, which is -1.

The factors of -6 are -6×1, 6×-1, -3×2, 3×-2.
Only $-3 + 2 = -1$

$$\therefore \quad x^2 - x - 6 = (x - 3)(x + 2)$$

Example 3

Factorise $x^2 + 6x$.

This is a much simpler quadratic expression to factorise than the standard type because it has a **common factor** which is x.

$$x^2 + 6x = x(x + 6)$$

Exercise 35E

Factorise the following:

1 $x^2 + 8x + 7$	**6** $x^2 - 2x - 15$	**11** $x^2 - 5x - 6$	**16** $x - 2x^2$
2 $x^2 + 7x + 10$	**7** $x^2 - 6x + 9$	**12** $x^2 - 6x - 16$	**17** $x^2 + 12x + 36$
3 $x^2 + x - 6$	**8** $x^2 - 4x - 12$	**13** $x^2 - 7x$	**18** $x^2 - 8x + 16$
4 $x^2 - 5x + 6$	**9** $x^2 - 7x + 12$	**14** $2x^2 + 6x$	
5 $x^2 + 2x - 8$	**10** $x^2 + 11x - 12$	**15** $6x^2 - 3x$	

---

*When the coefficient of x^2 is not unity

The coefficient of x^2 will not necessarily be 1.
In this section the method for factorising quadratic expressions is extended to include those where the coefficient of x^2 is 2 or 3.

Example

Factorise $2x^2 + 11x + 12$.

The method previously used must be adapted to take into account the coefficient 2.

As 2 will only factorise as 2×1, it can be seen that

$$2x^2 + 11x + 12 = (2x + a)(x + b)$$

As before, $a \times b = 12$, but now we require

$$a + 2b = 11 \text{ (not } a + b \text{ as before)}$$

Factors of 12 are $a \times b = 1 \times 12$ and $a + 2b = 1 + 24 = 25$
2×6 $\qquad = 2 + 12 = 14$
3×4 $\qquad = 3 + 8 = 11$
4×3 $\qquad = 4 + 6 = 10$
$\vdots$ $\qquad\qquad \vdots$

$$\therefore \quad 2x^2 + 11x + 12 = (2x + 3)(x + 4)$$

Alternatively list all possible solutions and select the one which 'works':

$$(2x + 1)(x + 12) = 2x^2 + 25x + 12$$
$$(2x + 2)(x + 6) = 2x^2 + 14x + 12$$
$$(2x + 3)(x + 4) = 2x^2 + 11x + 12$$

Exercise 35F

Factorise:

1 $2x^2 + 5x + 2$	**4** $2x^2 - 13x - 7$	**7** $3x^2 - 8x + 4$	**10** $3x^2 + 20x + 12$
2 $2x^2 + 7x + 6$	**5** $3x^2 + 8x + 5$	**8** $3x^2 - 13x - 10$	**11** $3x^2 - 22x - 16$
3 $2x^2 + 9x - 5$	**6** $3x^2 - 6x - 9$	**9** $2x^2 - 11x + 12$	**12** $2x^2 - 3x - 14$

*Difference of two squares

If you look back to Exercise 35D questions 9–12 on page 333, you will see that these brackets always multiplied out to give a quadratic expression with two terms and not the more usual three terms.

Moreover, the two terms are both perfect squares separated by a minus sign. This type of quadratic expression is known as **the difference of two squares**.

Once a quadratic has been recognised as the difference of squares, it is easily factorised:

$$a^2 - b^2 = (a - b)(a + b)$$

Example 1

Factorise $x^2 - 9$.
$$x^2 - 9 = (x - 3)(x + 3)$$

Example 2

Factorise $9x^2 - 16$.
$$9x^2 - 16 = (3x)^2 - (4)^2$$
$$= (3x - 4)(3x + 4)$$

Example 3

Factorise $8x^2 - 2$.

$8x^2$ and 2 are not perfect squares, but 2 is a **common factor** of the expression.

$$8x^2 - 2 = 2(4x^2 - 1)$$

$4x^2 - 1 = (2x)^2 - 1^2$ which is the difference of two squares

$$\therefore \quad 8x^2 - 2 = 2(2x - 1)(2x + 1)$$

(*Note.* If a quadratic has a common factor, always take out the common factor before factorising the quadratic into two brackets.)

Exercise 35G

Factorise:

1 $x^2 - 1$	**4** $4x^2 - 9$	**7** $49 - x^2$	**10** $2x^2 - 8$
2 $x^2 - 16$	**5** $9x^2 - 1$	**8** $36 - 25x^2$	**11** $9x^2 - 36$
3 $x^2 - 25$	**6** $16x^2 - 25$	**9** $25 - 49x^2$	**12** $12x^2 - 75$

35.4   Quadratic equations

An equation which can be arranged so that a quadratic expression is equal to zero is called a **quadratic equation**.

A quadratic equation is of the form

$$ax^2 + bx + c = 0$$

where b and c can be any integer and a can be any integer except 0.

Of course, the equation does not have to be in terms of x. It could be in terms of y or z or x^2, provided the powers of the variable are 2, 1 and 0.

Solution of quadratic equations by factorising

If the product of two numbers is zero, we know that one of the numbers must itself be zero.

$xy = 0$ mean that either $x = 0$ or $y = 0$ (or both $x = 0$ and $y = 0$)

The concept is very important for the solution of quadratic equations by factors.

Example 1

Solve the equation $x^2 - 5x - 14 = 0$.

$$x^2 - 5x - 14 = 0$$

Factorising gives: $\quad (x - 7)(x + 2) = 0$

$\therefore$ either $\quad\quad\quad\quad (x - 7) = 0$ or $(x + 2) = 0$

$$\therefore \quad \text{either} \qquad\qquad x = 7 \text{ or } x = -2$$

$$\therefore \quad \text{The solution is } x = 7 \text{ or } -2.$$

Note. A quadratic equation always has *two* possible solutions. If the quadratic is a perfect square, the two solutions will be the same and they are called repeated roots: see Example 2 below.

Example 2

Solve $x^2 - 6x + 9 = 0$.

$$x^2 - 6x + 9 = 0$$

Factorising gives: $\qquad (x - 3)(x - 3) = 0$

so either $\qquad\qquad x - 3 = 0 \text{ or } x - 3 = 0$

either $\qquad\qquad\qquad x = 3 \text{ or } x = 3$

$$\therefore \quad x = 3$$

(In an example like this we say that the equation has **repeated roots**.)

---

*Example 3

Solve the equation $2x^2 - 5x + 3 = 0$.

$$2x^2 - 5x + 3 = 0$$

Factorising gives: $\qquad (2x - 3)(x - 1) = 0$

so either $\qquad\qquad (2x - 3) = 0 \text{ and } x = 1\frac{1}{2}$

or $\qquad\qquad\qquad (x - 1) = 0 \text{ and } x = 1$

$$\therefore \quad x = 1\frac{1}{2} \text{ or } 1$$

---

*Example 4

Solve $6x^2 + 8x - 8 = 0$.

The LHS has a common factor of 2. Dividing the equation by 2 gives:

$$3x^2 + 4x - 4 = 0$$

Factorising gives: $\qquad (3x - 2)(x + 2) = 0$

so either $\qquad\qquad (3x - 2) = 0 \text{ and } x = \frac{2}{3}$

or $\qquad\qquad\qquad (x + 2) = 0 \text{ and } x = -2$

$$\therefore \quad x = \frac{2}{3} \text{ or } -2$$

Exercise 35H

Solve the following equations:

1 $\quad x^2 + 3x + 2 = 0$ 3 $\quad x^2 + x - 2 = 0$ 5 $\quad x^2 - 4 = 0$ 7 $\quad x^2 - 16 = 0$

2 $\quad x^2 - 3x + 2 = 0$ 4 $\quad x^2 - x - 2 = 0$ 6 $\quad x^2 - 9 = 0$ 8 $\quad x^2 - 2x = 0$

9   $x^2 + 3x = 0$ 13 $x^2 + x - 12 = 0$ *17 $2x^2 + 7x + 3 = 0$ *21 $2x^2 + 3x - 14 = 0$

10   $x^2 - 6x = 0$ 14 $x^2 - 8x + 12 = 0$ *18 $3x^2 - 2x - 8 = 0$ *22 $2x^2 - 2x - 40 = 0$

11   $x^2 - 5x + 4 = 0$ 15 $x^2 - 9x + 14 = 0$ *19 $5x^2 - 2x = 0$ *23 $6x^2 + 3x - 3 = 0$

12   $x^2 + 6x + 9 = 0$ 16 $x^2 + 11x + 28 = 0$ *20 $3x^2 - 20x + 12 = 0$

*Solving quadratic equations by the formula

The equation $2x^2 - 3x - 1 = 0$ does not have simple factors.

To solve this type of equation the **quadratic formula** must be used.

(The formula will be given to you in an exam, but you must know how to use it.)

For the equation of $ax^2 + bx + c = 0$

The solution is:

$$x = \frac{-b \pm \sqrt{b^2 - 4ac}}{2a}$$

Example 1

Solve $x^2 + 3x + 1 = 0$.

Compare $x^2 + 3x + 1 = 0$
with $ax^2 + bx + c = 0$

Then $a = +1$, $b = +3$, $c = +1$
and substituting for a, b and c in the formula gives:

$$x = \frac{-3 \pm \sqrt{9 - 4 \times 1 \times 1}}{2 \times 1}$$

$$= \frac{-3 \pm \sqrt{5}}{2} = \frac{-3 \pm 2.236}{2}$$

$$= -0.38 \text{ or } -2.62 \text{ (to 2 d.p.)}$$

Example 2

Solve $2x^2 - 3x - 1 = 0$.

In this equation, $a = 2$, $b = -3$, $c = -1$ and substituting in the formula gives:

$$x = \frac{+3 \pm \sqrt{9 - 4 \times 2 \times (-1)}}{2 \times 2}$$

$$= \frac{3 \pm \sqrt{17}}{4} = \frac{3 \pm 4.123}{4}$$

$$= 1.78 \text{ or } -0.28 \text{ (to 2 d.p.)}$$

Exercise 35I

Solve the following equations using the quadratic formula and give your answers correct to 2 d.p.

1 $2x^2 + 2x - 3 = 0$ 5 $x^2 + 6x - 10 = 0$ 9 $2x^2 + 5x = 6$

2 $2x^2 + 4x + 1 = 0$ 6 $x^2 - 7x + 9 = 0$ 10 $3x^2 - 10x = -5$

3 $x^2 + 2x - 2 = 0$ 7 $4x^2 - 8x - 16 = 0$ 11 $x(x + 2) = 5$

4 $3x^2 - x - 1 = 0$ 8 $3x^2 - 6x + 2 = 0$ 12 $x(x - 3) = -1$

*Problems involving quadratic equations

Example

One side of a rectangle is 3 centimetres longer than the other. The area (in cm^2) of the rectangle is twice its perimeter (in cm). Find the dimensions of the rectangle.

Let the length of one side $= x$ cm

The length of the other side $= (x + 3)$ cm

The area $= x(x + 3)$ cm^2

The perimeter $= (4x + 6)$ cm

Since Area $= 2 \times$ Perimeter

$$x(x + 3) = 2(4x + 6)$$
$$x^2 + 3x = 8x + 12$$
$$x^2 - 5x - 12 = 0$$
$$x = \frac{+5 \pm \sqrt{25 + 48}}{2}$$
$$= \frac{5 \pm \sqrt{73}}{2}$$

As x cannot be negative, $x = \dfrac{5 + 8.544}{2} = 6.77$ cm (to 2 d.p.)

$\therefore$ The dimensions are 6.77 cm and 9.77 cm.

Exercise 35J

1 A carpet is 3 metres too long for a square floor in one direction and 1 metre too short in the other. The area of the carpet is 21 square metres.

   a If the floor is x metres square, write down, in terms of x, the dimensions of the carpet.

   b Write down an equation in x for the area of the carpet.

   c Solve the equation and find the dimensions of the floor.

2 The number of diagonals D of a polygon with n sides is given by the formula:

   $$D = \tfrac{1}{2}n(n - 3)$$

   a Find D if $n = 7$.

   b Find n if $D = 20$.

3 In a tennis tournament each team plays every other team.

a If there are x number of teams, write down an expression, in terms of x, for the total number of games played.
(*Remember.* Team A plays team B is the same game as team B plays team A.)

b If a total of 28 games is played, write down an equation in x.

c Solve the equation to find how many teams played in the tournament.

4 A farmer has 40 metres of fencing with which he intends to make a rectangular enclosure for a few sheep and their lambs.

a If one side of the rectangle is x metres, what is the length of the other side, in terms of x?

b Write down an expression for the area in terms of x.

c The farmer calculates that he requires $96\,m^2$ of grass in the sheep pen.
What are the dimensions of the pen?

d If the farmer had required an enclosure of area $80\,m^2$, but not less, what would the dimensions of the rectangle have been?
(Use the quadratic formula to solve the equation and give your answer to the nearest 0.1 m.)

e State the dimensions of the pen which would enclose the largest area of grass, using the 40 m of fencing.

5 A rectangle has sides of 7 cm and 4 cm.
Two strips, each x cm wide, are cut from the rectangle.
One strip is cut parallel to the shorter side and the other is cut parallel to the longer side.

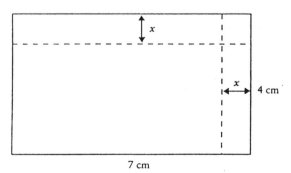

7 cm

The resulting rectangle is half the area of the original rectangle.

Form an equation in x and use the quadratic formula to find the value of x correct to 2 decimal places.

6 A photograph is 5 cm square. It is mounted on a piece of card so that there is a border which is x cm wide at the top and bottom and $\frac{1}{2}x$ cm wide at each side.
The area of the card is $80\,cm^2$.

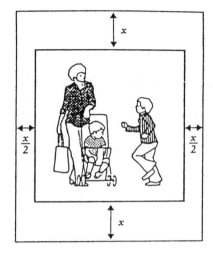

Write down an equation for the area of the card, in terms of x, and solve it to find the value of x correct to the nearest millimetre.

What are the dimensions of the card?

7 a A cube with dimensions $4 \times 4 \times 4$ is made of 64 small blocks, as shown in the diagram. The outside of the cube is painted. How many of the blocks will have just one side painted?

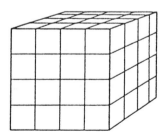

b A cube with dimensions $5 \times 5 \times 5$ would be made from 125 blocks. How many of these blocks would have one side painted?

c Find a formula for the number of blocks with one side painted (S) if the cube has dimensions $n \times n \times n$.

d Calculate the value of n if:
(i) $S = 24$, (ii) $S = 96$.

35.5 The transformation of formulae

Example

The profit (£P) made when an article is sold for a price £S with a percentage profit of $x\%$, can be calculated from the formula:

$$P = \frac{Sx}{100 + x}$$

a Find P when $x = 12$ and $S = 392$.

b Rearrange the formula to make x the subject.

c Find the percentage profit if a profit of £33 is made when an article is sold for £253.

a $P = \dfrac{392 \times 12}{100 + 12} = \dfrac{4704}{112} = £42$

b $P = \dfrac{Sx}{100 + x}$

   (i) Deal with the fraction by multiplying each side by $(100 + x)$:

$$P(100 + x) = Sx$$

  (ii) Multiply out the bracket:

$$100P + Px = Sx$$

 (iii) Collect x terms on one side, other terms on the opposite side (in this case, it is easier to collect x terms on RHS to avoid a negative sign for $100P$):

$$100P = Sx - Px$$

 (iv) Take out a common factor of x:

$$100P = x(S - P)$$

  (v) Divide each side by $(S - P)$ to find the value of x:

$$\frac{100P}{S - P} = x$$

c $P = £33$, $S = £253$

$$\text{Profit \%} = x = \frac{100P}{S - P} = \frac{100 \times 33}{253 - 33} = 15\%$$

Exercise 35K

1 The formula $s = \frac{1}{2}(u + v)t$ gives the distance travelled in time t by a vehicle, where u is its initial speed and v is the speed after travelling a distance s.

  a Rearrange the formula to give v in terms of s, u and t.

  b Calculate v if $s = 46$ m, $u = 5$ m/s, $t = 4$ s.

  c Calculate v if $s = 120$ m, $u = 24$ m/s, $t = 10$ s. What does this value of v tell you about the vehicle?

2 The focal length of a lens is given by the formula:

$$\frac{1}{f} = \frac{1}{u} + \frac{1}{v}$$

  a Calculate v if $f = 3.0$ cm and $u = 4.8$ cm.

  b Rearrange the formula to give u in terms of f and v.

  c Find u if $f = 2.6$ cm and $v = 6.5$ cm.

3 The time, in seconds, for a full swing, or oscillation, of a pendulum, length l metres, is given by:

$$T = 2\pi\sqrt{\frac{l}{10}}$$

  a Rearrange the formula to make l the subject.

  b The pendulum of a clock is to make a complete oscillation every one second. What length, to the nearest millimetre, must the pendulum be?

4 If £P is invested for 2 years at a compound interest rate of r%, the amount in the account at the end of the 2 years is given by:

$$A = P\left(1 + \frac{r}{100}\right)^2$$

  a Rearrange the formula to make r the subject.

  b £2500 was invested and at the end of 2 years had amounted to £3775. At what rate of compound interest was the money invested?

    The formula can be simplified if $\left(1 + \frac{r}{100}\right)$ is replaced by R.

  c Rewrite the formula for A in terms of P and R.

  d Rearrange the formula to make R the subject.

  e (i) Find R if $A = £4143.75$ and $P = £3750$.
     (ii) At what rate was the money invested?

5 A boy, standing near the end of a cliff, throws a stone up into the air. The velocity, v metres per second, of the stone, when it is a distance s metres from the boy's hand, is given by:

$$v^2 = 100 - 20s$$

  a Rearrange the formula to give s in terms of v.

  b Find s when
    (i) $v = 5$ m/s
    (ii) $v = 0$ m/s
    (iii) $v = 10$ m/s
    (iv) $v = 15$ m/s
    and describe the position of the stone in each case.

6 A window is made in the shape of a rectangle of width $2x$ and height $2h$ surmounted by a semicircle.

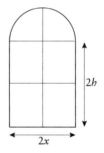

  a Write down an equation for the perimeter of the window, P, in terms of x and h.

  b Rearrange the above equation to give x in terms of P and h.

  c Find x if $P = 15.7$ and $h = 2$.

7 A rectangle has sides of length x and y.

  a Write down an expression for the perimeter of the rectangle in terms of x and y.

  b Write down an expression for the area of the rectangle in terms of x and y.

  c The area of the rectangle is three times the length of the perimeter. Write down an equation which describes this relationship.

  d Rearrange the equation to give a formula for x in terms of y.

  e Find a value for x if $y = 10$ cm.

Exercise 35L

1 **a** Solve the equations:

(i) $2x - y = 8$ (ii) $2x - 4y = 1$
    $3x + y = 7$ $3x - 5y = 2\frac{1}{2}$

b Solve, by drawing a graph, the equations:

$$x - y = -2$$
$$2y = 3x + 1$$

2 A school buys two types of calculator. One type costs £9 each, the other costs £30 each. The school ordered 74 calculators and paid £1170.
How many of each type were bought?

3 Expand:

a $(x + 5)(x + 2)$ **c** $(2x - 1)(x - 8)$

b $(x - 7)(x + 7)$ **d** $(2x - 3y)(x + y)$

4 Factorise:

a $x^2 - 8x - 9$ **d** $x^2 + x - 6$

b $3x^2 + 12x$ ***e** $3x^2 + 12x + 9$

c $3x^2 - 12$ ***f** $2x^2 + 11x - 6$

5 Solve:

a $x^2 - 4x - 5 = 0$ ***e** $4x^2 - 8x = 0$

b $x^2 - 3x = 0$ ***f** $9x^2 - 16 = 0$

c $x^2 - 25 = 0$ ***g** $2x^2 + 7x + 6 = 0$

d $x^2 + 7x + 12 = 0$

*6 Solve, giving your answers correct to 2 d.p.

a $x^2 + 2x - 7 = 0$ **b** $2x^2 - 5x + 1 = 0$

7 Use a trial and improvement method to solve $x^2 + 2x - 7 = 0$ correct to 1 d.p.

8 Solve the inequalities

a $2x + 4 \geqslant 16$

b $4x - 1 > x + 8$

c $x^2 \leqslant 9$

9 Find a formula for the nth term of the sequence 2, 6, 10, 14, ...

*10 A rectangle has a side of length x centimetres. The other side of the rectangle is 2 centimetres longer.

a Write down an expression, in terms of x, for the area of the rectangle.

b Write down an expression, in terms of x, for the perimeter of the rectangle.

c The area, in square centimetres, is equal to the perimeter, in centimetres. Write down an equation in x.

d Simplify your equation and solve it to find x, using the formula:

$$x = \frac{-b \pm \sqrt{b^2 - 4ac}}{2a}$$

(Give your answer correct to 3 s.f.)

*11 A firm makes novelties for Christmas crackers. Each novelty costs x pence to make and they are sold for £y per 1000.

a Write down an expression for the cost of manufacturing N novelties.

b Write down an expression for the selling price (in pence) of N novelties.

c Write down a formula for the profit P pence in terms of N, x and y.

d Calculate the profit if $N = 3000$, $x = 3$ and $y = 50$.

e Rearrange the formula to make N the subject.

f If the cost of manufacture and the selling price are unchanged, how many novelities must be sold in order to make a profit of £100?

36 Iteration and Convergence

36.1 Iteration

An **iterative process** is one which is repeated several times, following exactly the same process on each repeat.

One iterative process for finding the square root of a number is:

(i) Take a guess at the square root, i.e. make a first estimate.

(ii) Divide the number by your estimate then find the average of this number and your estimate.

(iii) This average is an improved estimate so repeat (ii) using this second estimate to find a third estimate.

(iv) Repeat (ii) with each new estimate of the square root of a number A. If the first estimate is called x_1, then the second estimate, x_2, is given by

$$x_2 = \frac{1}{2}\left(x_1 + \frac{A}{x_1}\right)$$

The third estimate is
$$x_3 = \frac{1}{2}\left(x_2 + \frac{A}{x_2}\right)$$

The fourth estimate is
$$x_4 = \frac{1}{2}\left(x_3 + \frac{A}{x_3}\right)$$

The $(n+1)$th estimate is
$$x_{n+1} = \frac{1}{2}\left(x_n + \frac{A}{x_n}\right)$$

and this is an iterative formula for the square root of A.

Example

a Using the iteration formula $x_{n+1} = \frac{1}{2}\left(x_n + \frac{A}{x_n}\right)$, find the square root of 85 correct to 5 decimal places.

b By writing x for x_{n+1} and x_n in the iteration formula, show that your answer is a solution of the equation $x^2 = 85$.

a
$$x_{n+1} = \frac{1}{2}\left(x_n + \frac{A}{x_n}\right)$$

One estimate of $\sqrt{85}$ is 9

$$x_1 = 9$$

$$x_2 = \frac{1}{2}\left(9 + \frac{85}{9}\right) = 9.2222222$$

(some calculators will display 10 digits)

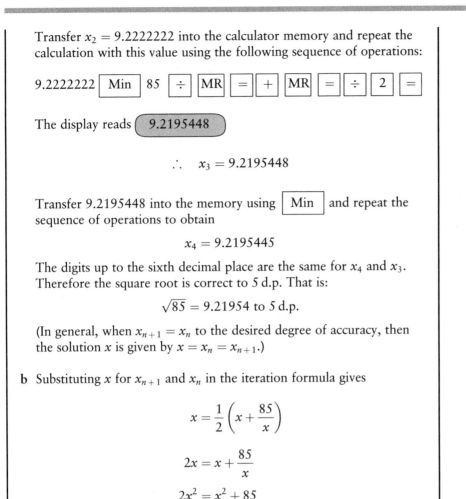

Transfer $x_2 = 9.2222222$ into the calculator memory and repeat the calculation with this value using the following sequence of operations:

9.2222222 | Min | 85 | ÷ | MR | = | + | MR | = | ÷ | 2 | =

The display reads (9.2195448)

$$\therefore \quad x_3 = 9.2195448$$

Transfer 9.2195448 into the memory using | Min | and repeat the sequence of operations to obtain

$$x_4 = 9.2195445$$

The digits up to the sixth decimal place are the same for x_4 and x_3. Therefore the square root is correct to 5 d.p. That is:

$$\sqrt{85} = 9.21954 \text{ to } 5 \text{ d.p.}$$

(In general, when $x_{n+1} = x_n$ to the desired degree of accuracy, then the solution x is given by $x = x_n = x_{n+1}$.)

b Substituting x for x_{n+1} and x_n in the iteration formula gives

$$x = \frac{1}{2}\left(x + \frac{85}{x}\right)$$

$$2x = x + \frac{85}{x}$$

$$2x^2 = x^2 + 85$$

$$x^2 = 85$$

This method of finding square roots was used before 1000 BC by the Babylonians.

Iterative methods are also used for solving equations.

Exercise 36A

1 Use the iterative formula $x_{n+1} = \frac{1}{2}\left(x_n + \frac{A}{x_n}\right)$

to find the square root of 60, correct to 5 d.p.

2 One root of the equation $x^2 - 8x + 2 = 0$ can be solved by using the iterative formula

$$x_{n+1} = 8 - \frac{2}{x_n}$$

a Take any value for x_1 and use the formula to find a solution to the equation correct to 3 d.p.

b Use the quadratic formula to solve the equation.

3 a The second root of the equation
$$x^2 - 8x + 2 = 0$$
can be found by using the iteration
$$x_{n+1} = \frac{x_n^2 + 2}{8}$$

(i) Take $x_1 = 2$, (ii) take $x_1 = 0.2$ and calculate:

x_2, x_3, x_4, x_5, x_6 and x_7

Hence find the second root of the equation correct to 3 d.p.

b How many iterations are required to find the root, correct to 3 d.p., when:

(i) $x_1 = 2$ (ii) $x_1 = 0.2$?

4 An iterative formula for solving a cubic equation is

$$x_{n+1} = \frac{3}{x_n^2} - 4$$

a Take $x_1 = 4$ and calculate $x_2, x_3, x_4, x_5, x_6, x_7$ and x_8.

For each iteration, write down the first five digits.

b What is the solution, correct to 3 d.p.?

c How many iterations are required to find this solution?

d By replacing x_{n+1} and x_n with x, show that this value is a solution of the equation

$$x^3 + 4x^2 - 3 = 0$$

5 A sequence of numbers $x_1, x_2, x_3, \ldots$ is defined by the relationship

$$x_{n+1} = 4.5 + \frac{5}{2x_n}$$

and by

$$x_1 = 4.$$

a Find x_2, x_3, x_4, giving all the figures on your calculator each time.

b Suggest the value that the sequence appears to be approaching.

c (i) Replace x_{n+1} and x_n in the relationship by x and show that the result is equivalent to the quadratic equation

$$2x^2 - 9x - 5 = 0$$

(ii) Show that the limit suggested in **b** is a solution of this equation.

d (i) Factorise $2x^2 - 9x - 5$.

(ii) What is the complete solution of the equation $2x^2 - 9x - 5 = 0$?

(SEG Sp88)

6 The equation $3x^2 - 11x - 4 = 0$ can be rearranged as

$$x = \frac{1}{3}\left(11 + \frac{4}{x}\right) \quad \text{or as} \quad x = \frac{3x^2 - 4}{11}$$

a Write down two possible iterative formulae for solving this equation.

b Take $x_1 = 3$ and calculate x_2, x_3, and x_4 for each of your formulae.

c What can you deduce from your results?

d Take $x_1 = 0.3$ and calculate x_2, x_3, x_4, x_5 and x_6 for each of your formulae.

e What do you deduce are the roots of the equation $3x^2 - 11x - 4 = 0$?

f Verify your answer to **e** by solving the equation $3x^2 - 11x - 4 = 0$ by factorising.

g Rearrange the equation $2x^2 - 4x + 1 = 0$ to find an iterative formula which will locate the solution to the equation which is greater than 1.

36.2   Convergence

The sequences generated by the iterative formulae above all **converge**, i.e. all the sequences become nearer and nearer a finite number. Many other sequences do not converge. For example:

1, 2, 4, 8, 16, ...
1, −1, 1, −1, 1, ...

are both sequences in which, however far you go, the consecutive numbers will not become close together.

Example

For the sequence $x_{n+1} = \dfrac{5}{(1 - x_n)}$, starting with $x_1 = 2$:

a Find x_2, x_3, x_4 and comment on your results.

b Find the quadratic equation which this iteration is trying to solve.

c By considering the solutions to this equation, explain why the iteration does not converge.

a $x_2 = \dfrac{5}{-1} = -5$

$x_3 = \dfrac{5}{6} = 0.8333$

$x_4 = \dfrac{5}{0.16667} = 30$

The iteration is diverging.

b The equation is:

$x = \dfrac{5}{(1 - x)}$

$x - x^2 = 5$

$x^2 - x + 5 = 0$

c The solutions of the quadratic equation are:

$x = \dfrac{1 \pm \sqrt{1 - 20}}{2}$

i.e. there are no real solutions, since there are no real solutions of the square root of a negative number. Therefore the iteration cannot converge to a finite number which would be its solution.

Exercise 36B

1 Taking $x_1 = 1.5$, find whether or not the following iterations converge, and if they do, give their limit:

a $x_{n+1} = \dfrac{3}{2 - x_n}$ **b** $x_{n+1} = 1 + \dfrac{2}{x_n}$

2 The equation $x^3 - 2x + 5 = 0$ may be written as:

a $x_{n+1} = \dfrac{2}{x_n} - \dfrac{5}{x_n^2}$ **b** $x_{n+1} = \sqrt[3]{2x_n - 5}$

Does either converge? Give the limit, if possible.

3 Consider the iteration $x_{n+1} = \dfrac{4}{x_n - 2}$ to 3 decimal points.

Does it matter which value you take for x_1?

Try $x = 1$, $x_1 = 2$, $x_1 = 3$, $x_1 = 4$, etc.

Shape, Space, and Measures

The following topics, covered in the Money Management and Number module, are also included in this module:

Measures	Metric and imperial conversions	see Unit 5	Section 5.2
Measures	Compound measures	see Unit 4	Section 4.5
Space	Discrete and continuous data	see Unit 7	Section 7.3
Space	Degrees of accuracy	see Unit 7	Section 7.1

37 Geometry

Geometry is one of the oldest branches of mathematics. The name is derived from the Greek words 'ge', meaning earth, and 'metrein', the verb to measure. The Ancient Babylonians, Egyptians and Greeks all contributed to the development of geometry.

37.1   Lines and angles

Lines

Lines can be curved or straight.

There are very few straight lines in nature, but much of geometry is concerned with straight lines.

On a flat surface – and most ancient civilisations believed the earth to be flat – a straight line is the shortest distance between two points.

A **point** marks a position and has no size.

The line below joins the two points A and B:

A   B

To indicate a point in a line, it is usual to mark the point with a small 'dash':

A

The next diagram shows two points A and B and a line passing between the two points.
The line extends infinitely in both directions.

A   B

If we wish to refer to the part of the line between the points A and B, we should, strictly speaking, refer to the **line segment AB**. It is, however, usual to use '**the line AB**' for the line segment.

Straight lines are drawn using a ruler, a rule or a straight edge.

Angles

When two straight lines meet they form an angle. The angle between AO and OB is called the angle AOB (or it could be called BOA).

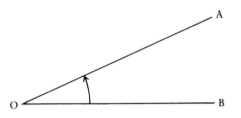

The size of the angle AOB above is a measure of how far the line OA has been rotated from a starting position along OB. If we continue to rotate OA in an anticlockwise direction until it once again lies along OB, it will have been rotated through a full circle or 360 degrees (also written as 360°).

The reason there are 360 degrees in a full circle is because the Babylonians believed there were 360 days in the year. An angular measure was derived in order to chart the movements of the stars, which took one year for a full rotation about the earth.

A rotation through half a circle, or half-turn, is a turn through 180°.

Angles which add together to make 180° form a straight line and are called **supplementary angles**. COB and AOC shown below are supplementary angles.

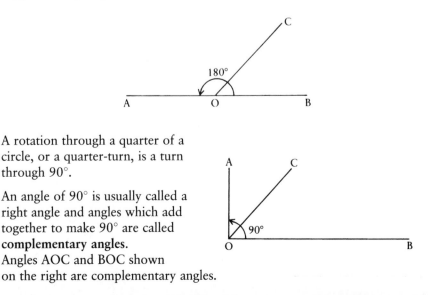

A rotation through a quarter of a circle, or a quarter-turn, is a turn through 90°.

An angle of 90° is usually called a right angle and angles which add together to make 90° are called **complementary angles**.
Angles AOC and BOC shown on the right are complementary angles.

Right angles occur a great deal in the man-made world. If you look around any room, you will see many examples of right angles.

Draughtsmen use a set square for drawing and checking right angles.

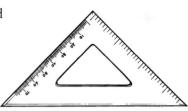

Two lines which intersect at right angles are called **perpendicular lines**. The diagram on the right shows two perpendicular lines.

The special symbol for a right angle has been used in the diagram.

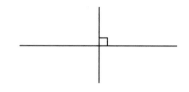

Other types of angle are:

acute angles, of size between 0° and 90°

obtuse angles, of size between 90° and 180°

reflex angles, of size greater than 180°

vertically opposite angles, which are formed when two straight lines cross.

In the diagram alongside, the angles marked x are vertically opposite angles and they are equal in size. The angles marked y are also equal.

From what you have read in this chapter you should be able to work out the angles in the following exercise.

Exercise 37A

1   Calculate the size of the missing angle in each case:

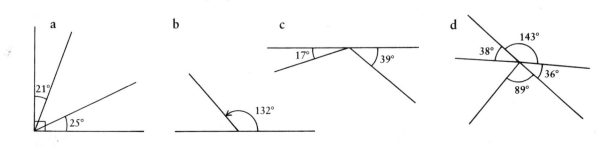

2 State whether the following angles are acute, obtuse or reflex:

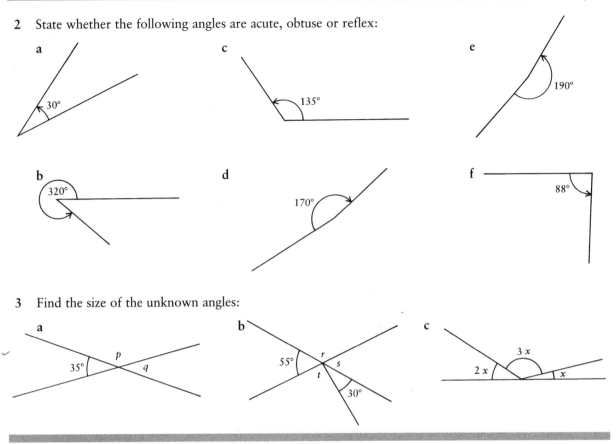

3 Find the size of the unknown angles:

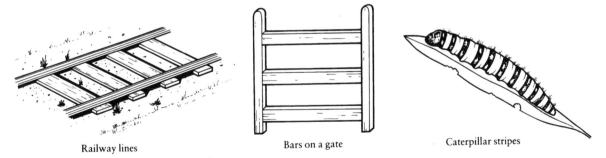

Angles and parallel lines

Lines which do not intersect, no matter how far they are extended, are called
parallel lines.

There are many examples of parallel lines around us.

Railway lines　　　　　　　　Bars on a gate　　　　　　　　Caterpillar stripes

On a wide gate, an extra bar is added at an angle to act as a brace or support.

Several angles are formed.

Angles such as a and b, which are on opposite sides of
the bracing bar, are called **alternate angles**.

Angles such as a and c, which are in similar positions
on the same side of the bar, are called **corresponding
angles**.

Angles a, b and c are all equal in size.

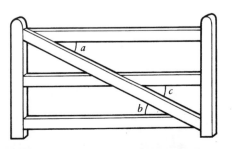

The arrows indicate that the lines are parallel.
All the angles marked *a* are equal in size.
All the angles marked *b* are equal in size.

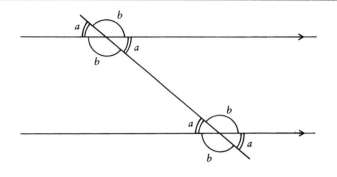

---Example---

Find the size of the angles lettered *a*, *b* and *c* in the diagram below:

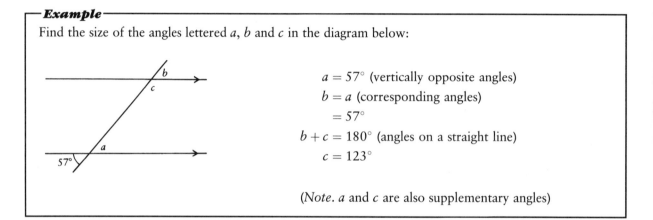

$a = 57°$ (vertically opposite angles)

$b = a$ (corresponding angles)

$= 57°$

$b + c = 180°$ (angles on a straight line)

$c = 123°$

(*Note.* *a* and *c* are also supplementary angles)

Exercise 37B

Find the size of each lettered angle in the following diagrams:

1

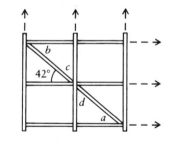

128°   *b*   *a*

c
d

5

b
42°   *c*
d
a

The diagram shows part of a fence panel.

2

s *t*   *r*
115°   *p* *q*

3

32°
s
r   *t*   *u*
w   *v*

6

120°
a
d   *b*   *c*

The diagram shows a cross-section through the roof of a barn.

4

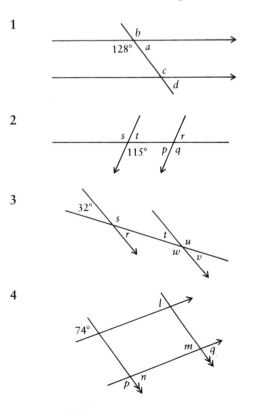

l
74°
m *q*
p *n*

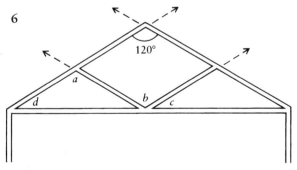

7   The diagram shows a set of steps used to board an aircraft.
    QS and the platform *PL* are both horizontal.

The bar *QP* is vertical.
QS = QL.
Angle *LQS* is 52°.

a   Explain why angle *PLQ* is 52°.

b   Calculate the size of angle *PLS*.

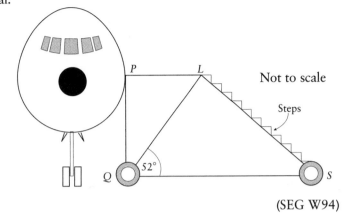

Not to scale

Steps

(SEG W94)

37.2   **Symmetry**

Symmetry implies balance and harmony.
Shapes which are symmetrical are balanced because there is a repeated
pattern. There are two types of symmetry, **reflective** (or line) symmetry and
rotational symmetry.

Reflective symmetry

A shape is said to have **reflective symmetry** if a line can be drawn which cuts
the shape into two halves and each half is the mirror image of the other.

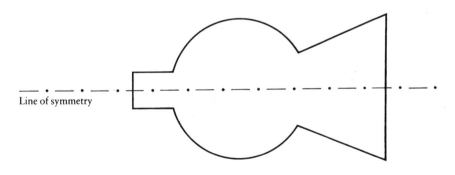

Line of symmetry

In the example below, AB is *not* a line of symmetry, even though the two
halves PQBA and ABRS are identical.

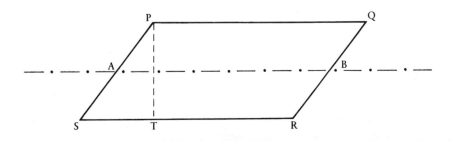

The mirror image of point P is point T, therefore ABRS cannot be the mirror image of PQBA.

Shapes may have several lines of symmetry.
For example, the hexagon on the right has six lines of symmetry, shown by the dotted lines.

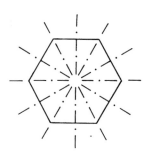

■ *Exercise 37C* ■

1   State the number of lines of symmetry which could be drawn through the following shapes:

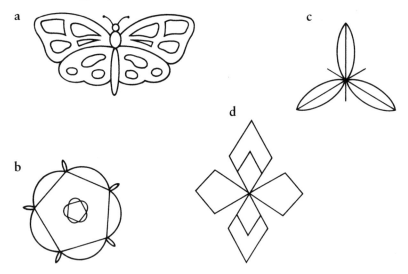

2   Copy the letters of the alphabet shown here and draw the maximum number of lines of symmetry on to each letter.

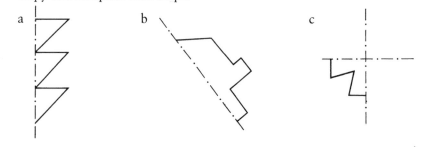

3   Lines of symmetry are indicated on the diagrams below, but the shapes are incomplete.
Copy and complete each shape.

Rotational symmetry

The parallelogram shown is not symmetrical about any line. If, however, the parallelogram is rotated about point O until PQ is in the position previously occupied by RS, the shape will look exactly the same as the original:

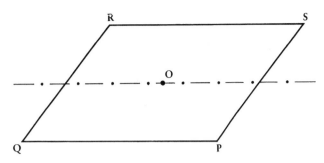

Parallelograms have **rotational symmetry**.

Because there are *two* positions in which the shape looks identical, the parallelogram has rotational symmetry of **order 2**.

The equilateral triangle ABC can be rotated about O, through 120° anticlockwise, to the position CAB, which is identical to ABC:

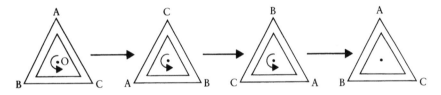

Repeating this rotation twice more will bring the triangle back to its original position.

The triangle has rotational symmetry of **order 3**.

Because all shapes fit into themselves, all shapes have rotational symmetry of at least 1. However, a shape with rotational symmetry order 1 is generally considered to have no rotational symmetry.

Exercise 37D

1   List the letters of the alphabet which have rotational symmetry and state the order of symmetry.

2   a   Which of the shapes in question 1 of Exercise 37C have rotational symmetry?

    b   What is the order of symmetry of each?

3   For each of the shapes shown below, state (where appropriate):
    (i) the number of lines of symmetry
    (ii) the order of rotational symmetry.

a   b

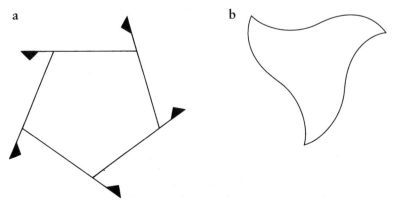

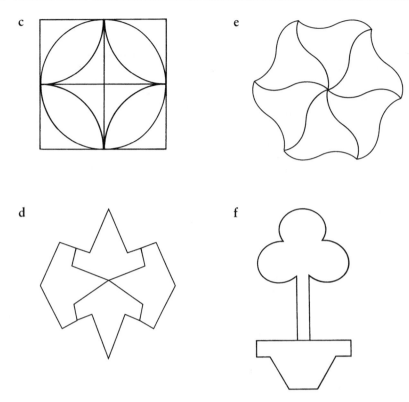

Historic note. Patterns **b**, **c** and **d** are Arabic. Symmetry played an important part in Arabic art and mathematics.

Pattern **e** was given to students in Babylonian times for the calculation of areas.

4   The following shapes are being considered for a company logo.

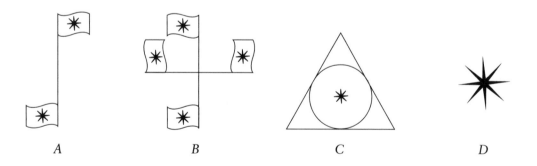

a   (i)   Which shape has no rotational symmetry?
    (ii) Which shape has rotational symmetry of order 2?
    (iii) Write down the order of rotational symmetry of shape C.

b   (i)   Draw all the lines of symmetry on the two dominoes.

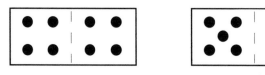

The two dominoes are put together in two arrangements as shown.

(ii) Write down the number of lines of symmetry for each arrangement.

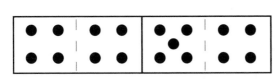

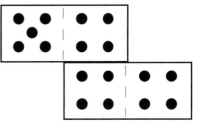

(SEG S94)

37.3 Planes of symmetry

Just as a two-dimensional shape may have one or more lines of symmetry, a three-dimensional shape may have one or more **planes of symmetry**.

Example 1

A wedge of cheese is cut from a round of Gouda, as shown.

How many planes of symmetry has the wedge?

If we 'slice' through the wedge of cheese vertically, as shown, with a very thin two-sided mirror, the wedge would appear unchanged from whichever side of the mirror we viewed it. The pieces are **mirror images** of each other.

The wedge has a plane of symmetry in the vertical direction, shown by the shaded area.

We could also cut the wedge into two pieces by 'slicing' it horizontally with the mirror, and the wedge would appear unchanged. Again, the two pieces are mirror images of each other.

The plane of symmetry is shown by the shaded area.
There is no other way we could form two pieces which are mirror images. Any other cut with the mirror would result in a change in the shape of the wedge.
The wedge has **two** planes of symmetry.

Example 2

Cuisennaire rods are used to teach young children about number. The rods all have the same square cross-section, but are of different lengths.

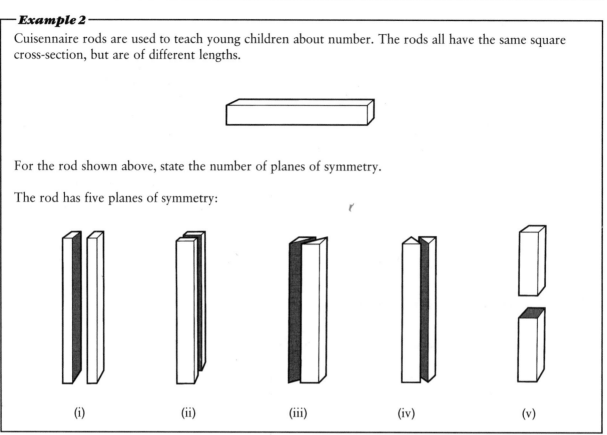

For the rod shown above, state the number of planes of symmetry.

The rod has five planes of symmetry:

| (i) | (ii) | (iii) | (iv) | (v) |

Exercise 37E

For each of the following solids, state the number of planes of symmetry.

1   the square-based pyramid:

2   the triangular prism:

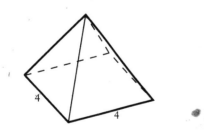

3   the triangular prism:

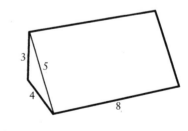

4   a cuboid with dimensions 4 cm × 5 cm × 6 cm

5   a cuboid with dimensions 4 cm × 4 cm × 6 cm

6   a cube

7   a tetrahedron with all edges 5 cm long:

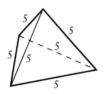

9   an hexagonal prism:

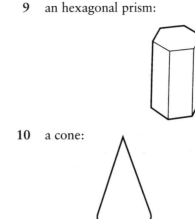

8   a cylinder:

10   a cone:

37.4   Nets

If you join together six identical squares edge to edge with adhesive tape, as shown, you will obtain a cube; or more precisely the outside surface of a cube.

A neater method of producing the outside of a cube is to keep most of the faces joined together.

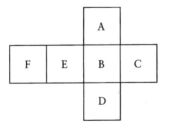

If you fold up the faces A, C, D, E upon base B, and then fold face F over the top, you will again obtain a cube.

A cross plan such as the one above, which can be folded to form a cube, is called a **net** of the cube.

Other nets of a cube include:

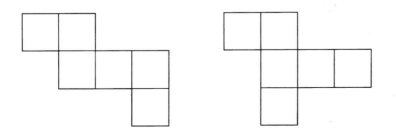

Can you find any more?

In many cases, when the six squares are joined together, it is not possible to form a cube. Here is one example:

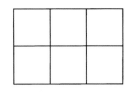

The net for an equilateral triangular prism (see pages 358, 448) would need these rectangles for the sides:

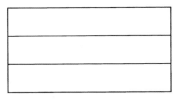

It would also need two equilateral triangles for the two ends. They can be added to opposite **ends** of any of the rectangles.

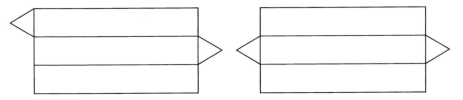

These are both nets of a triangular prism.

Constructing a net

When you have sketched a net, you must draw it accurately if you are to obtain a satisfactory result for the completed solid shape.

It is often simplest to draw a net on graph paper where lengths, and right angles, are easily checked.

If you are not using graph paper you must be able to draw the lengths and angles of the net accurately. Ensure you check lengths to within 1 mm, and angles to within 1°.

Forming a solid shape

If you were to create a cube from the first net on page 359, you would realise that you need to join F to face C. This can be done most easily by including a tab on face C (or F). Similarly, to strengthen the other joins, you would need more tabs. When you cut out your shape and use glue on all the tabs, you should be able to fold it to make a neat three-dimensional shape.

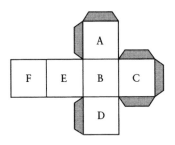

Example

Draw a net for a regular tetrahedron, each face of which is an equilateral triangle of side 3 cm.

One possible net is shown.
If necessary, please refer to the geometry section (page 388) for details of how to construct triangles.

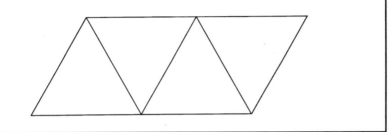

Exercise 37F

1   Draw a net of a cuboid 3 cm by 2 cm by 2 cm.

2   Draw a net for a regular tetrahedron with each side of length 4 cm.

3   Draw a net for a tetrahedron with an equilateral base of side 4 cm. Each slant edge is 6 cm.

4   Draw a net of a prism whose cross-section PQR is a right-angled triangle.

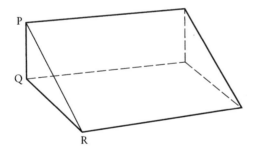

5   A plate of thickness 0.6 inches has radius 5 inches. Draw a net of a box which could contain four such plates.

6   A roll of adhesive tape is 2.5 cm in radius and 1.4 cm thick. Draw a net of a cuboid which will contain three such rolls.

7   The diagram is a net of a triangular prism.

   a Find the total area of the net.

   b Find the volume of the prism.

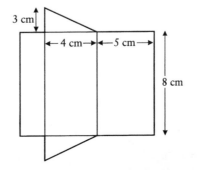

8   Which of the following are nets of a triangular prism?

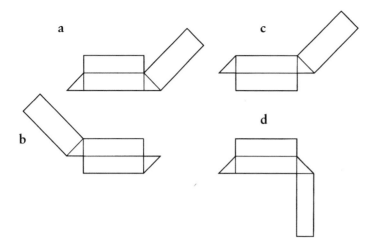

a

c

b

d

9   When the net below is folded to make a cube, which side is joined to side AB?

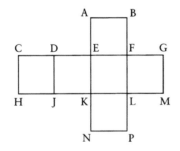

10   A bar of soap is packed in a closed cardboard prism with dimensions as shown.

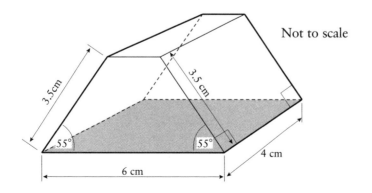

Not to scale

Draw an accurate net of the prism.   (SEG S95)

37.5 Triangles

A **triangle** is a three-sided polygon.

All triangles belong to one of three categories:

Scalene triangles

The three sides are of different length.

The three angles are of different size.

A scalene triangle has no reflective or rotational symmetry.

Isosceles triangles

Two sides are equal in length.

Two angles are equal in size.

An isosceles triangle has one line of symmetry.

Equilateral triangles

All three sides are equal in length.

All three angles are equal in size.

An equilateral triangle has three lines of symmetry and rotational symmetry of order 3.

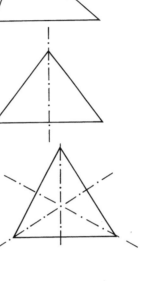

The triangles above are all **acute-angled** triangles, i.e. all three angles are less than 90°.

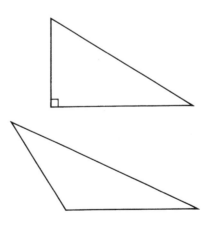

Triangles can also be:

Right-angled

if one angle is 90°.

Obtuse-angled

if one angle is greater than 90°.

Properties of triangles

The basic property of all triangles is this:

   The three interior angles of a triangle always total 180°.

$$a + b + c = 180°$$

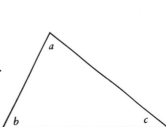

It follows from this that, in the triangle below, $d = a + b$

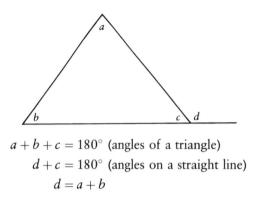

$$a + b + c = 180° \text{ (angles of a triangle)}$$
$$d + c = 180° \text{ (angles on a straight line)}$$
$$d = a + b$$

In other words:
The exterior angle of a triangle is equal to the sum of the opposite interior angles.

Equilateral triangles

Because all three angles are equal in size, the angles of an equilateral triangle are all 60°.

Isosceles triangles

An isosceles triangle has two sides equal in length and the angles opposite these sides are equal in size.
In the diagram below, the equal sides are marked with dashes and the equal angles with arcs.

$$AB = AC$$
$$\text{Angle B} = \text{Angle C}$$

which we can also write as

$$\text{Angle ABC} = \text{Angle ACB}$$
$$\text{or } \angle ABC = \angle ACB$$

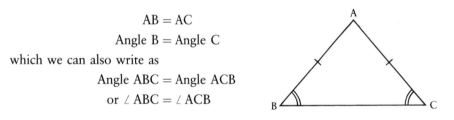

Example 1

In triangle PQR, PQ = PR and the angle PQR = 75°.

Find: **a** angle PRQ   **b** angle QPR.

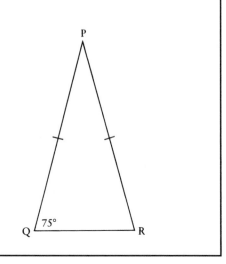

Angle PRQ = Angle PQR (angles of an isosceles triangle)
       = 75°
Angle QPR = 180° − (75° + 75°) (angles of a triangle)
       = 30°

Example 2

In the diagram, find angle XZY.

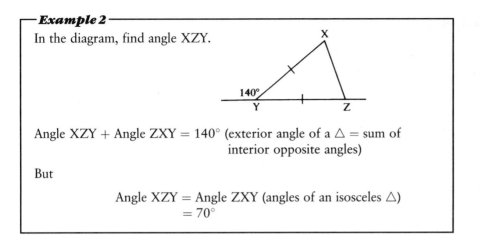

Angle XZY + Angle ZXY = 140° (exterior angle of a △ = sum of
interior opposite angles)

But

Angle XZY = Angle ZXY (angles of an isosceles △)
= 70°

Exercise 37G

1 Find angles P and R.

4 Find angles *x*, *y* and *z*.

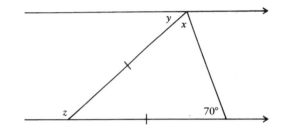

2 Find angles A and B.

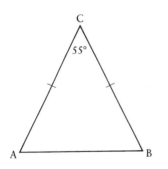

5 Find angles *x* and *y*.

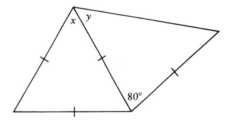

3 Find angles ACB, CAB, DCA, ADC.

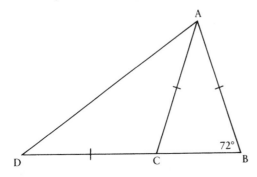

37.6   Quadrilaterals

As the name suggests, **quadrilaterals** are four-sided polygons.

A line joining two vertices of a polygon is called a **diagonal**.

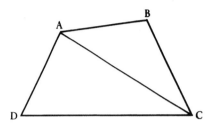

In the quadrilateral ABCD, above, the diagonal AC divides the quadrilateral into two triangles.

The sum of the angles of each triangle is 180°.

From this we deduce that:

The sum of the angles of a quadrilateral is 360.

Some special quadrilaterals are described below:

A trapezium is a quadrilateral which has one pair of opposite sides parallel.

A parallelogram is a quadrilateral which has both pairs of opposite sides parallel.

A kite is a quadrilateral which has two pairs of adjacent sides equal in length.

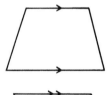

A rhombus is a parallelogram which has four sides of equal length.

A rectangle is a parallelogram which has angles of 90°.

A square is a rectangle which has four sides of equal length (or a rhombus with angles of 90°).

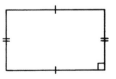

Properties of parallelograms

Applying a knowledge of symmetry, the properties of parallelograms can be listed.

All parallelograms. The parallelogram ABCD has rotational symmetry about O. Therefore:

1 AB = DC and AD = BC

   i.e. opposite sides are equal

2 ∠A = ∠C and ∠B = ∠D

   i.e. opposite angles are equal

3 AO = OC and BO = OD
   diagonals bisect each other,
   i.e. they meet at O, the
   midpoint of AC and of BD.

Rhombuses. A rhombus is a parallelogram and therefore all the properties listed above are true for a rhombus.
In addition, EG and FH are lines of symmetry. Therefore:

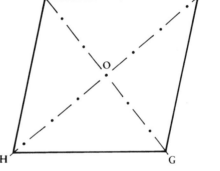

1 ∠HEG = ∠FEG,
   ∠EFH = ∠GFH,
   ∠FGE = ∠HGE,
   ∠GHF = ∠EHF

   i.e. Diagonals bisect opposite angles.

2 ∠EOF = ∠EOH = ∠GOH = ∠GOF = 90°.

   i.e. The diagonals bisect each other at 90°.

Rectangles. A rectangle is a parallelogram which has angles of 90°. Rectangles have two lines of symmetry and rotational symmetry.
Therefore:

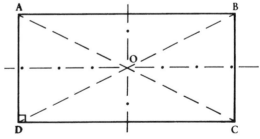

1 AO = OD = OC = OB

   i.e. The diagonals are equal.

Squares. A square is a rectangle which has all its sides equal. All the properties of a rectangle and a rhombus apply to a square.
Squares have four lines of symmetry and rotational symmetry of order 4.
Therefore:

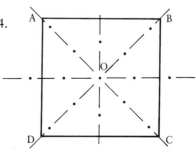

All the angles formed in the square ABCD are either 90° or 45°.

Example

ABCD is a parallelogram.

$\angle B = 63°$, $\angle BAC = 47°$

Find **a** $\angle A$ **b** $\angle ACB$

a $\angle A + \angle B = 180°$ (supplementary angles)
        $\angle A = 117°$

b $\angle DAC = 117° - 47°$
                $= 70°$
        $\angle ACB = \angle DAC$ (alternate angles)
                $= 70°$

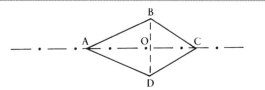

Exercise 37H

1   Use your knowledge of symmetry to list the
    properties of the kite ABCD.

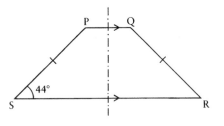

2   The trapezium PQRS is an isosceles trapezium,
    i.e. PS = QR.

    $\angle S = 44°$

    Find the size of angles P, Q and R.

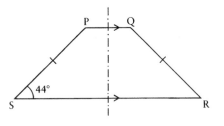

3   ABCD is a rectangle with BC = CX.
    Find the angle BXC.

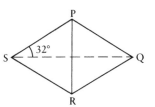

4   In rhombus PQRS, $\angle PSQ = 32°$.

    Find: **a** angle PQS   **b** angle PRQ.

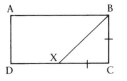

5   In parallelogram EFGH, $\angle EFH = 29°$,
    $\angle EOH = 57°$, $\angle EHG = 110°$.

    Find: **a** $\angle FOG$ **b** $\angle HFG$ **c** $\angle EGH$.

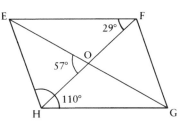

6 An open step ladder forms a triangle *RST* as shown.

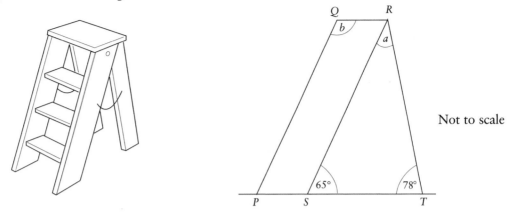

Not to scale

a The angles at the floor are 65° and 78°.
   (i) Calculate the size of the angle marked *a* in the diagram.
   (ii) Give a reason for your answer.

b *PQRS* is a parallelogram.
   (i) Calculate the size of the angle marked *b* in the diagram.
   (ii) Give a reason for your answer.

(SEG S95P)

37.7   Polygons and angles

A regular polygon has all its sides equal in length and all its angles
equal in size.

In the polygon below, the marked angles are the **exterior angles**
of the polygon.

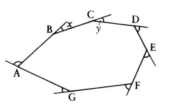

If you face along the direction AB and turn through the angle marked *x*, you
will face along direction BC.
Turning through *y* will face you in the direction CD.
When you have turned through all of the exterior angles, you will again be
facing the direction AB, having made a complete rotation of 360°.

The sum of the exterior angles of any polygon is 360°.

In the regular pentagon shown below, *x* is an exterior angle and *y* is an
interior angle.

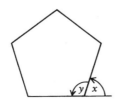

Interior angle + Exterior angle = 180° = 2 right angles

For a polygon with n sides,

$$\text{Interior angles} + \text{Exterior angles} = 180° \times n = 2n \text{ right angles}$$
$$\text{Exterior angles} = 360° = 4 \text{ right angles}$$
$$\therefore \quad \text{Interior angles} = (2n - 4) \text{ right angles}.$$

Sum of the interior angles of any polygon = $(2n - 4)$ right angles.

(*Note*. For regular polygons, it is sufficient to remember that the sum of the exterior angles is always 360°.)

Example 1

Find the size of an interior angle of a regular hexagon.

A regular hexagon has six exterior angles.

The sum of the exterior angles = 360°

$$\therefore \quad \text{Each exterior angle} = \frac{360°}{6} = 60°$$

Interior angle + Exterior angle = 180°

$$\therefore \quad \text{Each interior angle} = 180° - 60°$$
$$= 120°$$

Example 2

The sum of the interior angles of a regular polygon is 1260°.

How many sides has the polygon?

$$\text{Interior angles} + \text{Exterior angles} = 1260° + 360°$$
$$= 1620°$$
$$1 \text{ interior angle} + 1 \text{ exterior angle} = 180°$$

$$\therefore \quad \text{No. of sides of the polygon} = \frac{1620°}{180°} = 9$$

The polygon has 9 sides (a nonagon).

Exercise 37I

1 A regular polygon has 8 sides. What is the size of:

  a each exterior angle

  b each interior angle?

2 Repeat question 1 for a regular decagon, which is a polygon with 10 sides.

3 The exterior angle of a regular polygon is 20°. How many sides does it have?

4 The interior angle of a regular polygon is 140°. How many sides does it have?

5 What is the sum of the interior angles of a regular pentagon?

6 What is the sum of the interior angles of a regular decagon?

7 How many sides has a polygon if the sum of the interior angles is:

  a 1620°   b 2340°?

37.8 Circles: geometric properties

Definitions

A circle is a set of points which are a fixed distance from a given point.
The given point is the **centre** of the circle and the fixed distance is the **radius**.

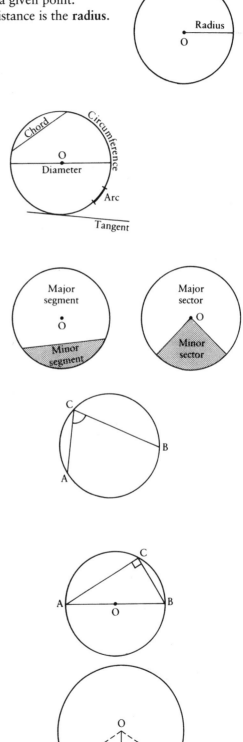

The **circumference** is the perimeter of the circle.

An **arc** is a section of the circumference.

A **chord** is a line joining two points on the circumference.

A **diameter** is a chord which passes through the centre.

A **tangent** is a line which touches the circle at one point only.

A **segment** is part of a circle cut off by a straight line.
A **sector** is part of a circle cut off by the radii.

Angle ACB is an angle subtended at a point, C, on the circumference by the arc AB.

Circles and right angles

(i) The angle subtended at the circumference by a diameter is a right angle (the angle in a semicircle).
$\angle ACB = 90°$

(ii) The line which is perpendicular to a chord and bisects the chord (i.e. the **perpendicular bisector** of a chord) passes through the centre of a circle.

Angle ODB = 90° and AD = DB.

Exercise 37J

1. Find the angles of triangle ABC if AB is the diameter of the circle:

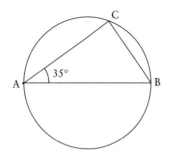

2. PQ is the diameter of the circle and RP = RQ. Find all the angles of the triangle.

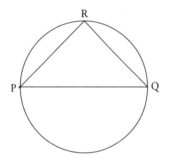

3. Find: **a** angle CBA   **b** angle ABD.

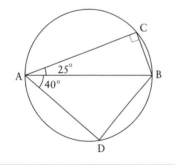

4. Find: **a** angle POR   **b** angle POQ.

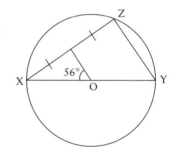

5. Find angle XYZ.

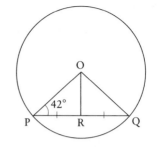

*Angle and tangent properties of circles

Circles and tangents

1. The angle formed by a radius and a tangent to the circle is a right angle:

∠ OAT = 90°

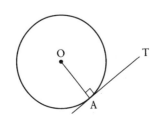

2 Tangents to a circle from an external point are equal.

PT = PS

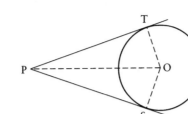

3 The angle between the tangent to a circle
and a chord is equal to the angle in the
alternate, or opposite, segment.

∠ QPT = ∠ PRQ

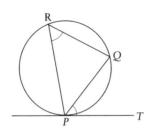

Circles and angles

There are several properties relating to angles in a circle which derive from
the fact that the angle subtended at the centre of the circle, by an arc, is twice
the angle subtended at the circumference, by the same arc. (In each of the
circles below the angle indicated by the double arc is twice the angle indicated
by the single arc.)

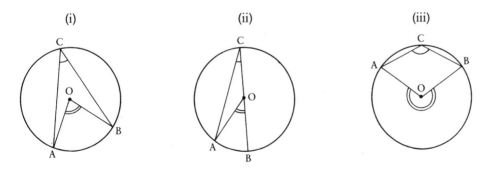

Theorems

1 The angle at the centre is twice the angle at the
circumference subtended by the same arc.

∠ AOB = 2 ∠ ACB

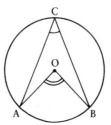

2 The angle in a semicircle is a right angle.

$$\angle ACB = 90°$$

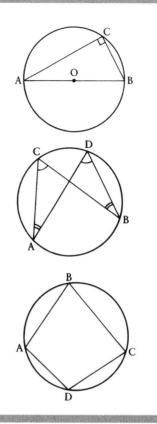

3 Angles in the same segment of a circle are equal
(i.e. angles subtended at the circumference by the
same arc).

$$\angle ACB = \angle ADB$$
$$\angle CAD = \angle CBD$$

4 The opposite angles of a cyclic quadrilateral are
supplementary.
(A cyclic quadrilateral is any quadrilateral whose
vertices are points on the circumference of a
circle.)

$$\angle ABC + \angle ADC = 180°$$
$$\angle BAD + \angle BCD = 180°$$

Exercise 37K

1 Find ∠PST.

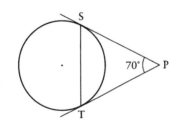

2 Find ∠SPT.

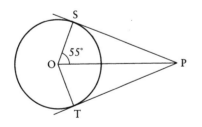

3 Find: **a** angle OPQ   **b** angle BCD.

4 Find the angles of triangle DEB.

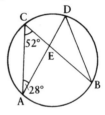

5 Find   **a** angle BCA   **b** angle CAD.

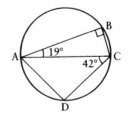

6 PQ and SR are parallel.
Angle PQS = 32°, angle SPQ = 74°.

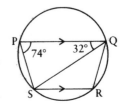

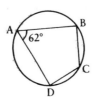

Find angles: **a** QSR   **b** QRS   **c** SQR.

7 AB and BC are equal in length.
Angle CAB = 41°.

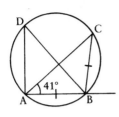

Find angles: **a** ACB **b** ADB.

8 Angle PTQ = 25°, RT is a diameter,
and TS = SR.

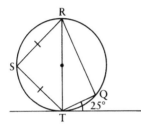

Find angles: **a** QTR **b** TRQ **c** SRQ.

9 Find: **a** ∠ APB **b** ∠ AOB **c** ∠ ACB

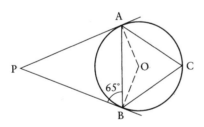

10 O is the centre of the circle.
∠ AOB = 108°, ∠ ACO = 33°

Find angle OBC.

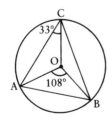

11 AB is a diameter of the circle.

Find:

a angle DCB

b angle DAB

c angle AXD.

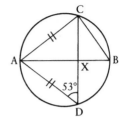

12 *BDEF* is a cyclic quadrilateral.
EFA is a straight line.
CBA is a tangent to the circle.

Angle *EDB* = 63° and angle *FBA* = 38°.

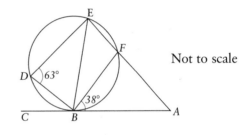

Not to scale

a Calculate the size of angle *BAF* showing
clearly how you found your answer.

b Calculate the size of angle *EBF* showing
clearly how you found your answer.

(SEG S96)

38 Congruence and Similarity

38.1   Congruent triangles

Two triangles are **congruent** if they are identical i.e. the second triangle is a translation, reflection or rotation of the first.

This means that:

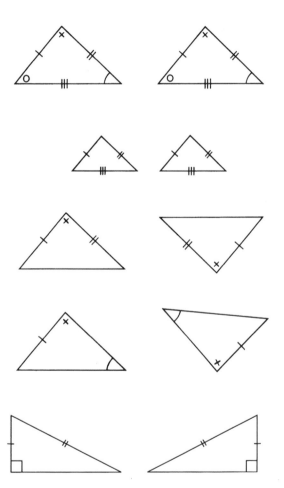

(i) the angles in the first triangle are equal to the corresponding angles in the second triangle

(ii) the lengths of the sides in the first triangle are equal to the corresponding lengths of the sides in the second triangle.

*Triangles are known to be congruent if:

(i) **three sides** in the first triangle are equal to three sides in the second triangle

(ii) **two sides and the angle between them** in the first triangle are equal to two sides and the angle between them in the second triangle (the only exception is a pair of right-angled triangles)

(iii) **two angles and one side** in the first triangle are equal to two angles and the corresponding side in the second triangle

(iv) **right angle, hypotenuse and side** in the first triangle are equal to the right angle, hypotenuse and the corresponding side in the second triangle.

Example

Name a pair of triangles in the following diagram which are definitely congruent.

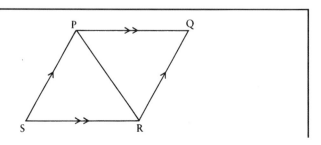

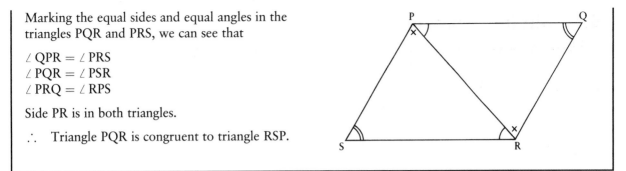

Marking the equal sides and equal angles in the triangles PQR and PRS, we can see that

∠ QPR = ∠ PRS
∠ PQR = ∠ PSR
∠ PRQ = ∠ RPS

Side PR is in both triangles.

∴   Triangle PQR is congruent to triangle RSP.

Exercise 38A

In each of the following questions, where possible, name a pair of triangles which are definitely congruent:

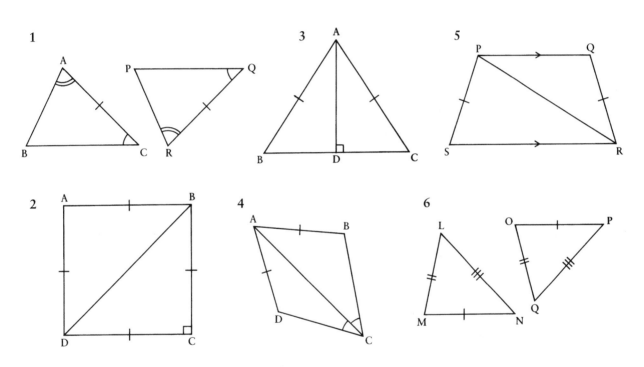

7   Two of the shapes below are congruent. Write down the letters of the **two** congruent shapes.

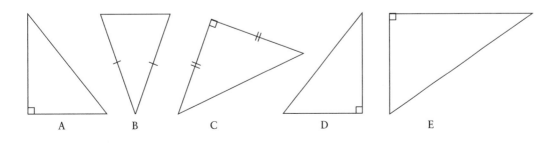

(SEG S94)

38.2   Similar triangles

Two triangles are **similar** if the angles in the first triangle are the same size as the angles in the second triangle, i.e. the second triangle is an enlargement of the first.

If two triangles are similar, they have the same shape but are of a different size, i.e. their corresponding sides are in the same ratio. This ratio is the factor of enlargement (see Unit 42).

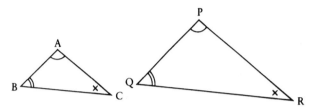

$\angle A = \angle P, \quad \angle B = \angle Q, \quad \angle C = \angle R$

∴   Triangles ABC and PQR are similar

and
$$\frac{AB}{PQ} = \frac{BC}{QR} = \frac{CA}{RP}$$

If AB = 2 cm and PQ = 5 cm,

$$\frac{PQ}{AB} = \frac{5}{2}$$

i.e. triangle PQR is an enlargement of triangle ABC and the scale factor of the enlargement is $\frac{5}{2}$ or 2.5.

Example

Triangles PQR and XYZ are similar.
QR = 2.5 cm, PR = 2.0 cm, PQ = 1.5 cm.
YZ = 1.5 cm

Find:

a the ratio between the sides of triangles PQR and XYZ

b the length of XZ

c the length of XY.

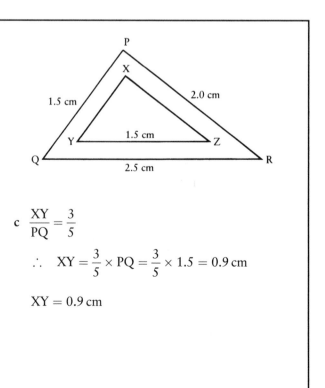

a $\dfrac{QR}{YZ} = \dfrac{2.5}{1.5} = \dfrac{5}{3}$

   ∴   The ratio is 5:3

b $\dfrac{XZ}{PR} = \dfrac{3}{5}$

   ∴ $XZ = \dfrac{3}{5} \times PR = \dfrac{3}{5} \times 2.0 = 1.2\,\text{cm}$

   XZ = 1.2 cm

c $\dfrac{XY}{PQ} = \dfrac{3}{5}$

   ∴ $XY = \dfrac{3}{5} \times PQ = \dfrac{3}{5} \times 1.5 = 0.9\,\text{cm}$

   XY = 0.9 cm

Exercise 38B

1 Triangles ABC and PQR are equiangular.

   Find PQ and PR.

2 Triangles LMN and OPQ are similar: $\dfrac{OP}{LM} = \dfrac{3}{2}$

   Find the lengths of the sides of triangle OPQ.

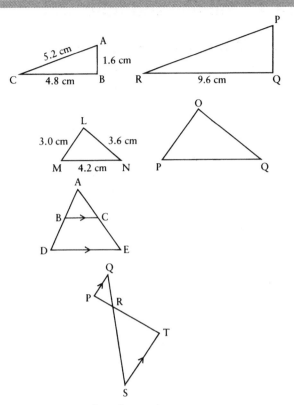

3 **a** AC = 2 cm, CE = 4 cm
      By what factor has △ ABC been enlarged to
      produce △ ADE?

   b If BC = 2.7 cm, what is the length of DE?

4 $\dfrac{PR}{RT} = \dfrac{2}{5}$

   PQ = 2.0 cm, QR = 2.5 cm, PR = 1.8 cm

   Find the lengths of:

   a RT   **b** TS   **c** SR

5 Pentagons *ABCDE* and *PQRST* are similar. All measurements are given in centimetres.

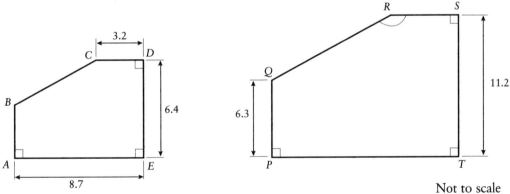

Not to scale

   a Calculate the length of *AB*.

   b Calculate the size of angle *QRS*.   (SEG W95)

6 State whether or not the triangles *ABC* and *XYZ* below are similar.
   Show working to support your answer.

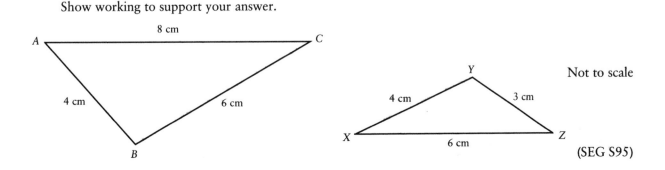

Not to scale

                                                               (SEG S95)

38.3   Models and map scales

Models and scale diagrams

A model is a replica, usually much smaller in size, of an original object. If the model is to be a faithful representation of the original, all the corresponding measurements of the original and the model must be in the same proportion.

The **scale** is the proportion by which each measurement has been reduced.

In order to fit a diagram on to paper, the measurements have to be 'scaled down', i.e. all reduced in the same proportion.
Once this has been done, you have a 'scale diagram'.

A scale is expressed either as a comparison between 2 lengths,
e.g. $1 \, \text{cm} : 2 \, \text{m}$
or as a ratio
e.g. $1 : 200$.

For example, a scale of 2 centimetres to 1 metre means that every length of 1 metre on the original is represented by a length of 2 centimetres on the model, i.e. the measurements are all in the ratio of $1 : 50$ and all the measurements on the model are $\frac{1}{50}$ of those of the original.

Example 1

The scale of a diagram is $1 : 5$. AB in the scaled diagram is $4 \, \text{cm}$. Find the original length AB.

$$\text{Original length} = 5 \times 4 \, \text{cm}$$
$$= 20 \, \text{cm}$$

Example 2

A model of an aircraft is made to a scale of $1 \, \text{cm} : 1.8 \, \text{m}$.

a   The length of the aircraft is $43 \, \text{m}$. What is the length of the model?

b   The width of the cargo hold on the model is $4.2 \, \text{cm}$.
What is the width of the cargo hold on the aircraft?

a $1.8 \, \text{m}$ is represented by $1 \, \text{cm}$

$1 \, \text{m}$ is represented by $\dfrac{1}{1.8} \, \text{cm}$

$43 \, \text{m}$ is represented by $\dfrac{1}{1.8} \times 43 \, \text{cm}$

$$= 23.9 \, \text{cm}$$

b $1 \, \text{cm}$ represents $1.8 \, \text{m}$
$4.2 \, \text{cm}$ represents $1.8 \times 4.2 \, \text{m}$
$$= 7.56 \, \text{m}$$

Example 3

The wing-span of a Boeing 747 is 59.6 metres. A model of the aircraft has a wing-span of 29.8 centimetres. To what scale was the model made?

$$29.8 \text{ cm represent } 59.6 \text{ m}$$

$$1 \text{ cm represents } \frac{59.6}{29.8} \text{ m} = 2 \text{ m} = 200 \text{ cm}$$

The scale is '1 cm represents 200 cm' or $1:200$

Provided both lengths have the same units,

$$\text{Scale} = \frac{\text{Original length}}{\text{Scaled length}}$$

Exercise 38C

1 The diagram below shows a rough plan (not drawn to scale) for a garden design.
The designer then draws an accurate scale diagram using a scale of $1:125$.

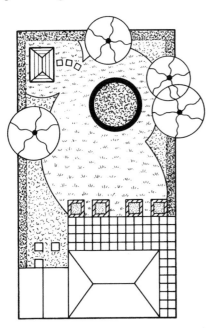

a What should the dimensions of the scale diagram be if the actual garden is 9.5 m wide by 20 m long?

b A patio is to be constructed which will measure 800 cm by 500 cm.
What dimensions should the patio have on the plan?

c The circular flower bed will have a diameter of 4 m.
What radius should the designer use to draw it on the plan?

2 A doll's house is made to a scale of 1 centimetre representing 30 centimetres.

a What are the lengths, in the doll's house, of the following:
  (i) a 2.1 m high door
  (ii) a carpet measuring 4.2 m × 3.6 m
  (iii) a plate of diameter 24 cm?

b What would be the full-scale sizes of the following doll's house furnishings:
  (i) a chair of height 3.1 cm
  (ii) a candlestick 4.5 mm high
  (iii) a picture 23 mm × 20 mm?

3 **a** When Alice ate a cake marked 'Eat me' she found herself enlarged on a scale of 1 to 1.9. If she was previously 4ft 10in tall, how tall was she after eating the cake?

b When Alice drank the contents of the bottle labelled 'Drink me', she shrank to a height of 10 inches.
If the miniature Alice had feet which were 3.6 cm long, how long were the feet on the full-sized Alice?

4 An 'N' gauge model railway is made to a scale of $\frac{1}{150}$. The roof of a model station is 136 mm long.

a How long would the actual station roof be?

b Taking 30 cm = 1 foot, calculate the actual length of the roof in feet.

5 Model railway scales are often quoted as the number of millimetres which represent 1 foot. The 00 gauge is a 4 mm scale, which means that 4 mm represents 1 foot.

a The wheel on a train has a diameter of 4ft 3in. How many millimetres will the diameter of the wheel of a 00 gauge train be?

b A 2 mm scale railway has passengers of height 11.5 mm. What is the actual height represented by these models:

(i) in feet and inches

(ii) in centimetres?

Map scales

Maps are two-dimensional representations of geographical areas. Map scales are usually given as a ratio or as a scaled line.

For example, a scale of $1:50\,000$ means that 1 cm on the map represents $50\,000$ cm actual distance, or 1 inch on the map represents $50\,000$ inches on the ground.

or   Scale: 0.5   0   0.5   1.0   1.5 kilometres

Example 1

The scale of a map is $1:50\,000$.

a What distance on the ground is represented by 6.3 cm on the map?

b What distance on the map represents 23 km on the ground?

a 1 cm represents $50\,000$ cm

$$6.3 \text{ cm represents } 50\,000 \times 6.3 \text{ cm} = \frac{315\,000}{100} \text{ m} = \frac{3150}{1000} \text{ km}$$

$$= 3.15 \text{ km}$$

Actual distance = Map scale × Length on map

b $50\,000 \text{ cm} = \dfrac{50\,000}{100} \text{ m} = 500 \text{ m} = \dfrac{500}{1000} \text{ km} = 0.5 \text{ km}$

0.5 km is represented by 1 cm
1 km is represented by 2 cm
23 km is represented by 2×23 cm = 46 cm

or   **Length on map** $= \dfrac{\textbf{Actual length}}{\textbf{Map scale}}$

$$= \frac{23 \times 1000 \times 100}{50\,000} \text{ cm}$$

$$= 46 \text{ cm}$$

Example 2

The side of a field on the ground is 400 m. The size on the map is 5 cm. Find the scale of the map.

$$\text{Scale} = \frac{\text{Original length}}{\text{Scaled length}}$$

Converting all units to centimetres gives:

$$\text{Scale} = \frac{400 \times 100}{5}$$

$$= 8000$$

The scale of the map is $1:8000$.

Exercise 38D

1 Convert the following actual lengths to map lengths using the scales given:

Actual length	Scale
a 5 metres	1:500
b 120 metres	1:2500
c 11 kilometres	1:10 000
d 14.4 kilometres	1:50 000

2 Convert the following map lengths to the original lengths using the scales given:

Map length	Scale
a 2 centimetres	1:1000
b 3.1 centimetres	1:4000
c 2.8 centimetres	1:6000
d 3 inches	1:3000
e 4.5 inches	1:2000

3 Calculate the scale of each of the following:

a a street map if 500 m is represented by 5 cm

b a small-scale road atlas if 16 km is represented by 32 cm

c an Ordnance Survey map if 3.2 km is represented by 12.8 cm

d a town plan if 5.6 cm represents 700 m

e a large-scale road atlas if 30 mm represents 30 km.

4 An Ordnance Survey map has a scale of 1:25 000.

a How many metres, on the ground, are represented by 1 cm on the map?

b What is the distance on the ground if the distance on the map is:

(i) 7.4 cm   (ii) 17.6 cm   (iii) 20 mm
(iv) 6.2 cm?

c What is the distance on the map if the distance on the ground is:

(i) 5 km   (ii) 1.8 km   (iii) 4.75 km
(iv) 10.6 km?

5 A road map has a scale of 1:50 000.

a How many centimetres on the map represent 1 kilometre on the ground?

b What is the distance on the ground if the distance on the map is:

(i) 4 cm   (ii) 48.8 cm   (iii) 16.3 cm
(iv) 36.9 cm?

c What is the distance on the map if the distance on the ground is:

(i) 27 km   (ii) 10.6 km   (iii) 58 m
(iv) 59.4 km?

6 On a map a reservoir covers an area of 0.6 cm². The scale of the map is 1:250 000.

a What distance, on the ground, is represented by 1 cm?

b What area, on the ground, is represented by 1 cm²?

c What is the actual surface area of the reservoir?

d A forest covers an area of 25 km². What area on the map does the forest cover?

7 The map below shows the centre of York.
(Scale: 1 cm represents 60 m)

a Mrs Hogan arrives at the station for a meeting at the careers office in Piccadilly:

   (i) Find the shortest route from the station to Piccadilly.

   (ii) What distance (to the nearest 50 m) will Mrs Hogan have to walk?

b By using a suitable measuring device, find the distance between the Castle and York Minster (to the nearest 50 m):

   (i) as the crow flies

   (ii) by road.

c Plan a guided walk around York City Centre starting and finishing at York Railway Station and taking in all the main places of interest.

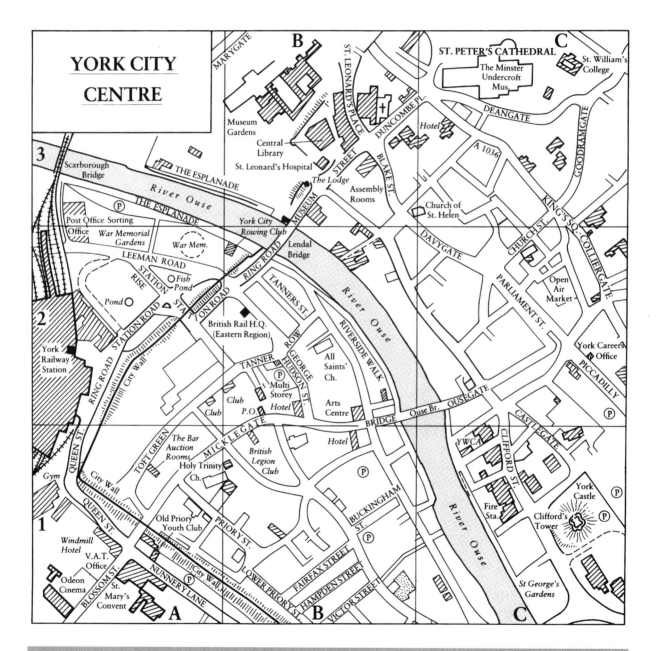

39 *Bearings*

An instrument for measuring the direction, on the ground, of one location from another, is called a **compass**.

The compass direction is called a **bearing**.

A compass rose, often shown on maps, gives the bearings in terms of the four cardinal points, north, south, east and west.

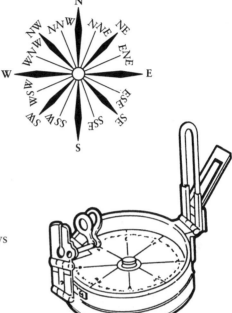

A mariner's compass, or a surveyor's prismatic compass, gives a bearing as an angle measured from north in a clockwise direction.

This is the most usual method of giving a bearing, which is always written as a three-digit number. For example:

SW is 225°
E is 090°
N is 000°

To find the bearing of a point B from a point A,

1 Join A and B.
2 Draw in the **north** line at A.
3 Find the angle between north and AB, measured **clockwise**.
4 Record the angle, in degrees, as a three-digit figure.

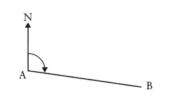

Example

In the diagram, the bearing of A from B is 250°.

Find the bearing of B from A.

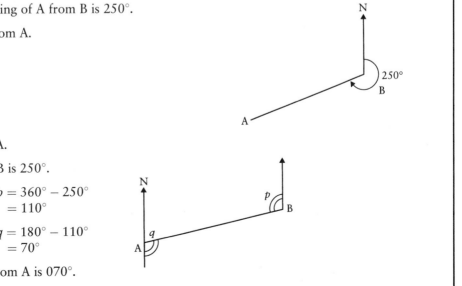

Draw the north line at A.

The bearing of A from B is 250°.

$$\therefore \quad \text{Angle } p = 360° - 250°$$
$$= 110°$$

$$\therefore \quad \text{Angle } q = 180° - 110°$$
$$= 70°$$

$\therefore$ The bearing of B from A is 070°.

Exercise 39A

1   Write the following directions as three-digit bearings:

a East

b SE

c NW

d N 30° W

e S 15° W

2   In each of the following, find the bearing of B from A.

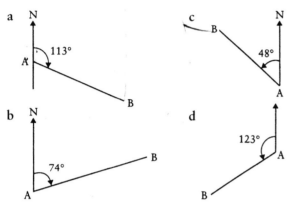

a 113°

b 74°

c 48°

d 123°

3   The bearing of Q from P is 192°. What is the bearing of P from Q?

4   The bearing of R from S is 061°. What is the bearing of S from R?

5   The bearing of D from C is 324°. What is the bearing of C from D?

6   Paris is 200 km due north of Bourges and 145 km due west of Chalons. What is the bearing of Bourges from Chalons?

PARIS

Chalons-s-Maine

Bourges

7   New Orleans is 360 miles due south of Memphis and Jacksonville is 500 miles due east of New Orleans.

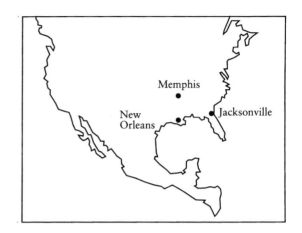

Memphis

New Orleans

Jacksonville

a On what bearing will a plane fly when travelling from Memphis to Jacksonville?

b On what bearing will a plane fly when travelling from Jacksonville to Memphis?

8   A ship leaves harbour, H, on a bearing of 310°. After sailing 3 miles, the ship is due north of a coastal town, T, which is due west of the harbour.
How far from the harbour is the town?

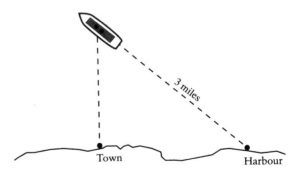

3 miles

Town

Harbour

9   A map of Pentland Firth is shown.
     A ferry is at the point marked **F**.

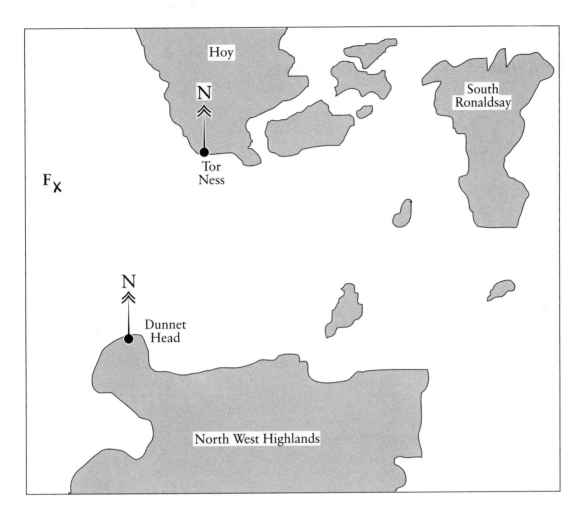

a Write down the bearing and distance in centimetres of the ferry from

   (i) Tor Ness,

   (ii) Dunnet Head.

b The scale of the map is 1 cm = 25 km.

   The ferry sails 250 kilometres on a bearing of 100° from **F**.

   Draw the ferry's path accurately on the map.   (SEG S95)

40 Geometrical Constructions

40.1 The use of drawing instruments

Equipment

You need:

a **ruler** or **rule** for measuring lengths, both in centimetres and inches

a **pencil** with a sharp point

a **protractor** for measuring angles (either circular or semicircular)

a **pair of compasses** for drawing arcs of circles

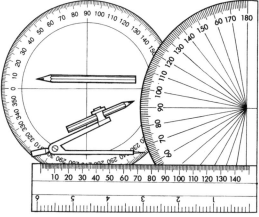

40.2 Construction of triangles

To construct a triangle, you need to know three facts about it, one of which must be the length of one side.

You need to know:
either   (i)   the lengths of all three sides
or   (ii)   the lengths of two sides, and the angle between them
or   (iii)   the length of one side, and the size of two angles.

It is conventional to represent the length of the side of a triangle by the lower case letter of the angle opposite to that side.

In triangle ABC on the right the lengths of the sides opposite to angles A, B and C are represented by a, b and c respectively.

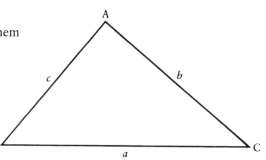

Example 1

Construct a triangle ABC given the three lengths:

$a = 7 \text{ cm}, \quad b = 9 \text{ cm}, \quad c = 5 \text{ cm}.$

Select the largest side to draw first.

Draw AC = 9 cm.

With centre A and radius 5 cm draw an arc.

Since AB = 5 cm, B must lie somewhere on this arc.
With centre C and radius 7 cm draw an arc on which B also must lie, since CB = 7 cm.

Therefore this arc cuts the first arc at B.

Join AB and BC.

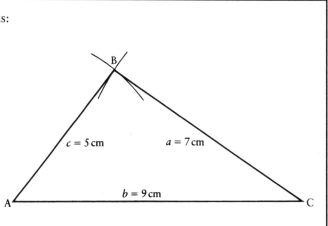

Example 2

Construct a triangle ABC given $a = 7$ cm, $b = 9$ cm, $\angle C = 70°$.

Once again, select the largest side first.
Draw AC = 9 cm.

With your protractor, centre at C, mark a point
at 70° with AC.

This point is on BC.

Draw BC = 7 cm.

Join AB.

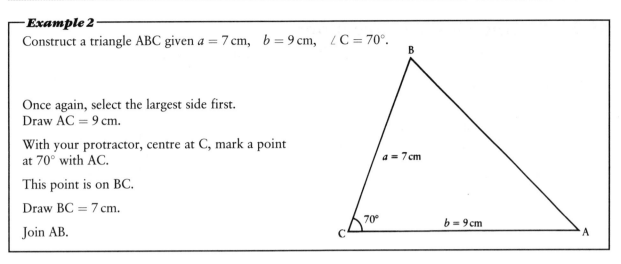

Example 3

Construct triangle ABC given $a = 7$ cm, $\angle B = 80°$, $\angle A = 48°$.

Draw BC = 7 cm.

With your protractor, centre at B, mark a point at 80° with BC.
This point is on AB.

Join B to this point and extend the line as appropriate.

($\angle C$ is required but was not given. However

$$\angle A + \angle B + \angle C = 180°$$
$$\therefore \quad \angle C = 52°.)$$

Repeat at C with 52°.

Where the two lines meet is point A.

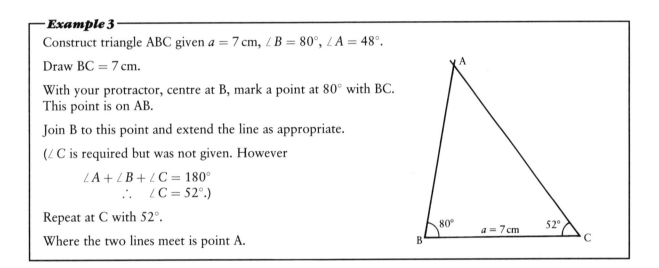

When you construct a diagram, you should try to ensure that your accuracy
is within 1 mm (for length) and 1° (for angles).

Exercise 40A

Draw the following triangles accurately and make the specified measurement.

1 AB = 6 cm   Angle A = 50°   Angle B = 70°   Measure AC

2 PQ = 4.3 cm   Angle P = 70°   Angle Q = 48°   Measure PR

3 PQ = 2.5 in   QR = 3.7 in   RP = 2.8 in   Measure angle P

4 XY = 6.4 cm   YZ = 7.1 cm   ZX = 4.9 cm   Measure angle Z

5 PQ = 4.9 cm   QR = 3.1 cm   Angle Q = 71°   Measure PR

40.3   Geometrical constructions

The following constructions require only the use of a ruler and a pair of compasses.

(i) To bisect the line AB

With A as centre, draw arcs above and below AB.
(The radius must be more than $\frac{1}{2}$AB.)
With B as centre, draw arcs to intersect the first arcs.
Join the points of intersection.
This line bisects AB.
(It is also perpendicular to AB.)

(ii) To bisect angle BAC
With A as centre, draw arcs to cut AB and AC.
With these two points as centres draw arcs to
intersect each other.
Join A to the point of intersection.
The line bisects ∠ BAC.

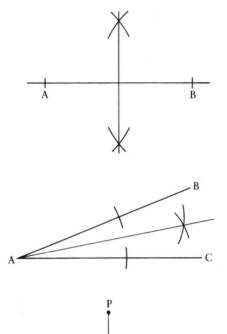

(iii) To draw a perpendicular from point P to line AB

With P as centre, draw an arc to cut AB in two
points.
With these points as centres, draw arcs to
intersect each other above, or below AB.
Join P to the point of intersection, producing the
line if necessary, to cut AB.
This line is perpendicular to AB.

Exercise 40B

1 Construct an equilateral triangle of side 5 cm,
using ruler and compass only.

2 Triangle ABC is isosceles with BC = 6.4 cm,
AB = AC = 5.2 cm.
Construct the perpendicular bisector of the
base BC.
Measure:

a angle ABC
b the altitude of triangle ABC.

(The **altitude** of the triangle is its height, i.e. the
perpendicular line from A to BC.)

3 Draw any triangle. Bisect the three angles of the
triangle to show that the angle bisectors are
concurrent (all meet in the same point).

4 Show, by construction, that the three altitudes
of a triangle are concurrent.

5 The position of a boat entering a harbour
protected by two piers with lighthouses is
shown in the diagram.

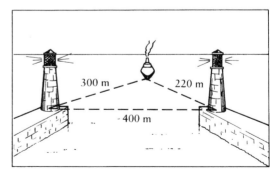

Construct the triangle formed by the boat and
the two lighthouses.

By constructing the perpendicular from the
position of the boat to the line between the
lighthouses, find the closest distance the boat
will sail to a pier.

6   A helicopter is on a mercy flight with food supplies to an isolated farm which has been cut off by snow drifts.

It flies a horizontal path at an altitude of 650 m above the ground.

Parcels dropped by the helicopter will be carried by the wind in a direction which is 30° to the vertical.

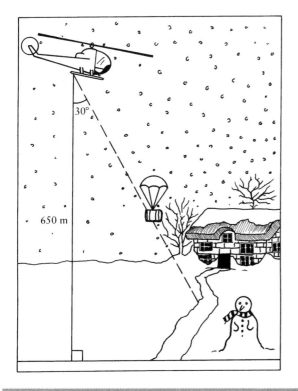

Using ruler and compasses only, construct a diagram to show the position of the helicopter, relative to the dropping point, when the parcels are released.

What is the horizontal distance of the helicopter from the dropping point?

7   A bridge, 200 metres long, spans a river. The bridge is suspended by steel cables from arches which are arcs of a circle, as shown in the diagram.

Using ruler and compasses only, find the height of an arch at its centre.

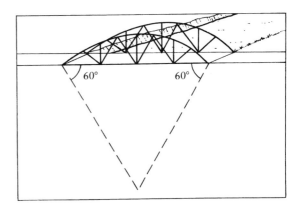

40.4   Locus

The word **locus** (plural is **loci**) means the 'path' an object traces out when it moves according to a given rule.

The 'path' may be a straight line, a curve or a region.

Example 1

The donkey in the picture below is always the same distance from the centre of the well.
What is its locus?

The locus of the donkey is a circle.

Example 2

The goat is tied to a post by a length of rope.
What is the locus of the goat?

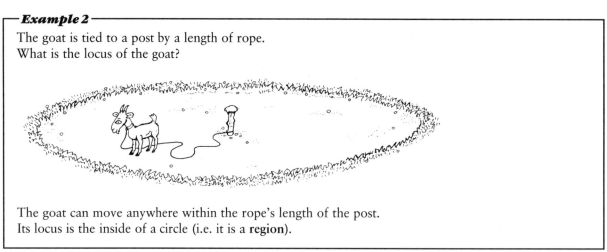

The goat can move anywhere within the rope's length of the post.
Its locus is the inside of a circle (i.e. it is a **region**).

Example 3

What is the locus of points which are always an equal distance from each
of two intersecting lines?

The locus is the bisector of the angle formed by the lines.

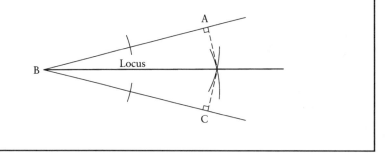

Exercise 40C

1 A boat sails between two buoys so that it is always an equal distance from each buoy.
Describe the locus of the boat.

2 What is the locus of P if:

   a P is always the same distance from a fixed line,

   b ∠APB is always 90°, where A and B are fixed points.

3 Sketch the loci of each of the points A, B, C
marked on the bicycle wheel.
A is at the centre of the hub, B is a point on a
spoke and C is a point on the rim.

   (These loci may be found by cutting out a circle of
cardboard and piercing holes to correspond with
A, B and C. As the circle is rolled along a straight
line, mark points at intervals and then join them
to show the loci.)

4 The diagram shows a simple pulley system for raising a beam (represented by the line ABC) from a horizontal position to a vertical one.
Describe the loci of points A, B and C (where B is the mid-point of AC) and draw the locus of B.

5 A dog (D) is fastened by a rope 10 m long to the corner (A) of a shed. AB is a wall which is more than 10 m long. Draw and shade the locus of the dog.

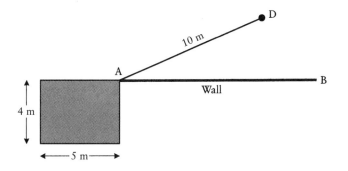

6 The diagram below shows the plan of a house and garden.
The house (rectangle ABCD) is 12 m by 15 m and the garden (rectangle GHCK) is 21 m by 45 m.
The back door of the house is in the middle of side DA and there is a garden gate centred at G.

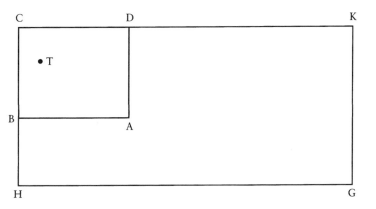

a A garden path is laid from the back door (E), initially at right angles to the house. The path then turns at a point F and finishes at gate G. FG is a straight line which is at an equal distance from fences GK and GH.
Copy the plan and construct the path EFG.

b Measure the angle EFG.

c A mobile telephone is sited at point T in the house, 3 m from BC and 4.5 m from CD. If the telephone is taken into the garden, it can be used up to a distance of 30 m from T. If someone is working in the garden, shade the area between the path and the fence GK in which the telephone cannot be heard.

7   The diagram shows a port (P)
    and two lighthouses (L and M).

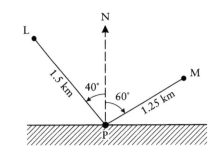

a   The captain of a ship sailing from P knows that there is always deeper
    water nearer to PL than to PM.
    Draw a diagram and shade the region of shallower water.

b   The captain also knows that there are rocks within an area of 660 m
    radius from L.
    Draw a line showing the direction in which the ship would sail if it
    just missed the rocks.

c   The ship sails in a straight line from P so that it remains in deeper
    water and misses the rocks. What is the angle between the limits of
    the ship's route?

41 Vectors

41.1   Vector representation

Some quantities in mathematics need to be specified by both their magnitude and direction. For example:

- The wind speed is 50 mph from the NE.
- Everyone take two paces forward.
- The lift accelerated downwards at 2 m/s^2.

These quantities are called **vectors**.

Velocity, displacement and acceleration are vectors. Force is also a vector.

Many quantities are completely specified by their magnitude. For example:

- I have been waiting for 40 minutes.
- The temperature today is 24°C.
- The car was travelling at 70 mph.

These quantities are called **scalars**.

Time, temperature, speed, mass and area are all scalar quantities.

A vector can be represented in several ways:

(i)   Geometrically

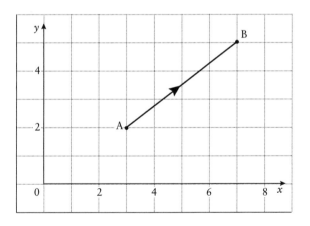

The vector is represented by the line from point A (3, 2) to point B (7, 5).

Its **magnitude** is the length of the line AB and its **direction** is the angle it makes with the x-axis.

(ii) As a column vector

The displacement from A at (3, 2) to B at (7, 5) is 4 along and 3 up.

To distinguish the vector from the coordinates of a point, it is written as $\begin{pmatrix} 4 \\ 3 \end{pmatrix}$ with the x-value on top of the y-value.

Because the vector goes from A to B, we write the vector as $\overrightarrow{AB} = \begin{pmatrix} 4 \\ 3 \end{pmatrix}$

The vector from B to A is $\overrightarrow{BA} = \begin{pmatrix} -4 \\ -3 \end{pmatrix}$.

The vector $\overrightarrow{AB}$ (or $\overrightarrow{BA}$) is called a **displacement vector**.

(iii) In unit base vectors

A **unit vector** is a vector of magnitude 1.

The unit vectors $\begin{pmatrix} 1 \\ 0 \end{pmatrix}$ and $\begin{pmatrix} 0 \\ 1 \end{pmatrix}$ are important because their directions are along the x-axis and y-axis respectively.

These vectors $\begin{pmatrix} 1 \\ 0 \end{pmatrix}$ and $\begin{pmatrix} 0 \\ 1 \end{pmatrix}$ are called the **unit base vectors**.

$\begin{pmatrix} 1 \\ 0 \end{pmatrix}$ is given the letter i and $\begin{pmatrix} 0 \\ 1 \end{pmatrix}$ is given the letter j.

i and j are always printed in bold type.
When they are written, they should be underlined: $\underline{i}$, $\underline{j}$.

All two-dimensional vectors can be written in terms of i and j.

For example, the vector $\begin{pmatrix} 4 \\ 3 \end{pmatrix}$, which is a displacement of 4 in the x-direction and 3 in the y-direction, is written as:

$$4i + 3j$$

Example

Find the displacement vector $\overrightarrow{PQ}$ which joins P(3, −2) and Q(7, −4) and write each vector (i) as a column vector, (ii) in terms of the unit base vectors i and j.

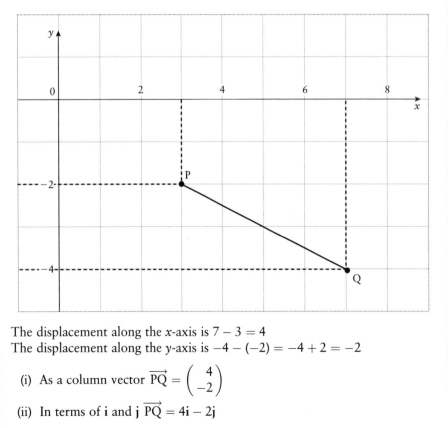

The displacement along the x-axis is $7 - 3 = 4$
The displacement along the y-axis is $-4 - (-2) = -4 + 2 = -2$

(i) As a column vector $\overrightarrow{PQ} = \begin{pmatrix} 4 \\ -2 \end{pmatrix}$

(ii) In terms of i and j $\overrightarrow{PQ} = 4i - 2j$

Exercise 41A

1 Find the displacement vectors in the following diagrams and write each vector (i) as a column vector, (ii) in terms of the base vectors **i** and **j**.

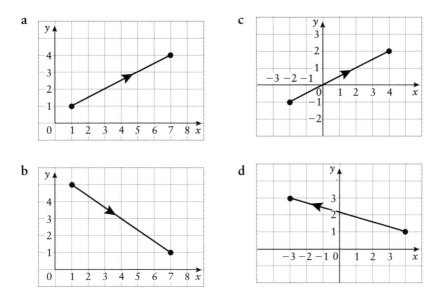

2 Plot the points A(1, 1), B(1.5, 3), C(6, 5) and D(8,1). Write, as column vectors:

a $\overrightarrow{AB}$ b $\overrightarrow{AD}$ c $\overrightarrow{CB}$ d $\overrightarrow{AC}$ e $\overrightarrow{DB}$

3 Find the displacement vector $\overrightarrow{PQ}$ which joins each of the following points, giving your answer (i) as a column vector, (ii) in terms of the vectors **i** and **j**.

a P(3, 5) Q(7, 6)   c P(−5, 2) Q(4, 1)   e P(6, −4) Q(−1, 3)

b P(0, 6) Q(4, 2)   d P(6, 3) Q(−4, −2)   f P(−2, −7) Q(−5, 0)

*41.2 The modulus of a vector

The **modulus** of a vector is another term for the magnitude.

The modulus of $\overrightarrow{AB}$ is written as AB or $|\overrightarrow{AB}|$.

To calculate the modulus, we use Pythagoras' theorem:

To find the length of AB:

$$AB^2 = 4^2 + 3^2$$
$$= 25$$
$$AB = 5 \text{ or } |\overrightarrow{AB}| = 5$$

If $\overrightarrow{AB} = \begin{pmatrix} x \\ y \end{pmatrix} = x\mathbf{i} + y\mathbf{j}$, then $|\overrightarrow{AB}| = \sqrt{(x^2 + y^2)}$

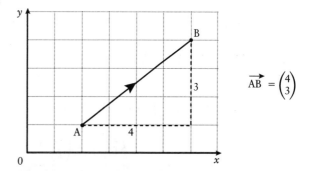

$\overrightarrow{AB} = \begin{pmatrix} 4 \\ 3 \end{pmatrix}$

Exercise 41B

1 Find the modulus of AB in each of the following:

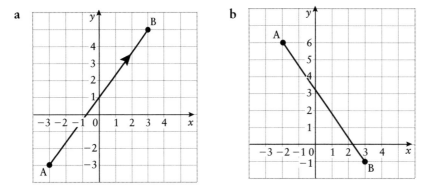

a   b

2 Calculate the modulus of the vectors:

a $\begin{pmatrix} 12 \\ 5 \end{pmatrix}$ b $\begin{pmatrix} -8 \\ 15 \end{pmatrix}$ c $3\mathbf{i} - 4\mathbf{j}$

3 Calculate the modulus of:

a the vector joining the points $(1, 3)$ and $(5, 5)$

b the vector joining the points $(-1, 3)$ and $(5, -5)$.

*41.3 Combining vectors

Multiplication by a scalar

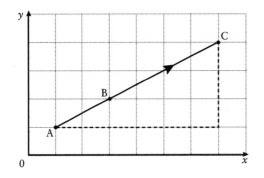

In the diagram above, the vector $\overrightarrow{AC}$ is three times longer than vector $\overrightarrow{AB}$.

$$\overrightarrow{AC} = 3\ \overrightarrow{AB} = 3\begin{pmatrix} 2 \\ 1 \end{pmatrix} = \begin{pmatrix} 6 \\ 3 \end{pmatrix}$$

Multiplying a vector $\begin{pmatrix} x \\ y \end{pmatrix}$ by a scalar k gives

$$k\begin{pmatrix} x \\ y \end{pmatrix} = \begin{pmatrix} kx \\ ky \end{pmatrix}$$

In the example above, both $\overrightarrow{AB}$ and $\overrightarrow{AC}$ can be written in terms of

the vector $\begin{pmatrix} 2 \\ 1 \end{pmatrix}$.

This is because, although the **magnitudes** of $\overrightarrow{AB}$ and $\overrightarrow{AC}$ are different, their **directions** are the same.

Example

Points P, Q, R and S have coordinates $(-1, 1)$, $(2, 3)$, $(4, 0)$ and $(1, -2)$ respectively.

a Find $\overrightarrow{QR}$, $\overrightarrow{QS}$, $\overrightarrow{PR}$ and $\overrightarrow{SP}$.

b State which pairs of vectors are equal in magnitude, parallel or both.

a $\overrightarrow{QR} = \begin{pmatrix} 4 - 2 \\ 0 - 3 \end{pmatrix} = \begin{pmatrix} 2 \\ -3 \end{pmatrix}$

$\overrightarrow{QS} = \begin{pmatrix} 1 - 2 \\ -2 - 3 \end{pmatrix} = \begin{pmatrix} -1 \\ -5 \end{pmatrix}$

$\overrightarrow{PR} = \begin{pmatrix} 4 - (-1) \\ 0 - 1 \end{pmatrix} = \begin{pmatrix} 5 \\ -1 \end{pmatrix}$

$\overrightarrow{SP} = \begin{pmatrix} -1 - 1 \\ 1 - (-2) \end{pmatrix} = \begin{pmatrix} -2 \\ 3 \end{pmatrix}$

b $\overrightarrow{QR} = -\overrightarrow{SP}$

∴ QR and PS are both equal in magnitude and parallel.

$|\overrightarrow{QS}| = |\overrightarrow{PR}|$

∴ QS and PR are equal in magnitude.

Exercise 41C

1 If $\mathbf{a} = \begin{pmatrix} 5 \\ -12 \end{pmatrix}$ calculate:

(i) $4\mathbf{a}$ (ii) $\frac{1}{2}\mathbf{a}$ (iii) $-2\mathbf{a}$ (iv) $-\mathbf{a}$ (v) $|\mathbf{a}|$ (vi) $|3\mathbf{a}|$

2 State which of the following pairs of vectors are parallel:

a $\begin{pmatrix} 3 \\ 6 \end{pmatrix}$ and $\begin{pmatrix} 1 \\ 3 \end{pmatrix}$

b $\begin{pmatrix} -2 \\ 8 \end{pmatrix}$ and $\begin{pmatrix} 1 \\ -4 \end{pmatrix}$

c $5\mathbf{i} + 10\mathbf{j}$ and $\frac{1}{4}\mathbf{i} + \frac{1}{2}\mathbf{j}$ **d** $4\mathbf{i} + -7\mathbf{j}$ and $2\mathbf{i} - 3\mathbf{j}$

e $\overrightarrow{PQ}$ and $\overrightarrow{RS}$ if P, Q, R and S have coordinates $(2, 5)$, $(4, 9)$, $(-1, 3)$ and $(-2, 1)$ respectively.

3   Calculate the displacement vectors $\overrightarrow{AB}$ and $\overrightarrow{CD}$ in each of the following and state whether AB and CD are equal, parallel or both.

a A(1, 3)   B(3, 7)   C(1, −1)   D(4, 5)

b A(2, 0)   B(6, 2)   C(4, 1)   D(8, −1)

c A($\frac{1}{2}$, 2)   B(2, 5)   C(−2, −7)   D($-\frac{1}{2}$, −4)

d A(−1, $3\frac{1}{2}$)   B(−2, 4)   C(0, −1)   D(3, $-2\frac{1}{2}$)

e A(3, −2)   B(0, 4)   C(3, 7)   D(6, 13)

Addition and subtraction

If we move from point A to point B and then from point B to point C, $\overrightarrow{AB}$ followed by $\overrightarrow{BC}$ produces the vector $\overrightarrow{AC}$, i.e.

$$\overrightarrow{AC} = \overrightarrow{AB} + \overrightarrow{BC}$$

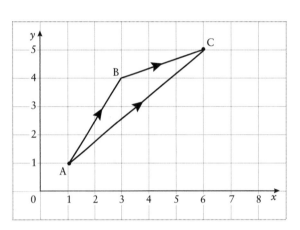

Addition

In the diagram above, $\overrightarrow{AB} = \begin{pmatrix} 2 \\ 3 \end{pmatrix}$ and $\overrightarrow{BC} = \begin{pmatrix} 3 \\ 1 \end{pmatrix}$

The displacement from A to C is 5 along and 4 up, i.e.

$$\overrightarrow{AC} = \begin{pmatrix} 2 \\ 3 \end{pmatrix} + \begin{pmatrix} 3 \\ 1 \end{pmatrix} = \begin{pmatrix} 5 \\ 4 \end{pmatrix}$$

Vectors are added by adding the corresponding x- and y-coordinates.

$$\begin{pmatrix} x_1 \\ y_1 \end{pmatrix} + \begin{pmatrix} x_2 \\ y_2 \end{pmatrix} = \begin{pmatrix} x_1 + x_2 \\ y_1 + y_2 \end{pmatrix}$$

Subtraction

$$\overrightarrow{BC} = \begin{pmatrix} 3 \\ 1 \end{pmatrix}$$

$$\therefore \quad \overrightarrow{CB} = \begin{pmatrix} -3 \\ -1 \end{pmatrix}$$

$$\overrightarrow{AB} - \overrightarrow{BC} = \overrightarrow{AB} + (-\overrightarrow{BC})$$
$$= \overrightarrow{AB} + \overrightarrow{CB}$$

$$= \begin{pmatrix} 2 \\ 3 \end{pmatrix} + \begin{pmatrix} -3 \\ -1 \end{pmatrix}$$

$$= \begin{pmatrix} 2-3 \\ 3-1 \end{pmatrix}$$

$$= \begin{pmatrix} -1 \\ 2 \end{pmatrix}$$

Vectors are subtracted by subtracting the corresponding x- and y-coordinates.

$$\begin{pmatrix} x_1 \\ y_1 \end{pmatrix} - \begin{pmatrix} x_2 \\ y_2 \end{pmatrix} = \begin{pmatrix} x_1 - x_2 \\ y_1 - y_2 \end{pmatrix}$$

Example

If $\mathbf{a} = \begin{pmatrix} 7 \\ -4 \end{pmatrix}$ and $\mathbf{b} = \begin{pmatrix} -3 \\ 6 \end{pmatrix}$, calculate

  (i) $\mathbf{a} - \mathbf{b}$ (ii) $4\mathbf{b} - 3\mathbf{a}$ (iii) $|4\mathbf{b} - 3\mathbf{a}|$

  (i) $\mathbf{a} - \mathbf{b} = \begin{pmatrix} 7 \\ -4 \end{pmatrix} - \begin{pmatrix} -3 \\ 6 \end{pmatrix} = \begin{pmatrix} 7 - (-3) \\ -4 - 6 \end{pmatrix} = \begin{pmatrix} 10 \\ -10 \end{pmatrix}$

  (ii) $4\mathbf{b} - 3\mathbf{a} = 4\begin{pmatrix} -3 \\ 6 \end{pmatrix} - 3\begin{pmatrix} 7 \\ -4 \end{pmatrix} = \begin{pmatrix} -12 \\ 24 \end{pmatrix} - \begin{pmatrix} 21 \\ -12 \end{pmatrix} = \begin{pmatrix} -33 \\ 36 \end{pmatrix}$

  (iii) $|4\mathbf{b} - 3\mathbf{a}| = \sqrt{(33^2 + 36^2)} = \sqrt{(1089 + 1296)} = \sqrt{2385} = 48.8$

Exercise 41D

1   If $\mathbf{a} = \begin{pmatrix} 4 \\ 5 \end{pmatrix}$ $\mathbf{b} = \begin{pmatrix} 2 \\ -1 \end{pmatrix}$ and $\mathbf{c} = \begin{pmatrix} 0 \\ -\frac{1}{2} \end{pmatrix}$ calculate:

   (i) $\mathbf{a} + \mathbf{b}$ (iv) $3\mathbf{a} - 2\mathbf{b}$ (vii) $\mathbf{c} + \frac{1}{4}(\mathbf{b} - \mathbf{a})$
   (ii) $\mathbf{b} - \mathbf{a}$ (v) $\frac{1}{2}\mathbf{b} - \mathbf{c}$ (viii) $\mathbf{a} - 2(\mathbf{b} - \mathbf{c})$
   (iii) $2\mathbf{a} + \mathbf{c}$ (vi) $\frac{1}{2}(\mathbf{a} - \mathbf{b})$

2   If $\mathbf{a} = 2\mathbf{i} + \mathbf{j}$ and $\mathbf{b} = 7\mathbf{i} - 2\mathbf{j}$, calculate:
   (i) $\mathbf{a} - \mathbf{b}$ (iii) $\mathbf{b} + 2\mathbf{a}$ (v) $\frac{1}{2}(\mathbf{a} + \mathbf{b})$
   (ii) $|\mathbf{a} - \mathbf{b}|$ (iv) $|\mathbf{b} + 2\mathbf{a}|$ (vi) $|\frac{1}{2}(\mathbf{a} + \mathbf{b})|$

3   $\mathbf{a} = \begin{pmatrix} 4 \\ 2 \end{pmatrix}$ $\mathbf{b} = \begin{pmatrix} 3 \\ 5 \end{pmatrix}$ $\mathbf{c} = \begin{pmatrix} -4 \\ 2 \end{pmatrix}$ $\mathbf{d} = \begin{pmatrix} -1 \\ -5 \end{pmatrix}$

   (i) Calculate $\mathbf{p} = 2\mathbf{a} + 3\mathbf{b}$.
   (ii) Calculate $\mathbf{q} = -3\mathbf{c} - 5\mathbf{d}$.
   (iii) What conclusion can you draw about $\mathbf{p}$ and $\mathbf{q}$?

*41.4 Vector geometry

In geometrical figures, the vectors are often represented by single letters and the questions are solved algebraically.

Example 1

In the triangle ABC, $\overrightarrow{AB} = \mathbf{a}$ and $\overrightarrow{BC} = \mathbf{b}$.
BD divides AC in the ratio $3:1$.

Find, in terms of $\mathbf{a}$ and $\mathbf{b}$:

(i) $\overrightarrow{AC}$ (ii) $\overrightarrow{AD}$ (iii) $\overrightarrow{BD}$

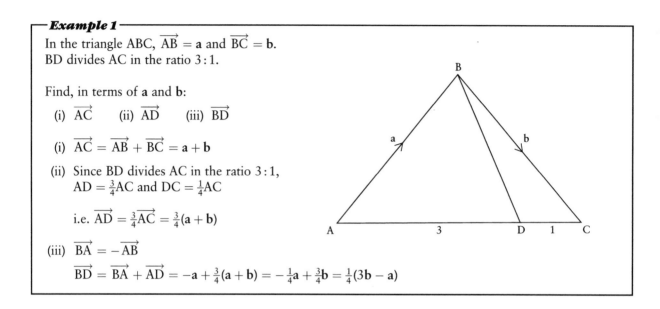

(i) $\overrightarrow{AC} = \overrightarrow{AB} + \overrightarrow{BC} = \mathbf{a} + \mathbf{b}$

(ii) Since BD divides AC in the ratio $3:1$,
$AD = \frac{3}{4}AC$ and $DC = \frac{1}{4}AC$

i.e. $\overrightarrow{AD} = \frac{3}{4}\overrightarrow{AC} = \frac{3}{4}(\mathbf{a} + \mathbf{b})$

(iii) $\overrightarrow{BA} = -\overrightarrow{AB}$

$\overrightarrow{BD} = \overrightarrow{BA} + \overrightarrow{AD} = -\mathbf{a} + \frac{3}{4}(\mathbf{a} + \mathbf{b}) = -\frac{1}{4}\mathbf{a} + \frac{3}{4}\mathbf{b} = \frac{1}{4}(3\mathbf{b} - \mathbf{a})$

Example 2

In quadrilateral OPQR, $\overrightarrow{OP} = \mathbf{p}$, $\overrightarrow{OR} = \mathbf{r}$, $\overrightarrow{PQ} = \mathbf{p} + \mathbf{r}$.
$RS = 2OR$.

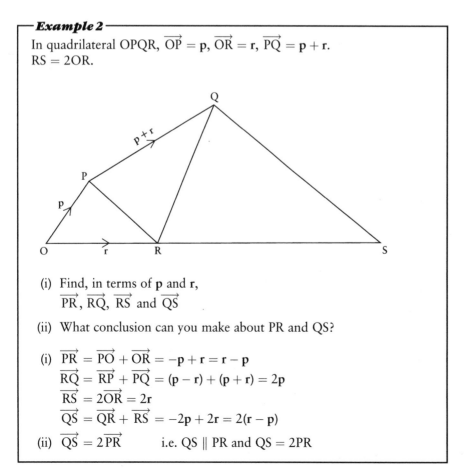

(i) Find, in terms of $\mathbf{p}$ and $\mathbf{r}$,
$\overrightarrow{PR}, \overrightarrow{RQ}, \overrightarrow{RS}$ and $\overrightarrow{QS}$

(ii) What conclusion can you make about PR and QS?

(i) $\overrightarrow{PR} = \overrightarrow{PO} + \overrightarrow{OR} = -\mathbf{p} + \mathbf{r} = \mathbf{r} - \mathbf{p}$
$\overrightarrow{RQ} = \overrightarrow{RP} + \overrightarrow{PQ} = (\mathbf{p} - \mathbf{r}) + (\mathbf{p} + \mathbf{r}) = 2\mathbf{p}$
$\overrightarrow{RS} = 2\overrightarrow{OR} = 2\mathbf{r}$
$\overrightarrow{QS} = \overrightarrow{QR} + \overrightarrow{RS} = -2\mathbf{p} + 2\mathbf{r} = 2(\mathbf{r} - \mathbf{p})$

(ii) $\overrightarrow{QS} = 2\overrightarrow{PR}$ i.e. QS ∥ PR and QS = 2PR

Exercise 41E

1   In triangle OPQ, OP = **p** and OQ = **q**.
R and S divide OP and OQ
respectively in the ratio of 3 : 1.

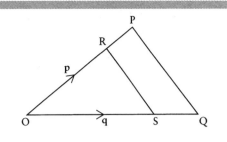

(i) Find, in terms of **p** and **q**: $\overrightarrow{PQ}$, $\overrightarrow{OR}$, $\overrightarrow{OS}$ and $\overrightarrow{RS}$.

(ii) What can you deduce about RS and PQ?

2   ABCD is a quadrilateral. P, Q, R and S are the mid-points of AB, BC, CD and DA respectively.
$\overrightarrow{AD}$ = **a**, $\overrightarrow{AB}$ = **b**, $\overrightarrow{BC}$ = **c**.

(i) Find, in terms of **a**, **b** and **c**.
$\overrightarrow{AP}$, $\overrightarrow{CD}$, $\overrightarrow{CR}$, $\overrightarrow{PS}$, $\overrightarrow{QR}$, $\overrightarrow{PQ}$, $\overrightarrow{SR}$.

(ii) What can you deduce about PQRS?

3   OPQRST is a regular hexagon. $\overrightarrow{OP}$ = **p** and $\overrightarrow{PQ}$ = **q**.

Find, in terms of **p** and **q**:
$\overrightarrow{RS}$, $\overrightarrow{ST}$, $\overrightarrow{QT}$, $\overrightarrow{QR}$, $\overrightarrow{PR}$ and $\overrightarrow{OR}$.

4   The diagram, not drawn to scale, shows a trapezium OABC.
X and Y are the mid-points of diagonals OB and AC respectively.
$\overrightarrow{OA}$ = **a**, $\overrightarrow{OC}$ = **c** and $\overrightarrow{CB}$ = $\frac{1}{2}$**a**

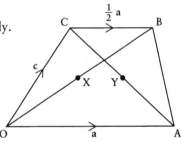

(i) Find each of the following vectors in terms of **a** and **c**, simplifying your answer where possible:
$\overrightarrow{OB}$, $\overrightarrow{OX}$, $\overrightarrow{AO}$, $\overrightarrow{AC}$, $\overrightarrow{AY}$, $\overrightarrow{OY}$, $\overrightarrow{XY}$.

(ii) What do you deduce about XY and OA?

5   In the diagram given below, $\overrightarrow{CA}$ = **a** and $\overrightarrow{CB}$ = **b**.
BE is parallel to CA, BE = $\frac{1}{2}$CA and AD = $\frac{2}{3}$AB.

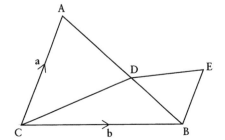

(i) Find, in terms of **a** and **b**:
$\overrightarrow{AB}$, $\overrightarrow{AD}$, $\overrightarrow{CD}$, $\overrightarrow{BE}$, $\overrightarrow{DB}$, $\overrightarrow{DE}$.

(ii) What can you deduce about CD and DE?

6　The diagram shows a triangle ABC.

P is a point on AC such that $\overrightarrow{AP} = \mathbf{a}$ and $\overrightarrow{AP} = \frac{1}{3}AC$.

Q is the mid point of AB and $\overrightarrow{AQ} = \mathbf{b}$.

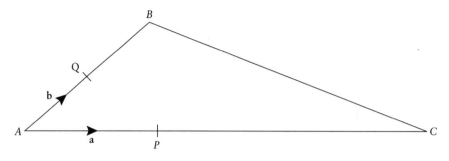

a　Write down, in terms of $\mathbf{a}$, the vector $\overrightarrow{AC}$.

b　Write down, in terms of $\mathbf{b}$, the vector $\overrightarrow{AB}$.

c　Write down, in terms of $\mathbf{a}$ and $\mathbf{b}$, the vector $\overrightarrow{BC}$.

(SEG S94)

42 Transformations

When the position, or size, of an object is changed, we call this a **transformation**.

There are four basic transformations.

The **position** of an object (e.g. the man below) can be changed by:

1 translation (i)   (ii)

2 reflection (i)   (ii)

3 rotation (i)   (ii)

and the size of an object can be changed by:

4 enlargement (i)   (ii)

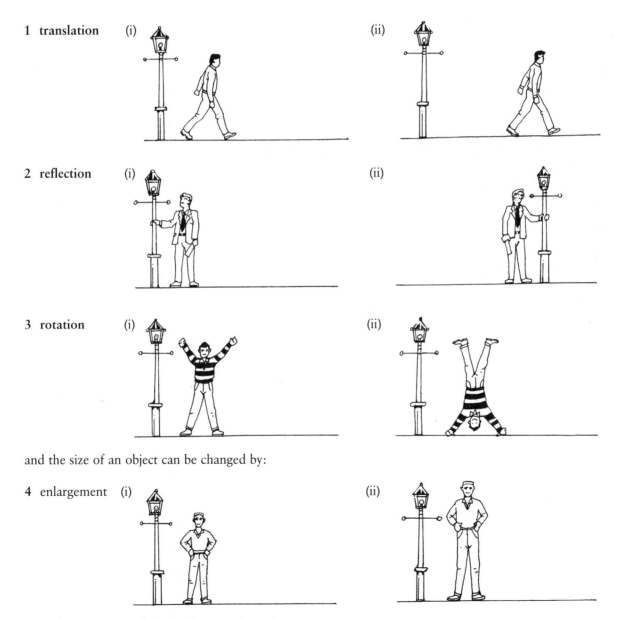

Transformations can be made in two dimensions or three dimensions, but all the work in this book will be in two dimensions.

42.1 Translation

A translation moves every point of an object in the same direction, through the same distance.

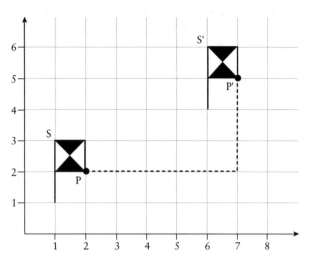

To find the translation, it is only necessary to look at one point in the object S as all the points move in the same way.

Consider the point P and its **image** P′ (i.e. the new position of P).

To change position from P to P′, the point is moved 5 units horizontally and 3 units vertically.

This translation is represented by the vector $\begin{pmatrix} 5 \\ 3 \end{pmatrix}$.

─Example─

Draw the rectangle ABCD with vertices A(3, 1), B(5, 1), C(5, 4) and D(3, 4).

The rectangle is transformed by the translation $\begin{pmatrix} -3 \\ 2 \end{pmatrix}$ into rectangle A′B′C′D′.

Draw rectangle A′B′C′D′.

Position of A′ = Position of A + Translation

$$= \begin{pmatrix} 3 \\ 1 \end{pmatrix} + \begin{pmatrix} -3 \\ 2 \end{pmatrix} = \begin{pmatrix} 0 \\ 3 \end{pmatrix}$$

i.e. A′ is the point (0, 3)

Similarly B′ is the point (2, 3)
C′ is the point (2, 6)
D′ is the point (0, 6)

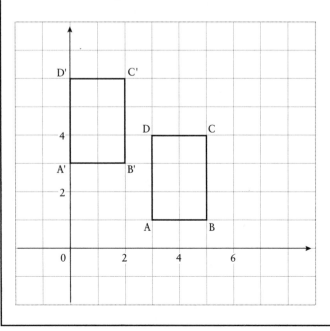

1 Write down, as vectors, the translations which move triangle ABC to each of the positions **a, b, c, d** and **e**.

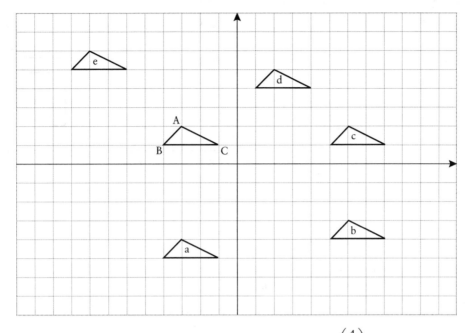

2 Triangle ABC is transformed into triangle A′B′C′ by the translation $\begin{pmatrix} 4 \\ 3 \end{pmatrix}$.

A is (2, 1), B is (3, 5) and C is (−1, −2). Find the coordinates of A′, B′ and C′.

3 Quadrilateral ABCD is transformed into A′B′C′D′ by a translation $\begin{pmatrix} 3 \\ -2 \end{pmatrix}$.

A is (1, 2), B is (3, 4), C is (7, 4) and D is (2, 5). Draw the quadrilaterals ABCD and A′B′C′D′ on graph paper.

42.2 Reflection

A reflection moves an object so that it becomes a **mirror image** of itself. Shape S′ is the image of S after it has been reflected in the line L, which acts like a mirror.

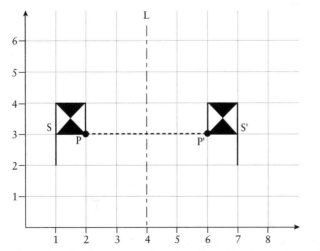

To find the mirror line L, choose a point P on S and its image P′ on S′. The mirror line is the perpendicular bisector of PP′.

Example 1

Draw the reflection of the given shape in the mirror line $x = 3$.

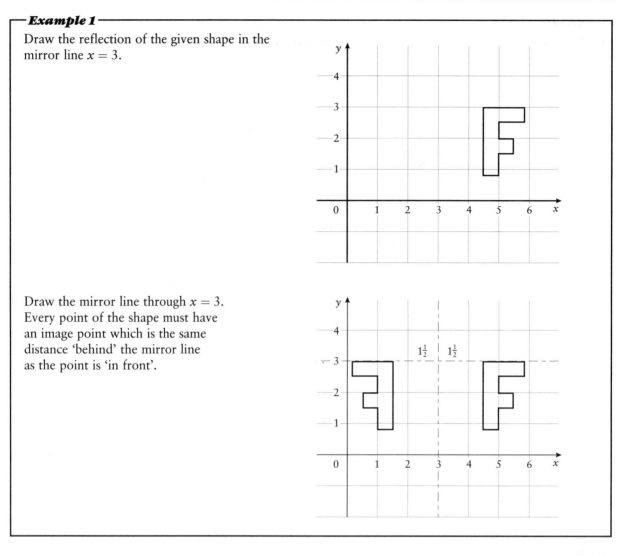

Draw the mirror line through $x = 3$.
Every point of the shape must have an image point which is the same distance 'behind' the mirror line as the point is 'in front'.

Example 2

In the diagram below (left), the shape A′B′C′D′ is a reflection of the shape ABCD.

Copy the diagram and draw the mirror line.

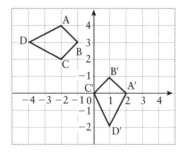

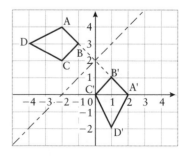

Join any point to its image: say BB′ (see above right).
Find the mid-point of this line: M(0, 2) is the mid-point of BB′.
Draw a line through the mid-point (M) which is perpendicular to BB′.
This is the mirror line.

Exercise 42B

1   Draw the reflections of each of the following shapes in the given mirror line.

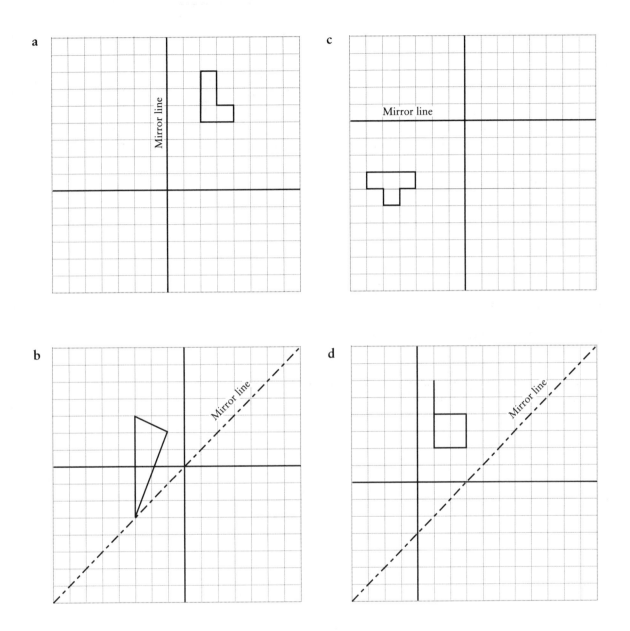

2   The vertices of triangle ABC are A(2, 1), B(3, −2) and C(5, −3). Give the coordinates of the vertices after:

a   a reflection in the x-axis

b   a reflection in the y-axis

c   a reflection in the line $x + y = 0$

d   a reflection in the line $y = x$.

3 The triangle PQR is reflected into triangle P'Q'R', as shown in the diagram.

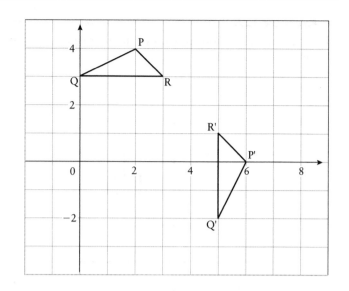

a Copy the diagram and draw the mirror line.

b Write down the equation of the mirror line.

42.3 Rotation

Rotation moves an object by turning it about a fixed point called the **centre of rotation**.

Shape S′ is the image of shape S after it has been rotated about point P, through an angle of 80° in a clockwise direction.

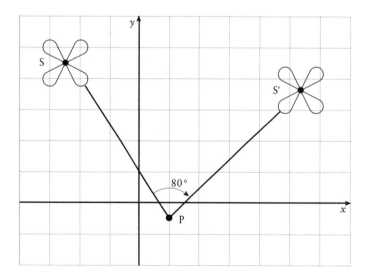

To define a rotation you need to state:

(i) the position of the centre of rotation

(ii) the angle of rotation

(iii) the direction (clockwise or anticlockwise) of rotation.

Example 1

Shape S is rotated about the origin through a three-quarter turn anticlockwise.
Draw the image of S (S′) after the rotation.

One way to find the position of S′ is to first trace the shape S, using the tracing paper.
Then, with a pin holding the tracing paper at the point (0, 0), turn it anticlockwise through a three-quarter turn (i.e. an angle of 270°).
The new position of the shape is the image S′.

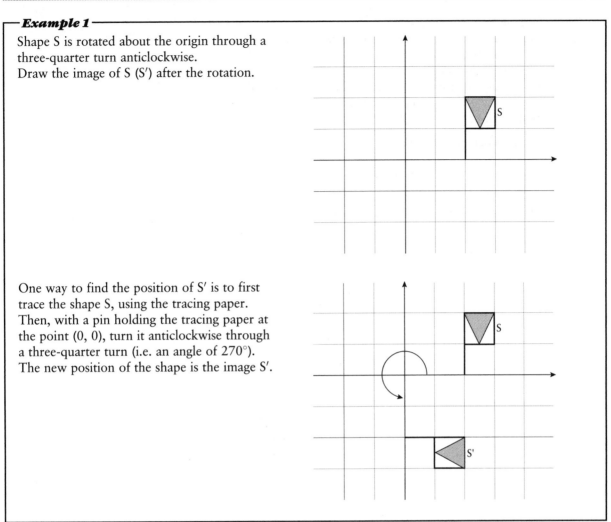

Example 2

Describe fully the rotation of shape S on to shape S′, shown in the diagram below.

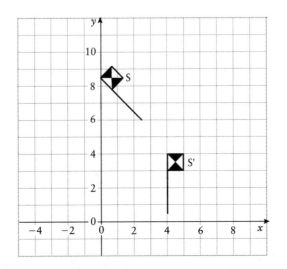

(i) To find the centre of rotation:
Join a pair of corresponding points on S and S', say A and A'.
Construct the perpendicular bisector of AA'.
The centre of rotation lies on this line.
Repeat for any other pair of corresponding points, say BB'.
The point at which the two perpendicular bisectors meet is the centre of rotation, i.e. point P.

(ii) To find the angle of rotation:

Measure angle APA', or angle BPB'.
     Angle APA' = 50°

(iii) The direction of rotation is always taken to be measured in an anticlockwise direction. This rotation is through 50° clockwise about point (−3, 2). This is an anticlockwise rotation of −50°. Thus the rotation is through −50° about point (−3, 2).

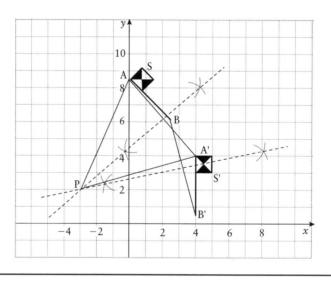

Exercise 42C

1 Copy the following diagrams and draw the image of each shape after rotation about the marked point, through the given angle.

a

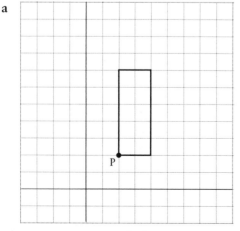

90° anticlockwise

b

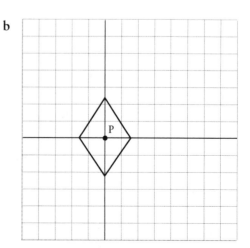

+135°

c

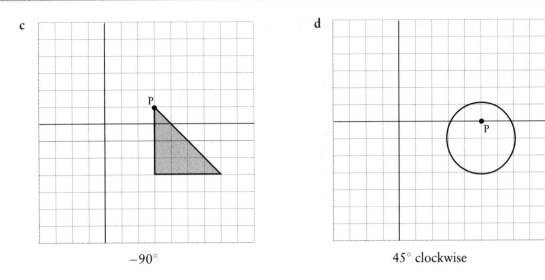

−90°

d

45° clockwise

2   Describe the rotation of each of the following:

a

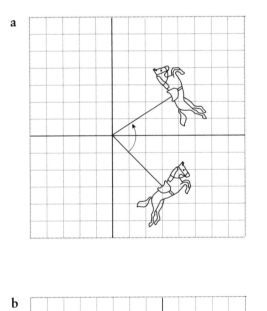

c

b

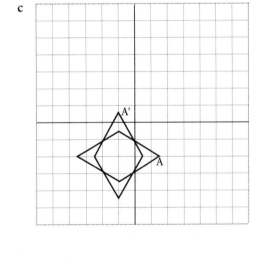

d

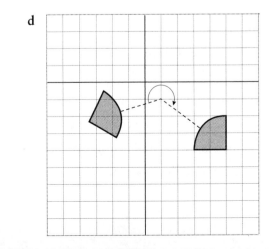

3   The triangle PQR is rotated into the triangle P'Q'R'. Define the rotation by finding:

a   the centre of rotation

b   the angle and direction
of rotation.

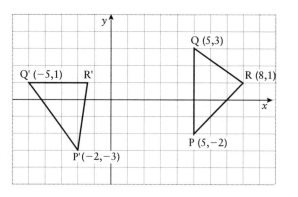

4   Square ABCD has vertices A(1, 2), B(4, 2), C(4, −1) and D(1, −1).
The square is rotated about (1, 0) through 180° on to A'B'C'D'.
Find the coordinates of A', B', C', D'

5   Rectangle ABCD is rotated
to the new position
A'B'C'D'.

Define the rotation.

42.4   Enlargement

Enlargements change the size of an object.
The image is similar to the original object (see
page 378),
i.e. the original lengths are all multiplied by a
scale factor to form the image.

Shape S' is the image of shape S after an
enlargement of scale factor 3, centre P.
That is, every length in S' is three times the
size of the corresponding length in S.

To define an enlargement you need to state:

(i)   the position of the centre of enlargement

(ii) the scale factor of the enlargement.

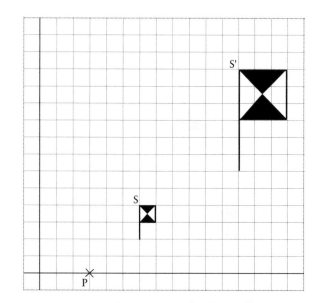

Example 1

Enlarge the following shape by a scale factor 2 with centre $(-2, -3)$.

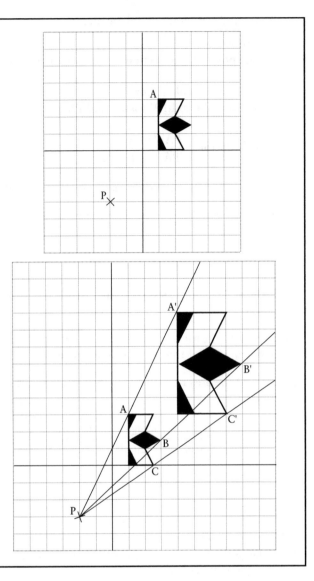

Draw a line from the centre of enlargement to a point on the given shape (PA) and produce the line (i.e. extend it).

Measure PA and multiply this length by the scale factor to find the position of A′ (PA′ = 2 × PA).

Repeat to find all the vertices of the image (B′, C′, etc.) which correspond to vertices of the original shape (B, C, etc.).

Join the image points to find the enlargement.

Example 2

Triangle A′B′C′ is the image of triangle ABC after enlargement.
Define the enlargement.

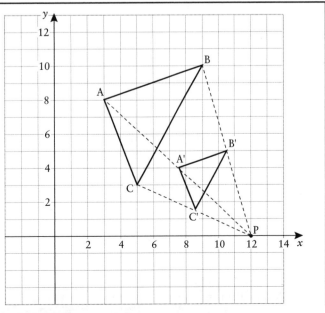

(i) To find the centre of enlargement:

Join AA′ and BB′.
The point of intersection of the two lines, P, is the centre of enlargement.

(ii) To find the scale factor:
Measure PA and PA′ (or PB and PB′)

$$\text{Scale factor} = \frac{PA'}{PA} = \frac{PB'}{PB} = \frac{1}{2}$$

The enlargement is scale factor $\frac{1}{2}$ with centre (0, 12).

Notes

(i) If the scale factor is less than one, the image is *smaller* than the original object.

(ii) If the scale factor is negative, the original object and its image are on the *opposite sides* of the centre of enlargement.

(iii) To find the area of the image after enlargement, use area of similar figures:

$$\frac{\text{Area of image}}{\text{Area of original}} = \frac{(\text{Length of side of image})^2}{(\text{Length of side of original})^2}$$

Exercise 42D

1 Enlarge the following shapes by the given scale factor with the centre of enlargement marked.

a

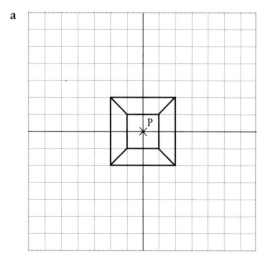

Scale factor = 2

b

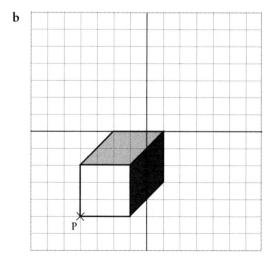

Scale factor = 3

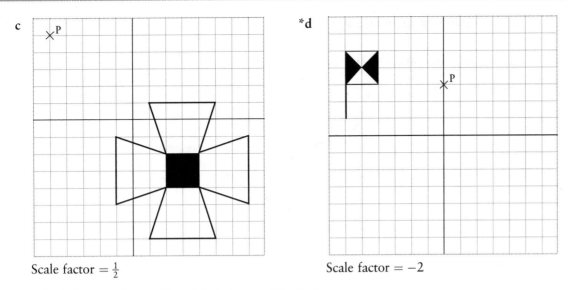

Scale factor $= \frac{1}{2}$

Scale factor $= -2$

2   For the following shapes (S) and their images (S'), find:

   (i) the scale factor of the enlargement (ii) the centre of the enlargement.

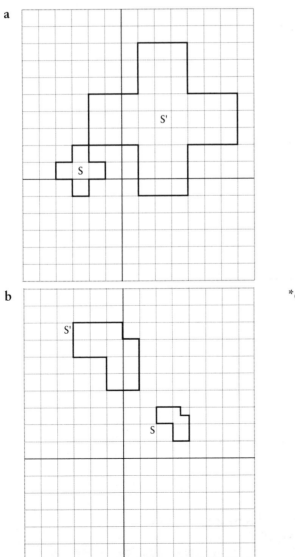

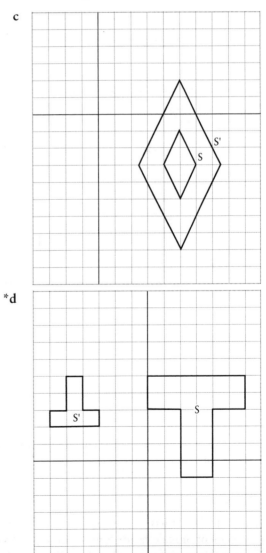

3 Triangle ABC is transformed into triangle A'B'C'.
Find:

   a the scale factor of the enlargement

   b the coordinates of A'.

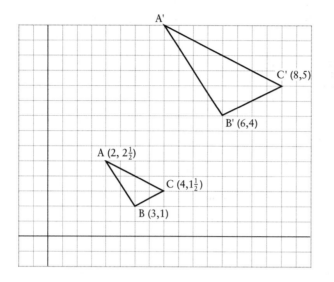

4 Rectangle ABCD is enlarged into A'B'C'D'.
Define the enlargement.

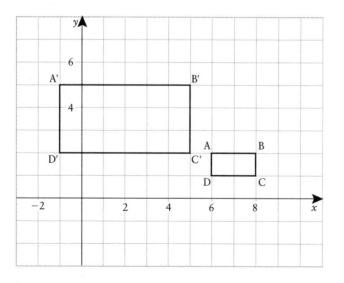

5 Triangle ABC has an area of 5 cm² and is transformed by an enlargement of scale factor 3 into triangle A'B'C'.
What is the area of triangle A'B'C'?

6 Rectangle ABCD has an area of 30 square inches and is transformed by an enlargement of scale factor 5 into rectangle A'B'C'D'.
What is the area of rectangle A'B'C'D'?

7

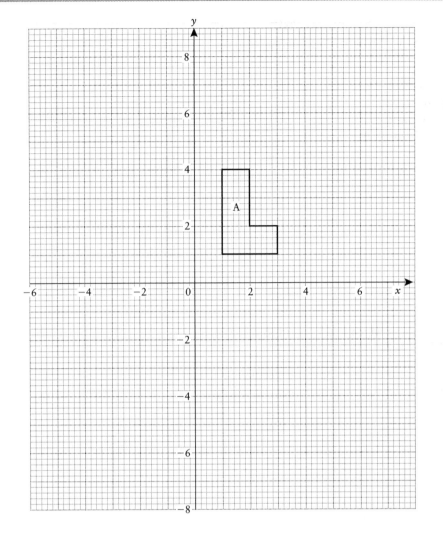

a Shape **A** is rotated through 90° in an anticlockwise direction about (0, 0). Draw its new position on the graph and label it **B**.

b Shape **A** is enlarged by scale factor 2, centre (0, 0). Draw the enlargement on the graph and label it **C**.

c Shape **A** is transformed by firstly reflecting it in the *y* axis and then by reflecting the result in the *x* axis. Draw the **final** position of shape **A** on the graph and label it **D**.

(SEG S94)

42.5 Combined transformations

When one transformation is followed by another, the resulting transformation can often be described by an equivalent, single transformation.

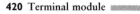

Example

For the shape S shown in the diagram, let M be the transformation:
reflection in the line $x = 4$,
   T be the transformation:
reflection in the y-axis,
   R be the transformation:
rotation of $-60°$ about $(0, 0)$.

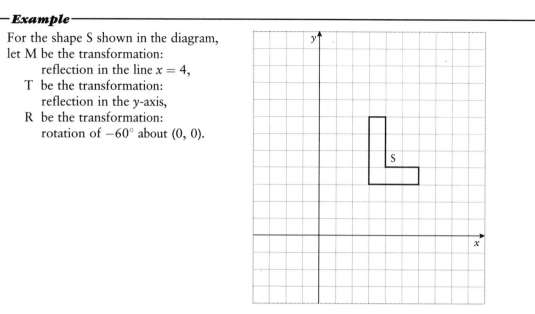

Find the single transformation which is equivalent to the combined transformation: **a** MT   **b** R².

a The transformation MT means first apply transformation T, then transformation M.

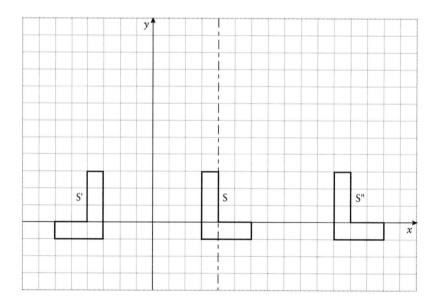

T reflects shape S in the y-axis into shape S′.
M reflects shape S′ in $x = 4$ into shape S″.
The two transformations are equivalent to the single transformation

$S \rightarrow S''$, which is a translation of $\begin{pmatrix} 8 \\ 0 \end{pmatrix}$

The transformation MT is a translation $\begin{pmatrix} 8 \\ 0 \end{pmatrix}$

Note. TM is a translation $\begin{pmatrix} -8 \\ 0 \end{pmatrix}$. You must ensure that the operations are carried out in the correct order.

b R^2 means RR, i.e. a rotation of $-60°$ followed by another rotation of $-60°$.

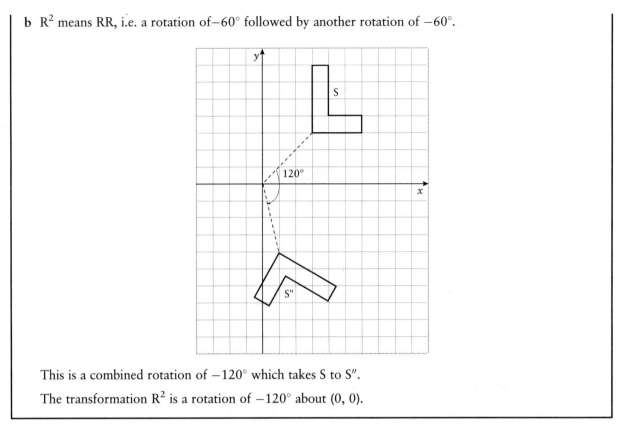

This is a combined rotation of $-120°$ which takes S to S″.

The transformation R^2 is a rotation of $-120°$ about $(0, 0)$.

Exercise 42E

1   R_1 is a rotation through $90°$ about the origin.
   R_2 is a rotation through $-90°$ (i.e. $90°$ clockwise) about point $(-5, 5)$.
   X is a reflection in the x-axis.

   a Draw the images of the square ABCD under the following
   transformations:

   (i) R_1R_2 (ii) R_2R_1 (iii) XR_1 (iv) R_1X

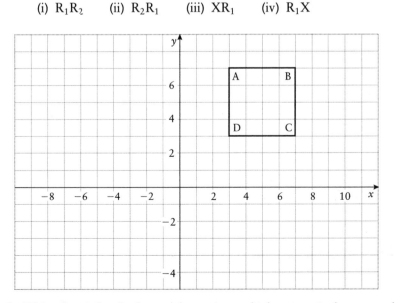

   b Write down the single transformations which are equivalent to each
   of the combined transformations in **a**.

2   Give the single transformation which is equivalent to:

    a R^2 **b** R^3 **c** X^2 **d** XY

    where
    R is a rotation about (0, 0) through 60° clockwise
    X is a reflection in the x-axis
    Y is a reflection in the y-axis.

3   Give the single transformation which is equivalent to

    a XH   **b** TW   **c** HT   **d** XW   **e** HW

    where
    H is a rotation about (0, 0) through 180°

    T is a translation of $\begin{pmatrix} 3 \\ 0 \end{pmatrix}$

    W is a translation of $\begin{pmatrix} 0 \\ 5 \end{pmatrix}$

    X is a reflection in the x-axis.

42.6   Inverse transformations

An inverse of a transformation is the transformation which will return an object to its original position:

(i)   The inverse of a translation $\begin{pmatrix} x \\ y \end{pmatrix}$ is a translation $\begin{pmatrix} -x \\ -y \end{pmatrix}$.

(ii)   The inverse of a reflection in a line is the same reflection in the same line, i.e. a reflection is a self-inverse.

(iii)   The inverse of a rotation about point P, through $\theta°$ anticlockwise is a rotation about point P, through $\theta°$ clockwise, i.e. through $-\theta°$.

(iv)   The inverse of an enlargement, centre P, scale factor k, is an enlargement, centre P, scale factor $\dfrac{1}{k}$.

Exercise 42F

Find the inverse transformations of:

1   a translation of: **a** $\begin{pmatrix} 2 \\ -3 \end{pmatrix}$ **b** $\begin{pmatrix} -5 \\ 6 \end{pmatrix}$

2   a reflection in **a** the y-axis **b** the line $y + 3x = 0$

3   a rotation about O, through 60° in a clockwise direction

4   an enlargement of scale factor 3, centre (2, −3).

43 Tessellations

43.1 Tessellating patterns

A tessellation is a pattern which when repeated will completely cover a plane without leaving any gaps. The pattern can either be one shape, or a combination of shapes.

Simple shapes which tessellate are:

Squares

Equilateral triangles

Regular hexagons

Parallelograms

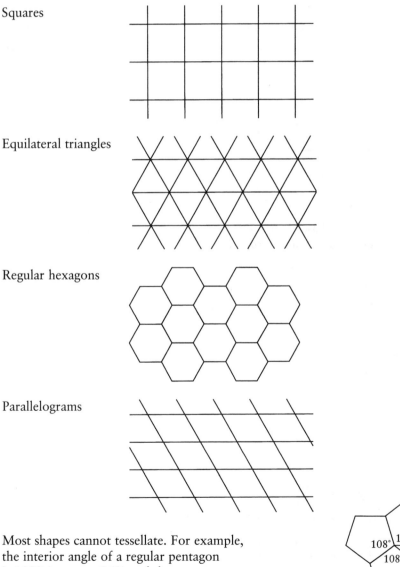

Most shapes cannot tessellate. For example, the interior angle of a regular pentagon is 1088 (see page 369) and these cannot be placed together to make 3608.

In 1891, a Russian crystallographer named Fedorov showed that there are only 17 different ways in which a basic pattern can be repeated. These repeating patterns are found frequently on dress materials, wallpaper, etc.

Any triangle can be placed with a triangle congruent to itself to form a parallelogram. Since parallelograms tessellate, any triangle can be the basis of a tessellation. Here is an example:

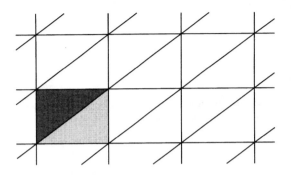

A shape which tessellates need not be regular. Here is an example:

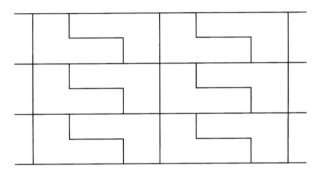

A tessellation can be formed by using two (or more) basic shapes. For example, regular octagons and squares form a tessellation. One such pattern is:

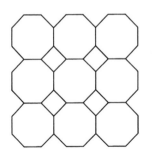

Example

Combine equilateral triangles and squares to form a tessellation.

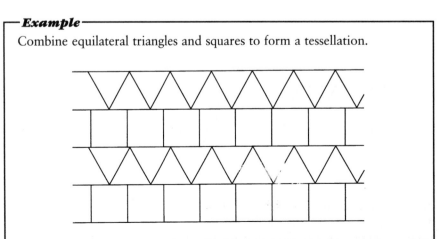

Exercise 43A

1   Draw a tessellation based on each of the following shapes:

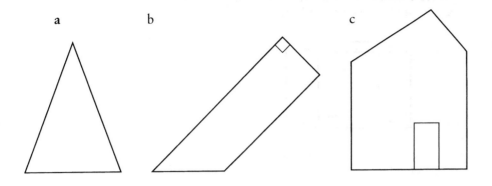

a   b   c

2   Square tiles of side 15 cm and rectangular tiles 5 cm by 15 cm are used to tile a bathroom wall, part of which is shown.

The wall measures 3.3 m by 2.4 m. How many of each type are needed to tile the wall?

3   Which additional shape is needed for this shape to tessellate?

4   Which of the shapes in question 3 of Exercise 37D on pages 355–6 tessellate?

43.2 2D representation of 3D objects

It is common to see pictures of three-dimensional objects, e.g. boxes, houses, people. These pictures are printed on paper and hence are in two dimensions.

The diagram on the right represents a cuboid.

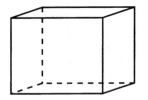

The cuboid has 8 **vertices**, 12 **edges** and 6 **faces**.

The edges you can see are shown as solid lines and the edges of the cuboid which you cannot see are shown as dotted lines.

There are also 3 faces and 1 vertex which you cannot see.

Diagrams of 3D objects are usually drawn with one of the horizontal directions in the *x* direction (as if it were a graph) with the vertical direction of the body being in the *y* direction. The second horizontal direction is shown at an angle.

Circles viewed obliquely are drawn as ovals and a cylinder has two oval ends.

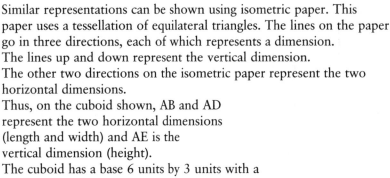

Similar representations can be shown using isometric paper. This paper uses a tessellation of equilateral triangles. The lines on the paper go in three directions, each of which represents a dimension.
The lines up and down represent the vertical dimension.
The other two directions on the isometric paper represent the two horizontal dimensions.
Thus, on the cuboid shown, AB and AD represent the two horizontal dimensions (length and width) and AE is the vertical dimension (height).
The cuboid has a base 6 units by 3 units with a height of 4 units.

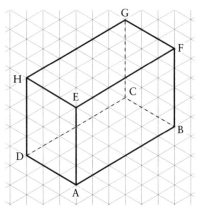

Example

Draw a triangular prism of length 8 units. The isosceles triangular face has a base of 4 units and a perpendicular height of 4 units.

Note. The base AB is 4 units in length, the length AD is 8 units and the height PQ of the triangular face is 4 units.

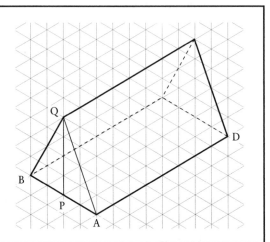

Exercise 43B

Draw two-dimensional representations of:

1   a triangular prism of length 6 units.
    The triangular face is equilateral of side 5 units.

2   a cube of 5 units.

3   a pyramid on a square base of side 4 units. The
    height is 6 units.

4   a house.

5   a triangular prism is drawn on the isometric dotted paper below.

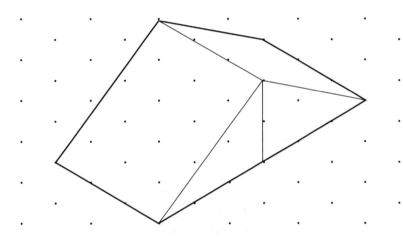

Using a scale factor of 2 draw an enlargement of the triangular prism.   (SEG W94)

44 Mensuration

44.1 Perimeters of polygons

A **plane figure** is a two-dimensional shape which is bounded by lines called **sides**.

Here are some examples:

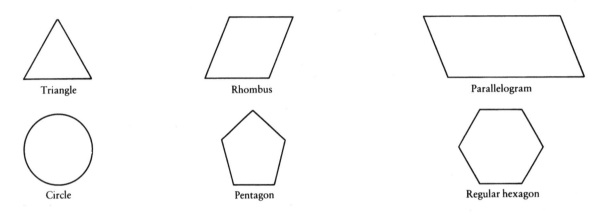

Plane figures which are bounded by straight lines are called **polygons**.

A **regular polygon** is one which has all its sides the same length. The regular three-sided polygon (like the one above) is an **equilateral triangle**.

The **perimeter** of a plane figure is the total length of the sides.

Example 1

A photograph frame has a metallic surround. The outer and inner perimeters of the surround are edged with gold.

Find the total length of the outer and inner perimeter and hence the length of the gold edging.

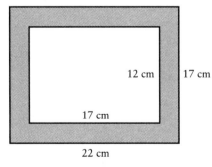

Length of gold edging = Outer perimeter + Inner perimeter
$$= 2 \times (22 + 17) + 2 \times (17 + 12) \, \text{cm}$$
$$= (78 + 58) \, \text{cm}$$
$$= 136 \, \text{cm}$$

Example 2

Find the perimeter of the shape given:

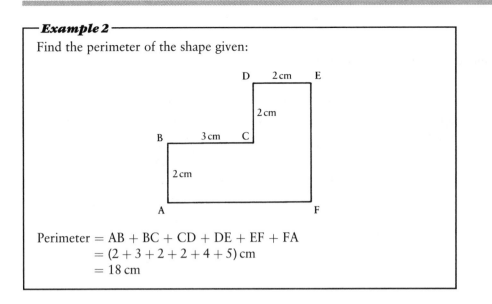

Perimeter = AB + BC + CD + DE + EF + FA
= (2 + 3 + 2 + 2 + 4 + 5) cm
= 18 cm

Exercise 44A

1 Find the perimeter of each of the following:

 a a rectangle of length 8 cm and width 3 cm   **b** a square of side 7 m

2 Find the perimeter of each of the following shapes:

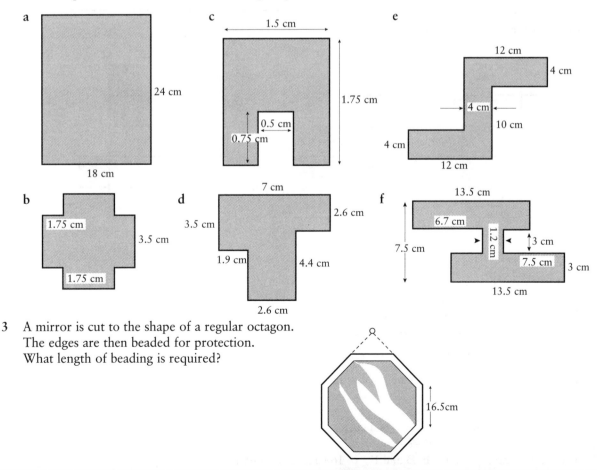

3 A mirror is cut to the shape of a regular octagon.
 The edges are then beaded for protection.
 What length of beading is required?

44.2   Area

Area is a measure of the surface covered by a given shape.

Example 1

Consider the shapes below and place them in ascending order according to their area.

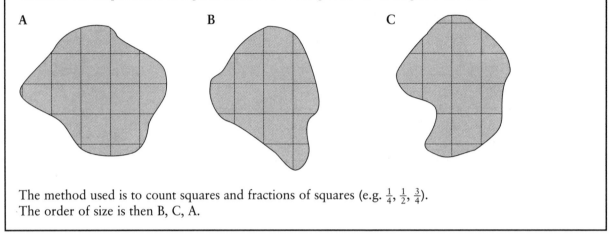

A   B   C

The method used is to count squares and fractions of squares (e.g. $\frac{1}{4}$, $\frac{1}{2}$, $\frac{3}{4}$).
The order of size is then B, C, A.

Figures which have straight sides (polygons) are much easier to compare.

Example 2

Consider the following polygons:

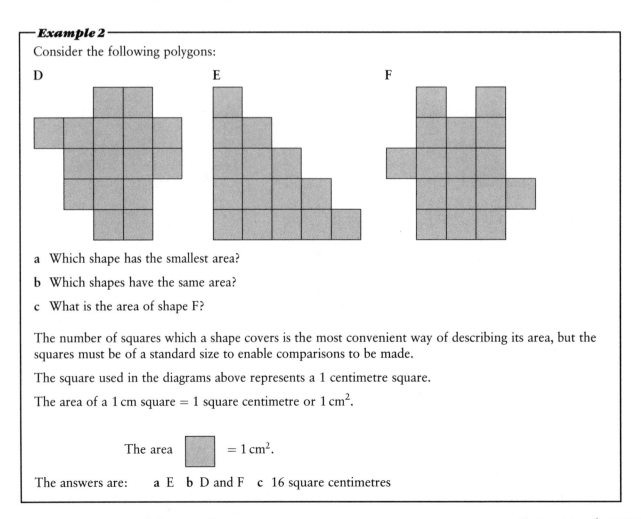

D   E   F

a Which shape has the smallest area?

b Which shapes have the same area?

c What is the area of shape F?

The number of squares which a shape covers is the most convenient way of describing its area, but the squares must be of a standard size to enable comparisons to be made.

The square used in the diagrams above represents a 1 centimetre square.

The area of a 1 cm square = 1 square centimetre or 1 cm².

The area ▢ = 1 cm².

The answers are: a E   b D and F   c 16 square centimetres

The most common measures of an area are based on:

$$\text{side} = 1\,\text{cm square}, \quad \text{area} = 1\,\text{square centimetre} = 1\,\text{cm}^2$$

$$\text{side} = 1\,\text{m square}, \quad \text{area} = 1\,\text{square metre} = 1\,\text{m}^2$$

$$\text{side} = 1\,\text{mm square}, \quad \text{area} = 1\,\text{square millimetre} = 1\,\text{mm}^2$$

The imperial units for area are, 1 square inch $= 1\,\text{in}^2$

$$1\,\text{square foot} = 1\,\text{ft}^2$$

$$1\,\text{square yard} = 1\,\text{yd}^2$$

Area of a rectangle

Example 1

What is the area of the rectangle shown?
(Each square represents $1\,\text{cm}^2$.)

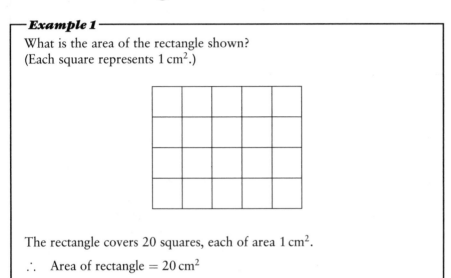

The rectangle covers 20 squares, each of area $1\,\text{cm}^2$.

∴   Area of rectangle $= 20\,\text{cm}^2$

Example 2

What is the area of this rectangle?

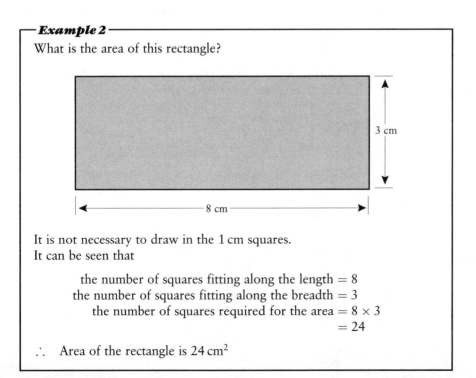

It is not necessary to draw in the 1 cm squares.
It can be seen that

the number of squares fitting along the length $= 8$
the number of squares fitting along the breadth $= 3$
the number of squares required for the area $= 8 \times 3$
$= 24$

∴   Area of the rectangle is $24\,\text{cm}^2$

For a rectangle: $\qquad$ **Area = Length × Breadth**
$$A = L \times B$$

For a square: $\qquad$ **Area = Length × Breadth**
$$A = L^2$$

Some shapes, which are more complicated, can be split up into rectangles in order to find the area.

Example 3

Find the area of this shape.

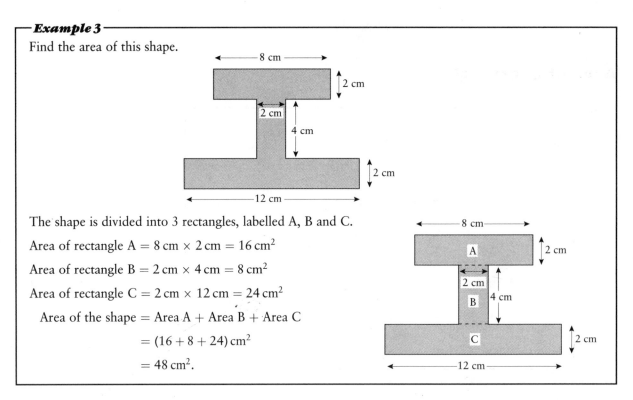

The shape is divided into 3 rectangles, labelled A, B and C.

Area of rectangle A = 8 cm × 2 cm = 16 cm²

Area of rectangle B = 2 cm × 4 cm = 8 cm²

Area of rectangle C = 2 cm × 12 cm = 24 cm²

$\quad$ Area of the shape = Area A + Area B + Area C

$$= (16 + 8 + 24) \text{ cm}^2$$

$$= 48 \text{ cm}^2.$$

In some cases it is quicker to subtract areas than to add.

Example 4

Find the area of this shape.

The shape could be divided into three rectangles:

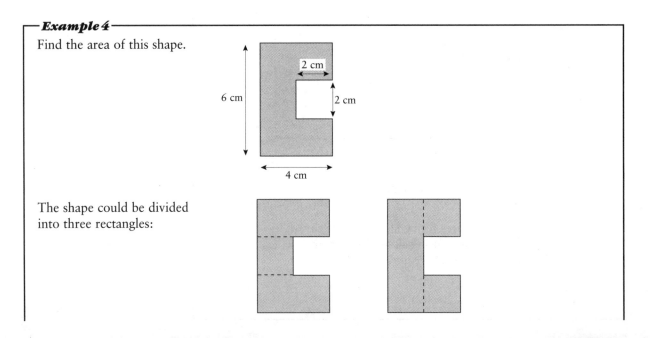

It is quicker, however, to consider two rectangles A and B:

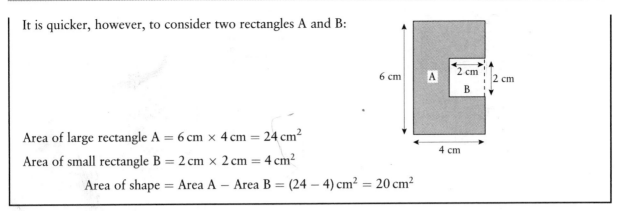

Area of large rectangle A = 6 cm × 4 cm = 24 cm²

Area of small rectangle B = 2 cm × 2 cm = 4 cm²

Area of shape = Area A − Area B = (24 − 4) cm² = 20 cm²

Exercise 44B

1 Write down the areas of the following rectangles. (Each square represents 1 cm².)

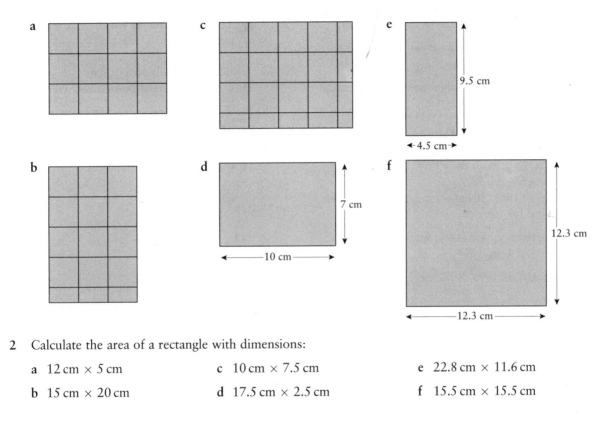

2 Calculate the area of a rectangle with dimensions:

 a 12 cm × 5 cm   c 10 cm × 7.5 cm   e 22.8 cm × 11.6 cm

 b 15 cm × 20 cm   d 17.5 cm × 2.5 cm   f 15.5 cm × 15.5 cm

3 Calculate the areas of the following shapes:

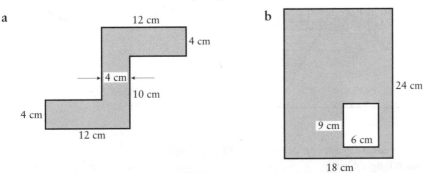

c

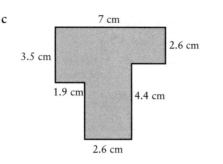

d

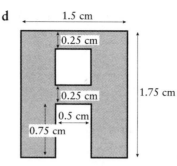

4   A window has 24 panes of glass, each measuring 26.5 cm by 21.5 cm.
What is the total area of glass in the window, correct to 3 significant figures?

5   Carpet is bought by the metre, but is priced by the square metre.

a   What is the area of a carpet which is 4 m wide and 7.5 m long?

b   What is the cost of the carpet if the price is £9.99 per square metre?

6   A path 1 m wide is laid all round the edge of a lawn which is 6 m by 5 m.
What is the area of the path?

7   A piece of card, 20 cm square, is used to make a surround for a picture.

a   What size square is cut out of the card if the frame is 4 cm wide?

b   What is the area of the card surround?

8   A rectangular piece of land covers an area of 156 m².

a   If its length is 13 m, what is its width?

b   What is its perimeter?

9   A square has a perimeter which is 22 cm long. What is the area of the square?

10   a   A rectangle has an area of 5 m².
How many square centimetres is the area?

b   How many square inches are there in an area of 6ft²?

c   A rectangular area is 12.6 cm².
How many square millimetres is this area?

Area of a triangle

Draw a rectangle ABCD and mark a point E anywhere along AB.
Join ED and EC.

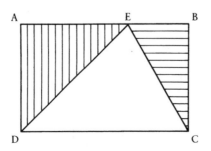

If triangles ADE and BCE are cut out, they can be placed on top of triangle DEC so that they exactly cover triangle DEC.

Area of triangle DEC = Half the area of rectangle ABCD
Area of rectangle ABCD = Length DC × Width BC
DC is the base of triangle DEC and BC equals its height.

Therefore: **Area of triangle DEC = $\frac{1}{2}$ Base × Height**

$$A = \tfrac{1}{2}\,b\,h$$

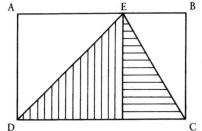

Example 1

Find the area of the triangle shown below.

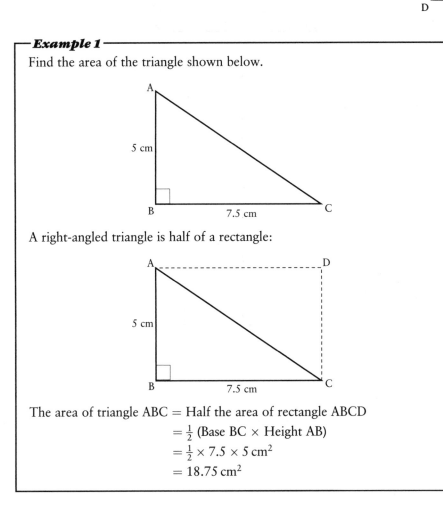

A right-angled triangle is half of a rectangle:

The area of triangle ABC = Half the area of rectangle ABCD
$= \frac{1}{2}$ (Base BC × Height AB)
$= \frac{1}{2} \times 7.5 \times 5$ cm²
$= 18.75$ cm²

Example 2

Find the area of the triangle shown below.

The area of triangle ABC

$= \frac{1}{2}$ (Base BC × Height AB)
$= \frac{1}{2} \times 6.2 \times 3$ cm²
$= 9.3$ cm²

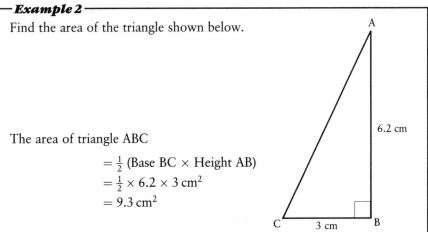

Example 3

Find the area of the following:

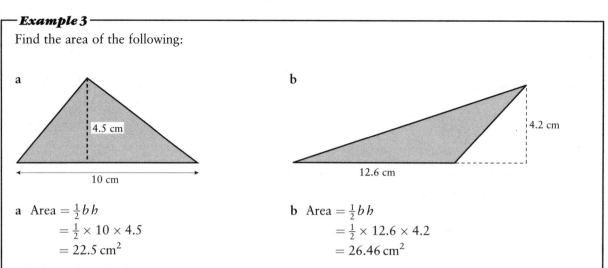

a

4.5 cm

10 cm

b

4.2 cm

12.6 cm

a Area $= \frac{1}{2}bh$

$= \frac{1}{2} \times 10 \times 4.5$

$= 22.5 \text{ cm}^2$

b Area $= \frac{1}{2}bh$

$= \frac{1}{2} \times 12.6 \times 4.2$

$= 26.46 \text{ cm}^2$

Example 4

A triangle has an area of 35 cm^2 and a height of 4 cm.
What is the length of its base?

$A = \frac{1}{2}bh$ or $A = \frac{1}{2}bh$

$35 = \frac{1}{2} \times b \times 4$ $b = \dfrac{2A}{h}$

$= 2b$ $= \dfrac{2 \times 35}{4} = 17.5$

$b = \dfrac{35}{2} = 17.5$

Base $= 17.5$ cm   Base $= 17.5$ cm

Exercise 44C

1   Find the areas of the following triangles:

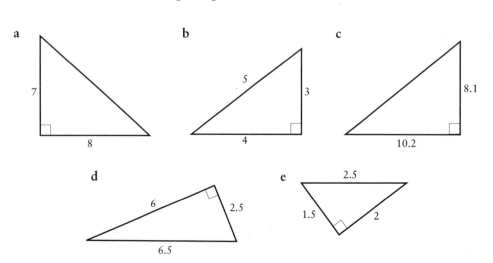

a

7

8

b

5

3

4

c

8.1

10.2

d

6

2.5

6.5

e

2.5

1.5

2

2 Find the areas of the following triangles:

a **b**

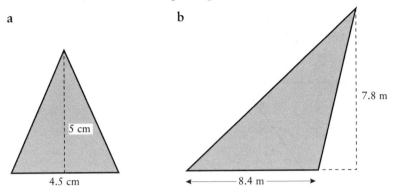

3 Find the area of a triangle with:

a Base = 16 cm, Height = 11 cm   **c** Base = 10.5 in, Height = 1 ft

b Base = 28 mm, Height = 17.5 mm   **d** Base = 4.6 m, Height = 8.4 m

4 **a** A triangle has an area of 40 m² and a height of 8 m.
  What is the length of its base?

b A triangle has a base of length 12 cm and an area of 96 cm².
  What is its height?

5 Calculate the missing dimension in the following triangles:

a Area = 144 cm²   Base = 18 cm   Height = ? cm

b Area = 52 cm²   Base = ? cm   Height = 13 cm

c Area = 45 mm²   Base = 7.5 mm   Height = ? mm

d Area = ? m²   Base = 7.2 m   Height = 3.4 m

e Area = ? in²   Base = 5.5 in   Height = 8.6 in

6 Find the area of each of the following shapes. Give the answers to 1 d.p.

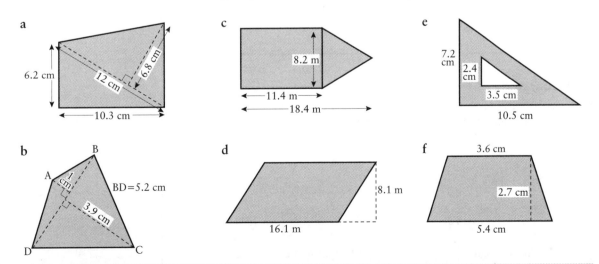

Area of a parallelogram

A parallelogram is a quadrilateral (four-sided figure) which has both pairs of opposite sides parallel.

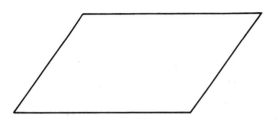

Any parallelogram can be divided into two identical triangles by drawing in a diagonal.

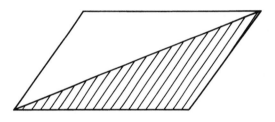

Area of the parallelogram = 2 × Area of one triangle

$$A = b \times h$$

Note. As on page 436 h is the perpendicular height of the parallelogram.

Example 1

Find the area of a parallelogram with base 12.5 cm and height 8.2 cm.

$$A = b \times h$$
$$= 12.5 \times 8.2 \, \text{cm}^2$$
$$\text{Area} = 102.5 \, \text{cm}^2$$

Example 2

A parallelogram has a base of 14 cm and an area of 168 cm^2.
What is the height of the parallelogram?

$$A = b \times h \qquad \text{or} \qquad A = bh$$

$$168 = 14 \times h \qquad\qquad h = \frac{A}{b}$$

$$h = \frac{168}{14} = 12 \qquad\qquad h = \frac{168}{14} = 12$$

Height = 12 cm   Height = 12 cm

Exercise 44D

1 Find the areas of the following parallelograms:

   a Base = 10 cm   Height = 12 cm

   b Base = 12.6 cm   Height = 6 cm

   c Base = 34.7 in   Height = 13 in

   d Base = 1.3 m   Height = 26 cm

   e Base = 19.2 mm   Height = 10.3 cm.

2 Find the missing dimension for the following parallelograms:

   a Area = 24 cm^2 Base = ?   Height = 10 cm

   b Area = 1.3 m^2 Base = 3.9 m   Height = ?

   c Area = ?   Base = 8.4 cm   Height = 16 mm

   d Area = 0.68 m^2 Base = 17 cm   Height = ?

   e Area = 2.4 yd^2 Base = ?   Height = 3 ft.

3 Calculate the area of the following shapes:

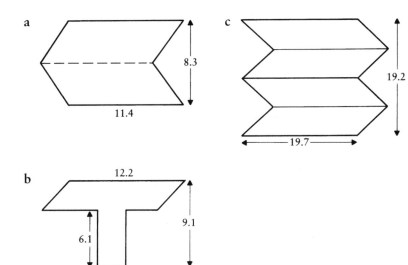

4 Find the height of a parallelogram which has:

   a area 32 cm^2, base 4 cm

   b area 36 cm^2, base 9 cm

   c area 42 in^2, base 8 in.

5 Find the base of a parallelogram which has:

   a area 70 cm^2, height 10 cm

   b area 20 cm^2, height 11.1 cm.

The area of a trapezium

The area of a trapezium is:

$$\frac{1}{2} \times \text{Sum of parallel sides} \times \text{Perpendicular height}$$

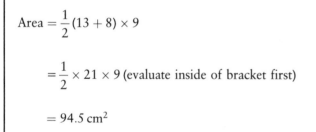

$$\text{Area} = \frac{1}{2}(a+b)h$$

Example

Trapezium PQRS has PQ parallel to SR. PQ=8 cm, SR=13 cm, and the perpendicular distance from S to PQ is 9 cm.

Find area PQRS.

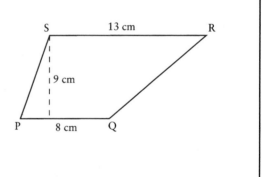

$$\text{Area} = \frac{1}{2}(13+8) \times 9$$

$$= \frac{1}{2} \times 21 \times 9 \text{ (evaluate inside of bracket first)}$$

$$= 94.5 \text{ cm}^2$$

Exercise 44E

Find the areas of the following trapeziums.

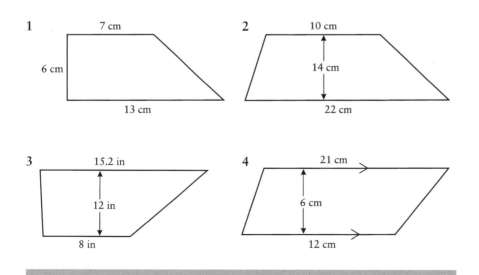

1   7 cm   6 cm   13 cm

2   10 cm   14 cm   22 cm

3   15.2 in   12 in   8 in

4   21 cm   6 cm   12 cm

44.3   The circumference and area of a circle

Circumference of a circle

The **perimeter** of a circle is called the **circumference**.

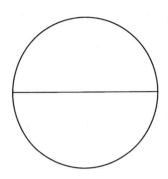

Take a long piece of thread or thin string and cut off
a piece equal to the length of the diameter of the
circle on the right.

Cut off a second length equal to the length of the circumference.
(You will need to take care when measuring the circumference.)

Compare the two lengths of thread and you should find that the longer piece
is just over three times the length of the shorter piece. That is:

$$\text{Circumference} \approx 3 \times \text{Diameter}$$

This formula was good enough for the mathematicians of ancient times, but
we now have an accurate value for this multiple of the diameter.

The multiple is given the symbol π, the Greek letter for 'p', pronounced 'pi'.

If you have a π button on your calculator, press it and you will see that
$\pi = 3.141\,592\,654$.

In fact, the number goes on for ever, but 9 decimal places are adequate for
most purposes! For all circles

Circumference = $\pi \times$ Diameter   or   **Circumference = 2 $\times$ π $\times$ Radius**

$$C = \pi d \qquad\qquad\qquad C = 2\pi r$$

In calculations, you may use the 'pi' button or an appropriate approximation,

e.g. $\pi = 3$ for a rough estimate,

$\pi = 3.14$ for a more accurate answer.

You will usually be told in an examination which value of π to use.

Example 1

Find the circumference of a circle with radius 6 cm. (Take $\pi = 3.14$.)

$$\begin{aligned}
\text{Circumference} &= 2 \times \pi \times r \\
&= 2 \times 3.14 \times 6\,\text{cm} \\
&= 37.68\,\text{cm}
\end{aligned}$$

Example 2

A circle has a circumference of length 30 m. Find:

a   a rough estimate of the diameter

a   For a rough estimate take

$$C \approx 3 \times d$$

$$\therefore \quad d \approx \frac{C}{3} = \frac{30}{3} = 10 \text{ and the diameter is } 10 \text{ m}$$

b an estimate of the diameter, correct to 2 d.p.

b $C = \pi \times d$

$$\therefore \quad d = \frac{C}{\pi} = \frac{30}{3.14} \text{ or } \frac{30}{\pi}$$

and use the π button on your calculator.

The diameter = 9.55 m (to 2 d.p.).

Exercise 44F

1 Use the π button on your calculator (or take $\pi = 3.14$) to calculate the missing lengths below. Give your answers correct to 1 d.p.

    a Diameter = 10 cm   Radius = ?   Circumference = ?

    b Diameter = ?   Radius = ?   Circumference = 27″

    c Diameter = ?   Radius = 3.5 mm   Circumference = ?

    d Diameter = ?   Radius = ?   Circumference = 51 cm

    e Diameter = ?   Radius = ?   Circumference = 6.12 m

2 The diameter of a £1 coin is 23 mm.
   What is the length of its circumference, correct to 2 d.p.?

3 The radius of a 1p coin is 5.5 mm.
   What is the length of its circumference in centimetres, correct to 2 d.p.?

4 A bicycle wheel has a radius of 34 cm.
   How far does it travel in 5 revolutions?

5 A gardener wishes to put an edging around his circular rose border.
   If the diameter of the border is 3.5 m, what length of edging will he need?

6 Find the perimeters of the following shapes:

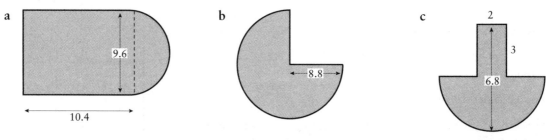

7   The wheel on a toy has a radius of 3 cm. The toy is pushed 5 metres along a floor. How many times does the wheel rotate?

Take π to be 3.14 or use the π key on your calculator.

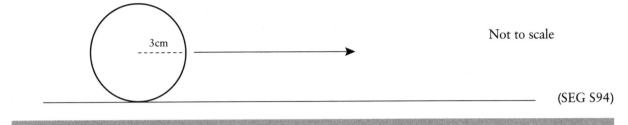

Not to scale

(SEG S94)

Area of a circle

A circle with radius 1 has an area $= \pi$

A circle with radius r has an area $= \pi \times r^2$.

For all circles:

$$A = \pi r^2$$

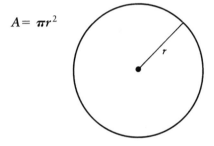

Example 1

A circle is drawn which has a diameter of 6 cm. What is the area of the circle?

$$\text{Diameter} = 6 \, \text{cm}$$
$$\text{Radius} = 3 \, \text{cm}$$
$$\text{Area} = \pi \times r^2$$
$$= \pi \times 3 \times 3 \, \text{cm}^2$$
$$= 28.3 \, \text{cm}^2 \text{ (to 3 s.f.)}$$

Example 2

A circle has an area of $10 \, \text{m}^2$. What is the radius of the circle?

$$\text{Area} = \pi r^2$$
$$\pi \times r^2 = 10$$
$$r^2 = \frac{10}{\pi}$$
$$r = \sqrt{\frac{10}{\pi}} = 1.78 \, \text{m} \text{ (to 3 s.f.)}$$

Exercise 44G

1   Use the π button on your calculator, or take $\pi = 3.14$, to calculate the area of:

  a   a circle of radius 7.1 cm

  b   a circle of radius 29.5 in

  c   a circle of diameter 13.6 mm

  d   a semicircle of radius 4.9 cm

  e   a semicircle of diameter 9.28 m

2   By taking an approximate value for π of 3, estimate the length of the radius of a circle with area:

  a $27 \, \text{cm}^2$ c $108 \, \text{cm}^2$ e $300 \, \text{ft}^2$

  b $150 \, \text{in}^2$ d $48 \, \text{m}^2$

3   Calculate the length of the radius, correct to 3 s.f., for each of the circles in question 2 above.

4   A discotheque has a circular dance floor which covers 50 m². What is the diameter of the dance floor, correct to 1 d.p.?

5   The Costains buy a semicircular rug to fit their hearth which is 1.2 m wide. If the carpet costs £12.99 per square metre, how much does the rug cost?

6   Mr Swain is making a circular fish pond. Each fish requires a surface area of 1000 square centimetres. If Mr Swain intends to keep ten fish, what is the smallest radius, correct to the nearest centimetre, he must use for his pond?

44.4   Composite areas

Example

A luggage label is made of thin cardboard with the dimensions shown in the figure.
Calculate the area of cardboard which the label covers.

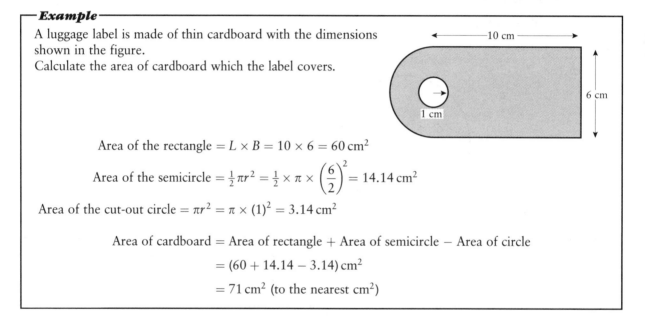

Area of the rectangle $= L \times B = 10 \times 6 = 60 \text{ cm}^2$

Area of the semicircle $= \frac{1}{2}\pi r^2 = \frac{1}{2} \times \pi \times \left(\frac{6}{2}\right)^2 = 14.14 \text{ cm}^2$

Area of the cut-out circle $= \pi r^2 = \pi \times (1)^2 = 3.14 \text{ cm}^2$

Area of cardboard = Area of rectangle + Area of semicircle − Area of circle

$= (60 + 14.14 - 3.14) \text{ cm}^2$

$= 71 \text{ cm}^2$ (to the nearest cm²)

Exercise 44H

1   A running track of length 400 m is the shape of a rectangle with 2 semicircular ends of radius 40 m.

   a   What is the length of each straight?

   b   What is the total area enclosed by the running track?

2   Find (i) the perimeter and (ii) the area of each of the following shapes:

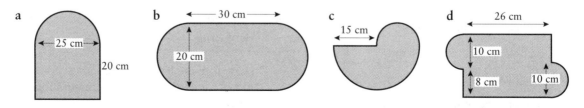

3 Calculate the shaded area in each of the following:

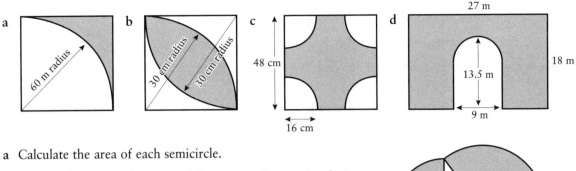

a 60 m radius

b 30 cm radius   30 cm radius

c 48 cm   16 cm

d 27 m   18 m   13.5 m   9 m

4 a Calculate the area of each semicircle.

b What is the sum of the areas of the two smaller semicircles?

c How are the areas of the three semicircles connected?

d What is the total area of the figure?

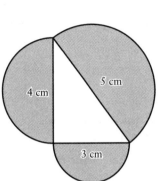

5 cm   4 cm   3 cm

5 The diagram shows the cross-section of a drainage pipe.

The radius of the outer circle
is *R* and of the inner circle is *r*.
(The shaded area is called an annulus.)

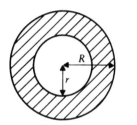

R   *r*

a Write down an expression for the area of the outer circle in terms of *R*.

b Write down an expression for the area of the inner circle in terms of *r*.

c Write down an expression for the shaded area.

d If $R = 13.2$ cm and $r = 10.2$ cm, find the area of the cross-section of the pipe.

6 Find the area of the cross-section of a pipe with an outer diameter of 16 mm and an inner diameter of 12 mm.

7 This millstone is a circle of radius 24 inches.
A square of side length 12 inches is cut out
of the centre of the circle as shown.

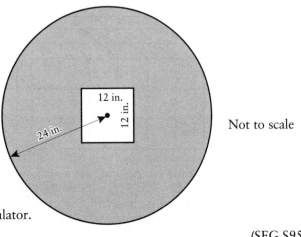

12 in.   12 in.   24 in.   Not to scale

a Calculate the circumference of the
millstone.

Take $\pi = 3.14$ or use the π key on your calculator.

b Calculate the shaded area.   (SEG S95)

44.5 Volume

Volume is a measure of the amount of space which is taken up by a solid shape.

Solids are three dimensional shapes, i.e. they have length, breadth and height.

A **cube** is a solid with six square faces (or sides), hence the length, breadth and height of a cube are all equal.

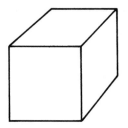

A cube which is 1 cm long, 1 cm wide and 1 cm high has a volume of 1 cubic centimetre ($= 1\,\text{cm}^3$).

Similarly, the volume of a cube of side 1 metre is 1 cubic metre ($= 1\,\text{m}^3$).

In imperial units the measures of volume are cubic feet, cubic inches, etc.

A **cuboid** is a solid with six rectangular faces.

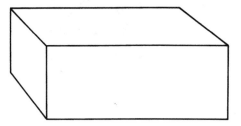

To find the volume of a cuboid we need to find out how many unit cubes it contains.

Example

A cuboid which is 6 cm × 4 cm × 3 cm can be divided as shown:

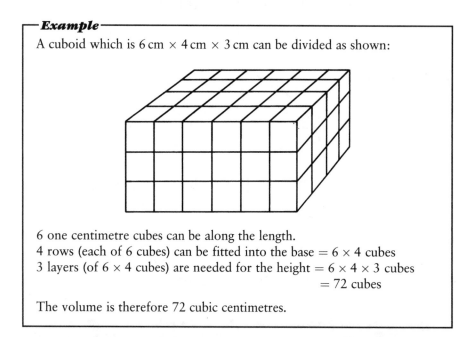

6 one centimetre cubes can be along the length.
4 rows (each of 6 cubes) can be fitted into the base = 6 × 4 cubes
3 layers (of 6 × 4 cubes) are needed for the height = 6 × 4 × 3 cubes
                                            = 72 cubes

The volume is therefore 72 cubic centimetres.

Volume of a cuboid = Length × Breadth × Height

$$V = l \times b \times h$$

For a cube, $Length = Breadth = Height$, and

$$V = l^3$$

—Example—

Find the volume of the solid shown below:

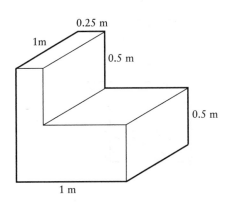

This solid may be divided into two cuboids, A and B, as shown:

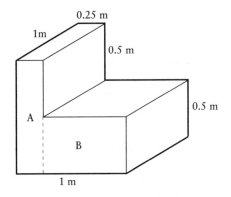

Volume of solid = Volume of cuboid A + Volume of cuboid B

$$= (1 \times 0.25 \times 1) + (0.5 \times 0.75 \times 1)$$

$$= 0.25 + 0.375$$

$$= 0.625 \text{ m}^3$$

Alternatively:

Volume of solid = Area of cross-section × Length

$$= (\text{Area A} + \text{Area B}) \times 1$$

$$= (1 \times 0.25 + 0.5 \times 0.75) \times 1$$

$$= 0.625 \text{ m}^3$$

Prisms

A **prism** is a solid which has a constant cross-section (i.e. the cross-section of the top is exactly the same as the cross-section of the base).

Some common prisms are:

(i) a **cylinder** (with a cross-section which is a circle)

(ii) a **triangular prism**

(iii) a **rectangular prism** or cuboid

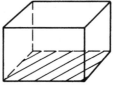

For a prism:

Volume = Area of cross-section × Height

For a cylinder:

Volume = Area of circle × Height

$$V = \pi r^2 h$$

Example 1

A Toblerone packet is a triangular prism.
The triangular cross-section has an area of 2.25 cm² and the length of the packet is 16.8 cm.
What is the volume of the Toblerone packet?

Volume = Area of cross-section × Length

$\quad\quad = 2.25 \text{ cm}^2 \times 16.8 \text{ cm}$

$\quad\quad = 37.8 \text{ cm}^3$

Example 2

Find the volume of a cylinder with a base radius of 3.5 cm and a height of 8 cm.

Volume of cylinder = Area of base × Height

$\quad\quad\quad\quad = \pi r^2 h$

$\quad\quad\quad\quad = \pi \times 3.5 \times 3.5 \times 8 \text{ cm}^3$

$\quad\quad\quad\quad = 307.9 \text{ cm}^3$

Example 3

Find the volume of a block of wood of length 12 cm which has a constant cross-section as shown in the diagram.

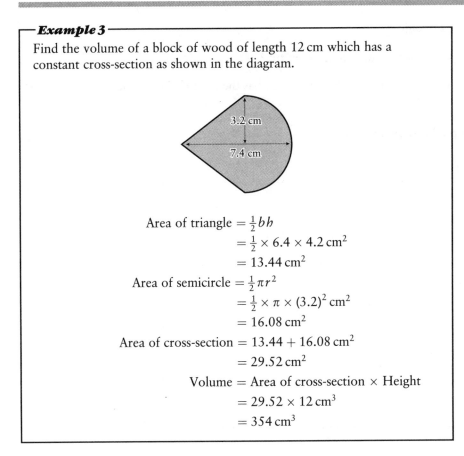

Area of triangle $= \frac{1}{2}bh$

$= \frac{1}{2} \times 6.4 \times 4.2 \, \text{cm}^2$

$= 13.44 \, \text{cm}^2$

Area of semicircle $= \frac{1}{2}\pi r^2$

$= \frac{1}{2} \times \pi \times (3.2)^2 \, \text{cm}^2$

$= 16.08 \, \text{cm}^2$

Area of cross-section $= 13.44 + 16.08 \, \text{cm}^2$

$= 29.52 \, \text{cm}^2$

Volume $=$ Area of cross-section $\times$ Height

$= 29.52 \times 12 \, \text{cm}^3$

$= 354 \, \text{cm}^3$

Exercise 44I

1 Calculate the volume of each of the following cuboids:

  a Length $= 6$ cm   Breadth $= 5$ cm   Height $= 3$ cm

  b Length $= 3.4$ m   Breadth $= 2.6$ m   Height $= 5.8$ m

  c Length $= 0.7$ m   Breadth $= 0.6$ m   Height $= 0.8$ m

  d Length $= 30$ cm   Breadth $= 22$ cm   Height $= 22$ mm

  e Length $= 2$ m   Breadth $= 60$ cm   Height $= 1$ m

2 The base of a cuboid has an area of $12 \, \text{cm}^2$. The volume of the cuboid is $40 \, \text{cm}^3$. What is the height of the cuboid?

3 Calculate the volume of a cube of side:

  a 8 cm   b 11 m   c 4.1 cm

4 The volume of a cube is $63 \, \text{cm}^3$.
What is the length of a side, correct to 1 d.p.?

5 A child has 27 wooden blocks, each a cube of edge 1 inch. She builds one large cube using all the blocks.

  a What is the volume of the large cube?

  b What is the length of an edge of the large cube?

6 An oil tank is 6 ft × 4 ft × 4 ft high.

   a How many cubic feet of oil does the tank hold when full?

   b When 54.5 cubic feet of oil has been used, by how much has the level in the tank fallen?

7 The plan of a swimming pool is shown below

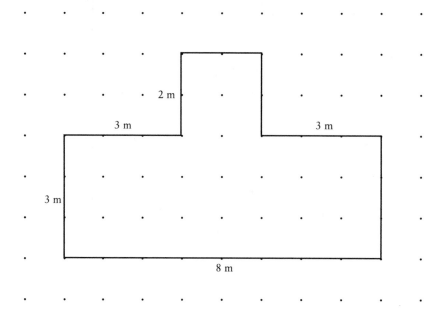

   a Calculate the perimeter of the swimming pool.

   b Calculate the area of the surface of the pool.

   The pool is to be bordered with square paving slabs of side length 0.5 m.

   c How many slabs are needed to put a border round the pool? Include all the corners in your calculations.

   d The pool is filled to a depth of 1.5 m.
   Calculate the volume of water in the pool.   (SEG W94)

*8 Find the volume of the following solids which have a length as given and a uniform cross-section as shown in the diagram.

   a Length = 15 cm   b Length = 5 cm   c Length = 10.5 cm

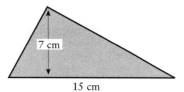

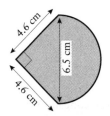

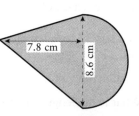

*9   A greenhouse is 10 m long and has a cross-section which is a square of
     side 5 m on top of which is a triangle, as shown in the diagram.

     The overall height of the greenhouse is 7.5 m.
     Calculate the volume of air inside the greenhouse.

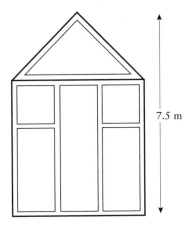

7.5 m

*10 Calculate the volume of a cylinder with:

     a Radius = 4 cm   Height = 12 cm

     b Radius = 9 cm   Length = 12 cm

     c Diameter = 14 in   Height = 11 in

     d Diameter = 8.4 cm   Length = 7 cm.

*11 Calculate the radius of a cylinder with

     a Volume = 32 cm³   Height = 5 cm

     b Volume = 49 cm³   Length = 2.3 cm

     c Volume = 84 in³   Height = 9 in.

*12 A tunnel is excavated from a hillside. The length of the tunnel is 300 m
     and its cross-section is as shown:

     Calculate the volume of earth which is removed to make the tunnel.

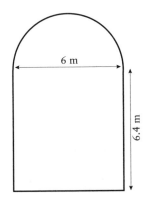

6 m

6.4 m

*13 A section of metal pipe has an outer diameter of 5 cm and the metal is
     3 mm thick. The pipe is 18 cm long.
     What volume of metal is used in making the section of pipe?

*14 The diagram shows triangle *ABC* with dimensions as shown.

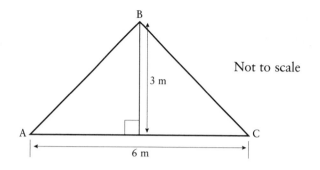

a Calculate the area of triangle *ABC*.

The roof of a warehouse is a triangular prism with dimensions as shown. The roof space is used for storage.

b Calculate the volume of storage space in the roof.

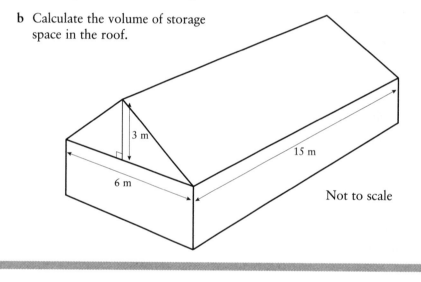

(SEG W94)

44.6 Dimensions in formulae

Length has one dimension.
Area is (length × length) and therefore has two dimensions.

Formulae representing area must involve the product of 2 lengths, for example r^2 or $2rh$ or $2r^2 + 2\pi rh$.

$3r$, $7r$, $7r + \pi h$ only involve one length and hence cannot represent areas. The 7 in the $7r$ is a constant and does not result in an extra dimension.

$3r + 2r^2$ is the sum of two terms, one representing length ($3r$) and one area ($2r^2$), which means that $3r + 2r^2$ is not a formula identifying length or area.

Volume is (length × length × length) and hence has three dimensions. Therefore formulae representing volume must involve the product of three lengths, e.g.

r^3, $\dfrac{4\pi r^3}{3}$, lbh, r^2h.

Example

State whether $\frac{1}{3}\pi r^2 h$ is a formula identifying perimeter, area or volume.

$\frac{1}{3}$ and π are numbers. r and h are lengths.

$\therefore$ $\frac{1}{3}\pi r^2 h$ is the product of 3 lengths.

$\therefore$ The formula identifies a volume.

Exercise 44J

State with a reason whether the following identify perimeter, area, volume or none of these.

1 $2r + 7h$

2 $3r^2 + 7rh$

3 $\pi r^2 h$

4 $3r^2 h + 7r$

5 $2r^2 h + 3r^3 + 4\pi h^3$

6 $2r^2 + 3rh + 7h$

7 $2r(r + h) + 3\pi r^2$

8 A biscuit tin is a cylinder of radius 8.4 cm.

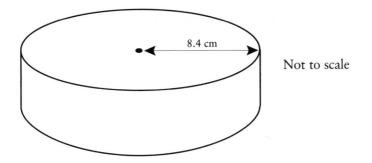

8.4 cm

Not to scale

a Calculate the area of the top of the biscuit tin.

b Sanjit uses the formula $V = 2\pi rh$ to calculate the volume of the biscuit tin.

By considering the dimensions, explain why this formula cannot be used to calculate volume.

(SEG S96)

45 Pythagoras and Trigonometry

45.1 Pythagoras' theorem

The name of Pythagoras must be the most widely recognised of all the famous Greek mathematicians.

The theorem attributed to his name appears in all school geometry and trigonometry textbooks.

He was born in 582BC on the island of Samos. He later lived in Crotona where he founded a secret society known as the Pythagoreans. The main purpose of this 'brotherhood' was political, but they were also interested in mathematics and astronomy. They believed that everything could be explained in terms of numbers and number patterns.

The theorem for which they are best known, concerning the sides of a right-angled triangle, had been known in a limited form for many hundreds of years.

Egyptian surveyors, or rope-stretchers, knew that a triangle with sides of lengths 3, 4 and 5 units always contained a right angle.
They used a rope with 12 equally spaced knots.

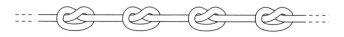

The ends could then be joined together and the rope stretched to form a right-angled triangle which could be used for marking out fields after the Nile floods or for building.

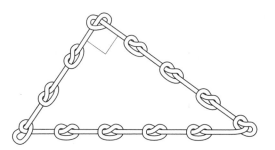

Even before the Egyptians, the Babylonians knew of hundreds more of these triangles.

For example, triangles with sides of 5, 12 and 13 units or 8, 15 and 17 units are always right-angled triangles.

It was Pythagoras, or one of the Pythagoreans, however, who discovered the relationship which connects the sides of these triangles.

Pythagoras' theorem states that in any right-angled triangle:

The square of the hypotenuse is equal to the sum of the squares of the other two sides.

(The **hypotenuse** is the side of a triangle which is opposite to a right angle.)

The theorem can easily be illustrated for the Egyptians' 3, 4, 5 triangle:

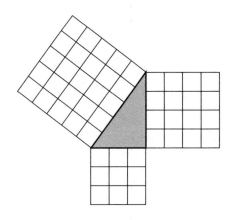

It was about two hundred years later that the theorem was proved, by Euclid when he was writing his textbook of geometry theorems.

(The theorem is in fact true for any shapes which are constructed on the three sides of a right-angled triangle: see Exercise 44H question 4 on page 446.)

For a triangle ABC with sides of length a, b and c, as shown:

$$c^2 = a^2 + b^2$$

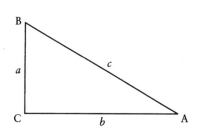

Example 1

John is building a garage with a rectangular base, 6.5 m long and 3.4 m wide.
John knows that to ensure he marks out right angles at each corner he must also mark out one of the diagonals.[‡]
Calculate the length of this diagonal.

By Pythagoras' theorem:

$$x^2 = 6.5^2 + 3.4^2$$
$$= 42.25 + 11.56$$
$$= 53.81$$
$$\therefore \quad x = 7.34$$

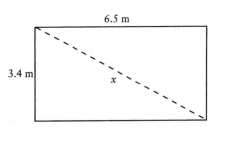

$\therefore$ The diagonal has a length of 7.34 m.

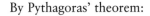

[‡]If the three sides of a triangle satisfy Pythagoras' theorem ($a^2 + b^2 = c^2$) the angle opposite the largest side is a right angle. Thus when John measures the diagonal in his garage base he knows that he will create a right angle when the diagonal is of the correct length.

Example 2

In PQR, R $= 90°$, PQ $= 17$ in, QR $= 10$ in. Find PR.

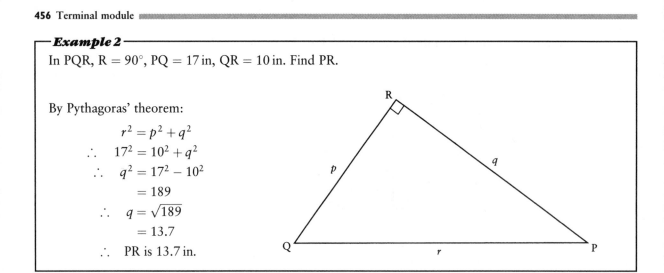

By Pythagoras' theorem:

$$r^2 = p^2 + q^2$$
$$\therefore \quad 17^2 = 10^2 + q^2$$
$$\therefore \quad q^2 = 17^2 - 10^2$$
$$= 189$$
$$\therefore \quad q = \sqrt{189}$$
$$= 13.7$$
$$\therefore \quad \text{PR is } 13.7 \text{ in.}$$

Example 3

In the diagram below, $\angle$ CAB $= \angle$ CBD $= 90°$. AC $= 8$ cm, AB $= 6$ cm, and CD $= 12$ cm.

Find BC and BD.

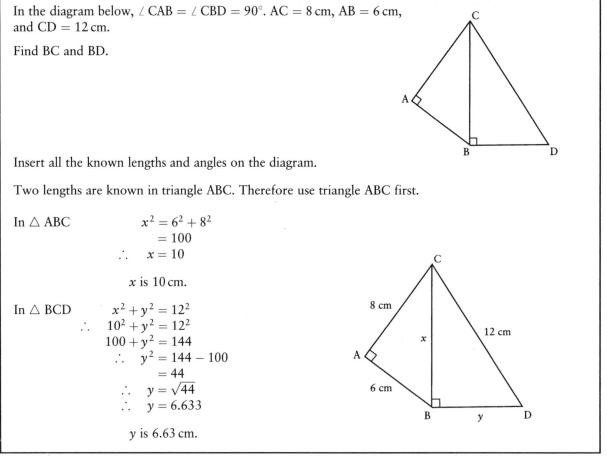

Insert all the known lengths and angles on the diagram.

Two lengths are known in triangle ABC. Therefore use triangle ABC first.

In △ ABC
$$x^2 = 6^2 + 8^2$$
$$= 100$$
$$\therefore \quad x = 10$$

x is 10 cm.

In △ BCD
$$x^2 + y^2 = 12^2$$
$$\therefore \quad 10^2 + y^2 = 12^2$$
$$100 + y^2 = 144$$
$$\therefore \quad y^2 = 144 - 100$$
$$= 44$$
$$\therefore \quad y = \sqrt{44}$$
$$\therefore \quad y = 6.633$$

y is 6.63 cm.

Exercise 45A

1 **a** Construct a triangle with sides 6 cm, 8 cm, 10 cm and check that it is right-angled by measuring the largest angle.
An accuracy of between 89° and 91° should be obtained.

   b Measure the two remaining acute angles in the triangle.

2 The following questions refer to the diagram opposite:

   a $b = 5$ cm, $c = 12$ cm. Find a.

   b $b = 2.7$ mm, $a = 3.8$ mm. Find c.

   c $a = 3.5$ in, $c = 2.1$ in. Find b.

3 A triangle has sides $p = 10$ cm, $q = 17$ cm, $r = 13$ cm.
Is it a right-angled triangle?

4 Find the unknown lengths:

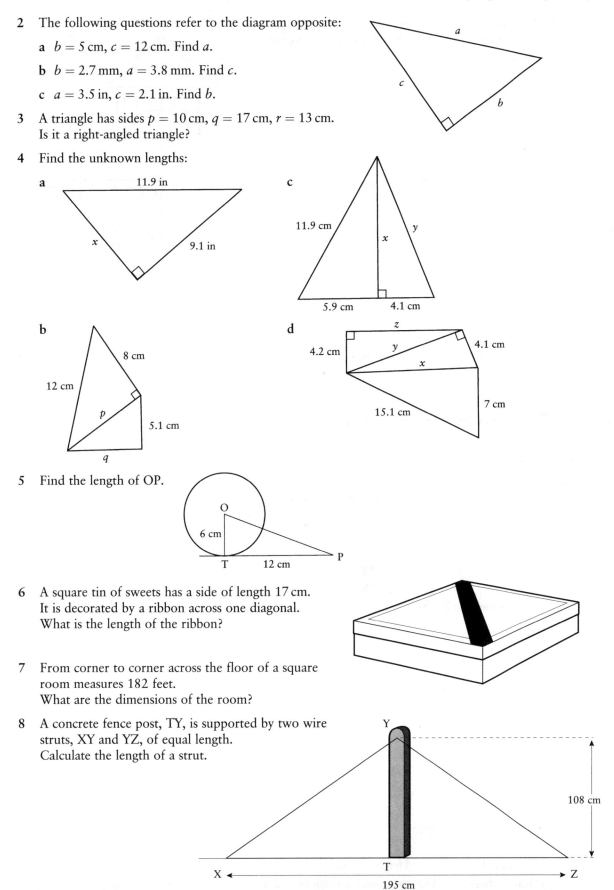

5 Find the length of OP.

6 A square tin of sweets has a side of length 17 cm.
It is decorated by a ribbon across one diagonal.
What is the length of the ribbon?

7 From corner to corner across the floor of a square
room measures 182 feet.
What are the dimensions of the room?

8 A concrete fence post, TY, is supported by two wire
struts, XY and YZ, of equal length.
Calculate the length of a strut.

9   A house has a dormer window, as shown in the diagram. Calculate the length of the roof, over the dormer window, from the ridge to the gutter edge.

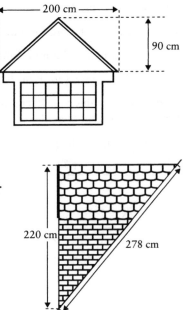

10 A side-view of the dormer window in question 9 is shown here. Calculate how far the window projects from the roof of the house.

45.2   Trigonometrical ratios

The tangent ratio

The word **trigonometry** means triangle measurement.
The surveyors of ancient times were able to find the height of a tall object by comparing its shadow with the shadow of an object (e.g. a stick) of known height.

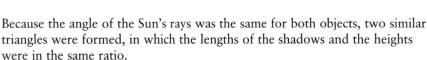

Because the angle of the Sun's rays was the same for both objects, two similar triangles were formed, in which the lengths of the shadows and the heights were in the same ratio.

$$\frac{\text{Height of pyramid}}{\text{Length of pyramid's shadow}} = \frac{\text{Height of stick}}{\text{Length of stick's shadow}} = \frac{4.0}{2.5}$$

The unknown height could then be calculated.

$$\text{Height of pyramid} = \frac{4.0}{2.5} \times \text{Length of its shadow}$$

$$= 1.6 \times 300 \, \text{ft}$$

$$= 480 \, \text{ft}$$

(The original height of the Great Pyramid was 481.4 ft, although 31 feet are now missing.)

Any other triangle with equal angles will also have its sides in the same ratio and a height can quickly be calculated if the ratio is known.

For example, in the triangles below, the angle at the base is 58°. When the base of the triangle is 2.5 units, the height of the triangle is 4.0 units. Therefore, when the base is 1 unit the height is 1.6 units.

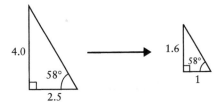

The height of any similar triangle can be found by multiplying the length of its base by 1.6.

Height of tree = 5.5 m × 1.6
    = 8.8 m

The ratio 1.6 is called the **tangent of 58°**.

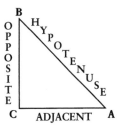

In about 150 BC, a Greek mathematician working in Alexandria, called Hipparchus, realized that a great deal of time could be saved if a table of values for different angles was constructed.

In a triangle ABC, if A is the angle involved in a calculation, the sides are named as shown in the diagram on the right:

The tangent ratio is therefore:

$$\text{tangent of angle A} = \frac{\text{Side opposite to angle A}}{\text{Side adjacent to angle A}} = \frac{BC}{CA}$$

or $$\tan A = \frac{\text{Opposite}}{\text{Adjacent}}$$

$$\tan A = x$$

implies $A = \tan^{-1} x$

To use your calculator to find $\tan^{-1} x$ key in x | INV | | tan |

Example 1

In the triangle shown, find angle A.

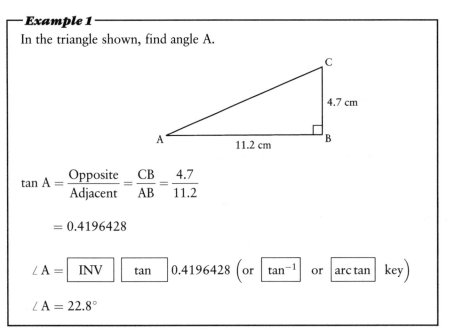

$$\tan A = \frac{\text{Opposite}}{\text{Adjacent}} = \frac{CB}{AB} = \frac{4.7}{11.2}$$

$$= 0.4196428$$

$\angle A = \boxed{\text{INV}} \; \boxed{\tan} \; 0.4196428 \; \left(\text{or} \; \boxed{\tan^{-1}} \; \text{or} \; \boxed{\text{arc tan}} \; \text{key} \right)$

$\angle A = 22.8°$

Example 2

In the triangle below, find PQ.

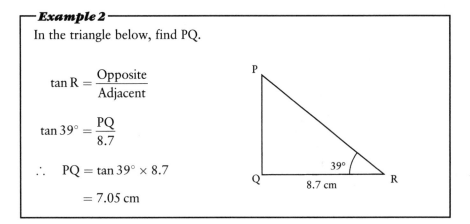

$$\tan R = \frac{\text{Opposite}}{\text{Adjacent}}$$

$$\tan 39° = \frac{PQ}{8.7}$$

$$\therefore \quad PQ = \tan 39° \times 8.7$$

$$= 7.05 \text{ cm}$$

Example 3

In the triangle shown, find XY.

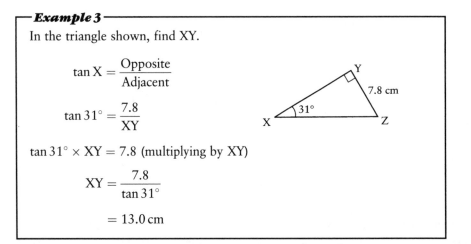

$$\tan X = \frac{\text{Opposite}}{\text{Adjacent}}$$

$$\tan 31° = \frac{7.8}{XY}$$

$$\tan 31° \times XY = 7.8 \quad \text{(multiplying by XY)}$$

$$XY = \frac{7.8}{\tan 31°}$$

$$= 13.0 \text{ cm}$$

Exercise 45B

1   Find AB.

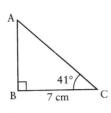

2   Find angle R.

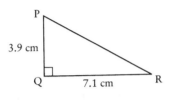

3   Find XY.

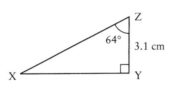

4   Find *x* and *y*.

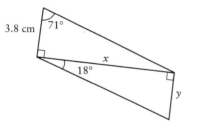

5   ABC is an isosceles triangle with AB = AC.

   a   Find the altitude of triangle ABC.
   [The altitude of triangle ABC is the line
   drawn from A perpendicular to BC.]

   b   Calculate the area of triangle ABC.

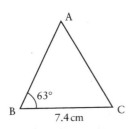

6   In the diagram, PT is a tangent to the circle,
centre O, which has a radius of 6.5 cm.
PT = 20 cm.

Find angle RST.

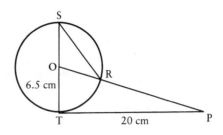

7   A chord of length 4 cm subtends an angle of
40° at O, the centre of the circle.
Find:

   a   the perpendicular distance of the chord from
   the centre

   b   the radius of the circle.

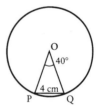

8   CA and CB are tangents to the circle.
Find ∠ ACB.

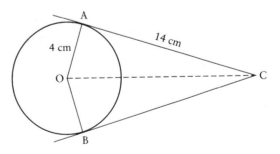

9   Find the length of the chord RQ.

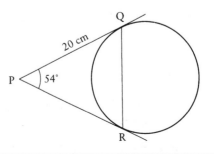

Sine and cosine ratios

The tangent ratio does not involve the hypotenuse, but the sides of a triangle can be paired in two more ways to give ratios which do involve the hypotenuse.

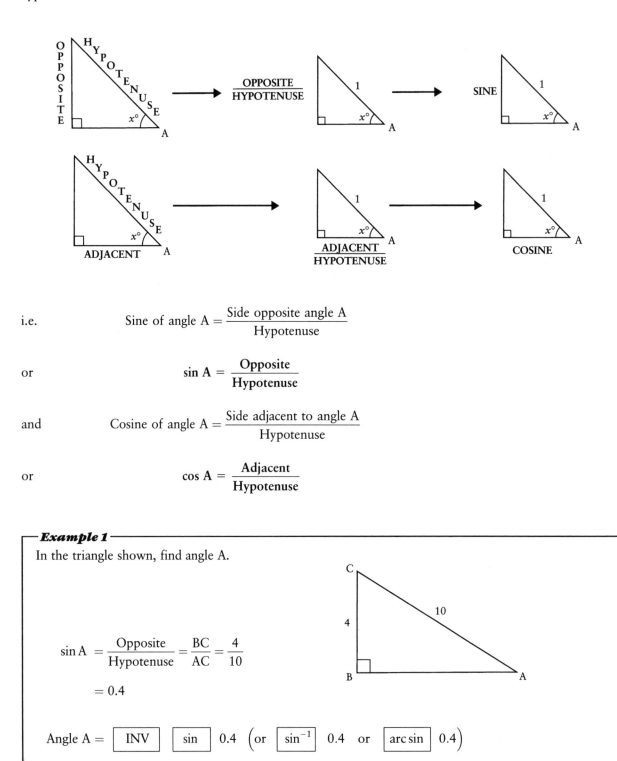

i.e.   Sine of angle A $= \dfrac{\text{Side opposite angle A}}{\text{Hypotenuse}}$

or $\sin A = \dfrac{\text{Opposite}}{\text{Hypotenuse}}$

and   Cosine of angle A $= \dfrac{\text{Side adjacent to angle A}}{\text{Hypotenuse}}$

or $\cos A = \dfrac{\text{Adjacent}}{\text{Hypotenuse}}$

Example 1

In the triangle shown, find angle A.

$\sin A = \dfrac{\text{Opposite}}{\text{Hypotenuse}} = \dfrac{BC}{AC} = \dfrac{4}{10}$

$= 0.4$

Angle A = $\boxed{\text{INV}}$ $\boxed{\text{sin}}$ 0.4 $\left(\text{or } \boxed{\sin^{-1}} \text{ 0.4 or } \boxed{\text{arc sin}} \text{ 0.4}\right)$

$= 23.6°$

Example 2

In the triangle shown, find BC.

$$\cos C = \frac{\text{Adjacent}}{\text{Hypotenuse}}$$

$$\cos 52° = \frac{a}{7}$$

$$a = \cos 52° \times 7$$

$$BC = 4.31 \text{ cm}$$

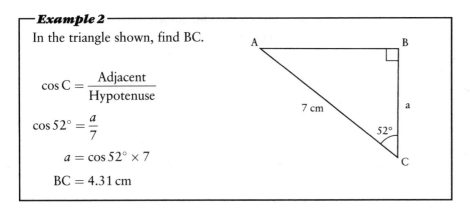

Example 3

Find the length of the radius OQ.

Draw in OR perpendicular to PQ, then ∠ ROQ = 20° and RQ = 2 cm.

$$\sin Y = \frac{\text{Opposite}}{\text{Hypotenuse}}$$

$$\sin 20° = \frac{2}{OQ}$$

$$\sin 20° \times OQ = 2$$

$$OQ = \frac{2}{\sin 20°} = 5.85 \text{ cm}$$

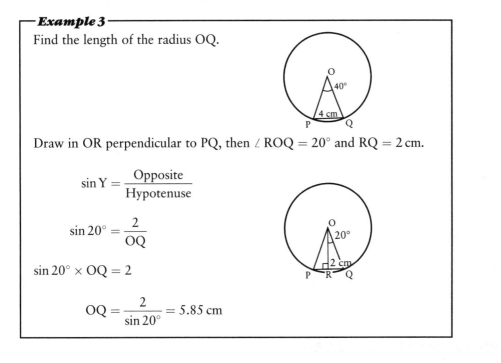

Exercise 45C

1   **a**   Find PQ.   **b**   Find XY.   3   **a**   Find angle C.   **b**   Find angle R.

2   **a**   Find AB.   **b**   Find PR.   4   **a**   Find angle A.   **b**   Find angle P.

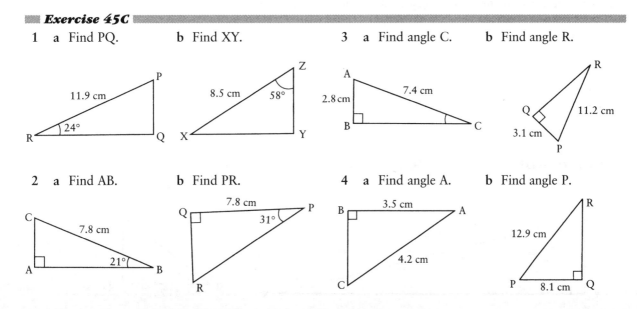

5   Find x and y.

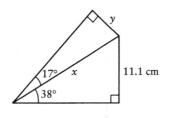

6   Find p and q.

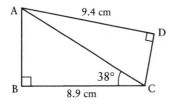

7   Find AC and angle CAD.

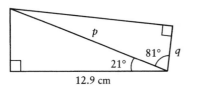

8   Draw in the altitude AD.
    Hence find angle BAC.

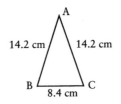

9   PQ and PR are tangents to the circle.
    Calculate the length of the chord QR.

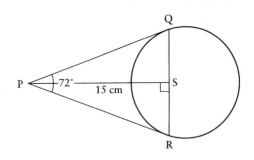

Choosing the ratio to solve a problem

When an angle is to be found in a right-angled triangle, first decide which
two sides have been given. Then use the trigonometric ratio which includes
these two sides.

When a side is to be found in a right-angled triangle and one angle (other
than the right angle) is given, consider which side is given (opposite, adjacent
or hypotenuse), and which is the side required. Then use the trigonometric
ratio which includes these two sides.

Exercise 45D

1   Find angle X.

2   Find PQ.

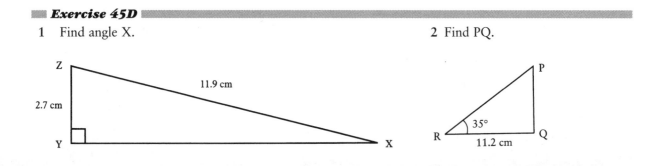

3 Sides AB and AC are equal.
Draw in the altitude AD.
Find AB.

4 Sides AB and DC are parallel.
Find p and q.

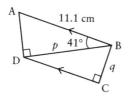

5 A ladder, 3.6 m long, rests against a window sill which is 3.3 m above the ground.
How far from the wall is the foot of the ladder?

6 A ramp is to be constructed to allow a wheelchair access to a building.
The height of the step into the building is 15 cm. The slope of the ramp is to have an angle of 10°.

a What is the length of the top of the ramp?

b How far from the doorstep will the ramp protrude?

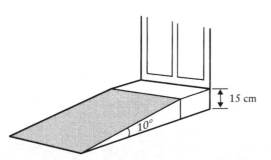

7 A dormer window is shown in the diagram below.

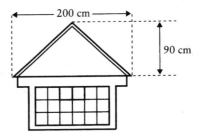

Calculate the pitch of the dormer roof (i.e. the angle the roof makes with the vertical).

8 The diagram shows the side view of the dormer window.
Calculate the pitch of the roof of the house.

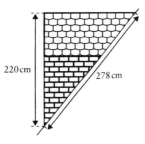

9 A step ladder is 125 cm long. When fully opened the distance between the foot of the ladder and its support is 82 cm.

a How high above the ground is the top of the ladder?

b What is the angle between the ladder and its support when it is fully opened?

10 A pawnbroker's sign consists of a right-angled triangular frame attached to a vertical wall as shown. The horizontal bar of the frame is 3 m above the ground and is 0.75 m long. The hypotenuse of the triangle is 2 m long.

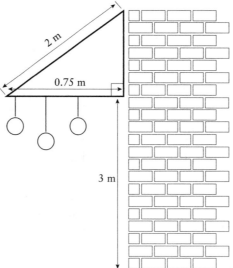

a Calculate the height of the wall.

b Calculate the angle between the hypotenuse and the horizontal side of the triangle.

(SEG S94)

45.3   Angles of elevation and depression

If you look at an object, the angle formed between the horizontal and your line of sight is called

the angle of elevation, if it is above the horizontal,
the angle of depression if it is below the horizontal.

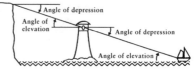

Example 1

Bill stands 200 m from a church and measures the angles of elevation of the top and the bottom of the steeple. These are 19° and 15° as shown below.

Calculate the height of the steeple.

$$\frac{PR}{200} = \tan 19°$$

$$\therefore \quad PR = 200 \tan 19°$$

$$= 68.86$$

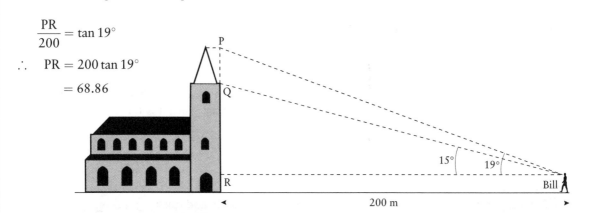

$$\frac{QR}{200} = \tan 15°$$

$$\therefore \quad QR = 200 \tan 15°$$

$$= 53.59$$

$$PQ = PR - PQ = (68.86 - 53.59) = 15.27 = 15.3 \text{ (to 3 s.f.)}$$

$\therefore$ The height of the steeple is 15.3 metres.

Example 2

A coastguard watching a yacht race sees a yacht at an angle of depression of 4°. He knows that he is 300 ft above sea level.
How far from him is the yacht?

Draw a diagram.

From the diagram:

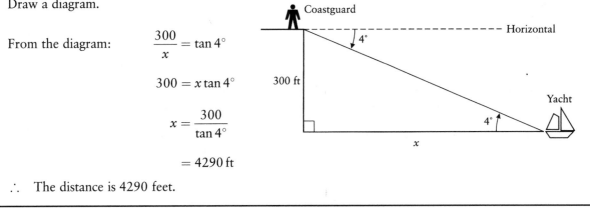

$$\frac{300}{x} = \tan 4°$$

$$300 = x \tan 4°$$

$$x = \frac{300}{\tan 4°}$$

$$= 4290 \text{ ft}$$

$\therefore$ The distance is 4290 feet.

Exercise 45E

1 From a boat, the angle of elevation of the top of a cliff is 74°.
The height of the cliff is known to be 800ft.
How far is the cliff from the boat?

2 The height of a tree is 58 m. Jacky, who is 2 m tall, is 200 m away from the tree on horizontal ground.
What is the angle of elevation of the top of the tree from Jacky's eyes?

3 John sees a plane at an angle of elevation of 3.8°. The plane is 2.1 miles horizontally from John.
What is its height?
Give your answer in feet.
(5280 ft = 1 mile.)

(*Note.* In this type of question, where the distances are large, the height of a person is negligible and can be ignored.)

4   A plane flies horizontally over Sue at a height of 2 miles. When she first sees it, the plane is at an angle of elevation of 74° and it takes 9.2 seconds to pass over her and once again to be at an elevation of 74°.

What is its speed? (Speed = $\dfrac{\text{Distance}}{\text{Time}}$)

5   Ken is 30 m from a flagpole. The angle of elevation of its top is 12.4°, and the angle of depression of the bottom is 2.1°.

How high is the flagpole?

6   In a penalty shoot-out, the player is 36 ft away from the goal mouth, and the goal has height 8 ft. What is the maximum angle of elevation at which the ball can be kicked to go into the goal if it travels in a straight line?

7   A space shuttle is rising at 2000 ft per second. Avril is 2 miles away from the launch pad. What is the angle of elevation of the space shuttle after:

   a 2 minutes   b 5 minutes?

8   A balloon is positioned 420 metres above a showground.

Angus is 3 miles away from the showground. What is the angle of elevation of the balloon? (1 mile = 1.6 kilometres.)

9   Joan is standing on a church tower 15 m high. She is looking at the top of a lamp post which has height 8 m. She finds that the angle of depression of the top of the lamp post is 8°.

The church tower and the lamp post stand on level ground.

   a Draw a diagram to show this information.

   b Calculate the distance from the church tower to the lamp post.   (SEG S89)

---

10 A group of geography students needs to calculate the width of a river. They measure the angle of elevation of the top of an electricity pylon from opposite banks of the river, P and L. The pylon is 70 metres high.

70m

Not to scale

70m

43°   36°
      P   L
      | River |

43°

P   36°

River

L

Calculate the width of the river PL.   (SEG S96)

11 A frog sits on one side of a road. Directly opposite the frog is a vertical lamp post of height 15 feet. The angle of elevation of the top of the lamp post from the frog is 31°.

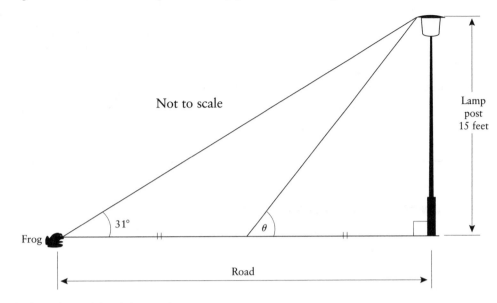

Not to scale

Lamp post 15 feet

Frog

31°

θ

Road

a Calculate the width of the road.

b Calculate the angle of elevation, θ, of the top of the lamp post when the frog is exactly half way across the road.   (SEG W94)

*45.4   Three-dimensional problems

Use of Pythagoras' theorem and trigometry in three-dimensional problems

Problems set in three-dimensions usually involve finding right-angled triangles within the three-dimensional structure.

It is advisable to show in separate clear diagrams every right-angled triangle to be used.

┌─ **Example** ───

The diagram shows a cuboid ABCDWXYZ.

AB = 7 cm, BC = 5 m, AW = 2.3 m.
Find the length of the diagonal AY.

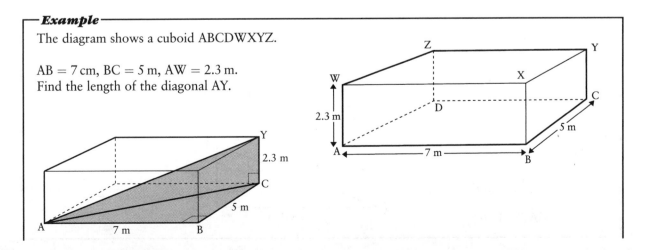

The right-angled triangles to be used are ABC, to find AC, and ACY to find AY.

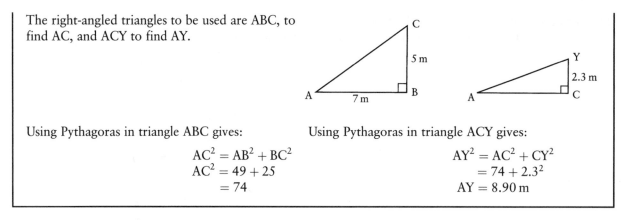

Using Pythagoras in triangle ABC gives:

$$AC^2 = AB^2 + BC^2$$
$$AC^2 = 49 + 25$$
$$= 74$$

Using Pythagoras in triangle ACY gives:

$$AY^2 = AC^2 + CY^2$$
$$= 74 + 2.3^2$$
$$AY = 8.90 \text{ m}$$

The angle between a line and a plane

Find a right-angled triangle containing the line and a line in the plane (which is a projection of the line on to the plane).

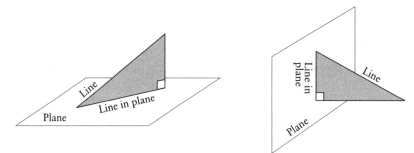

Example

Find the angle between:

a the line AZ and the plane ABCD

b the line AY and the plane ABCD.

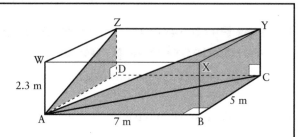

a Triangle ADZ contains the line AZ and the line AD which is in the plane ABCD. The angle ADZ is 90°.

Using ADZ gives:

$$\tan ZAD = \frac{2.3}{5.0} = 0.46$$

$$\angle ZAD = 24.7°$$

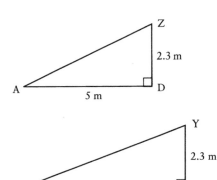

b Using CAY gives:

$$\tan CAY = \frac{2.3}{8.602}$$

$$\angle CAY = 15.0°.$$

The angle between two planes

Two planes which are not parallel, meet in a line. To find the angle between two planes which meet in a line:

1 identify this common line;

2 find two lines which are perpendicular to this common line, one in each plane.

The angle between these two lines is the angle between the two planes.

Example

ABCDV is a pyramid on a square base ABCD, with the apex V vertically above the centre of the square.

The side of the square is 12 cm and each slant edge is 10 cm, as shown in the diagram.
Find the angle between planes ABCD and ABV.

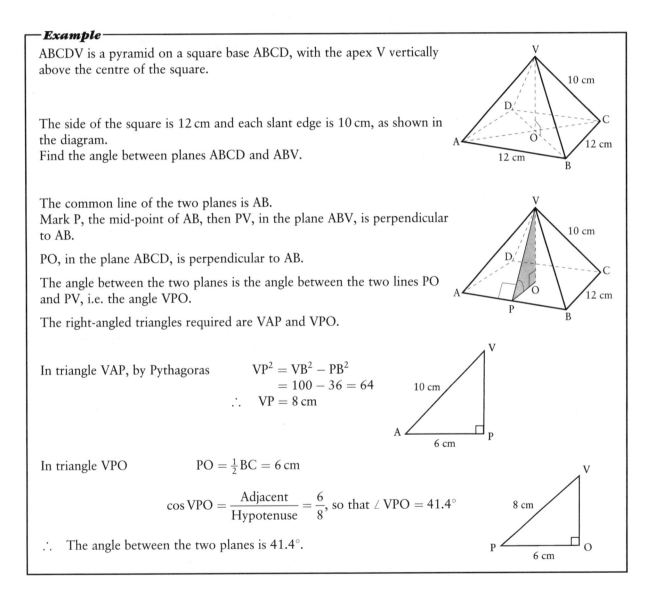

The common line of the two planes is AB.
Mark P, the mid-point of AB, then PV, in the plane ABV, is perpendicular to AB.

PO, in the plane ABCD, is perpendicular to AB.

The angle between the two planes is the angle between the two lines PO and PV, i.e. the angle VPO.

The right-angled triangles required are VAP and VPO.

In triangle VAP, by Pythagoras
$$VP^2 = VB^2 - PB^2$$
$$= 100 - 36 = 64$$
$$\therefore \quad VP = 8 \text{ cm}$$

In triangle VPO
$$PO = \tfrac{1}{2} BC = 6 \text{ cm}$$

$$\cos VPO = \frac{\text{Adjacent}}{\text{Hypotenuse}} = \frac{6}{8}, \text{ so that } \angle VPO = 41.4°$$

$\therefore$ The angle between the two planes is 41.4°.

Exercise 45F

1 A cardboard box, used for packing a television, is 26 inches long, 12 inches wide and 19 inches high.
What is the length of the longest diagonal of the box?

2 A room is 6.2 m by 4.3 m. Its height is 2.4 m. What is the length of the longest diagonal?

3 The volume of a cuboid container is 1254 cubic feet. Its length is 21 feet and its width is 8 feet. What is its height?
What is the length of the longest diagonal?

4 The volume of a water butt, in the shape of a cuboid, is 2200 litres. The width is 0.8 m and the height is 1.4 m.
What is the length of the diagonal?

5 A brick has length 10 inches and width 6 inches. The diagonal through the centre of the brick is 12.9 inches.
What is its height?

6 OABCD is a pyramid on a rectangular base of dimensions 16 cm and 12 cm. O is 15 cm vertically above the centre of the base.
What is the angle which OA makes with the base?

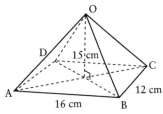

7 The diagram shows a pyramid on a square base, of side 7.8 cm, with
VP = VQ = VR = VS = 8 cm.
Calculate:

a the height of the pyramid,

b the angle which VQ makes with the base.

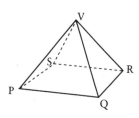

8 The barn shown in the diagram has a horizontal rectangular floor ABCD, with AB = 12 m and BC = 8 m. AP, BQ, CR and DS are all vertical and of length 5 m. X and Y are the highest points of semicircular arcs PXS and QYR. (Take π as 3.14 or use the π button on your calculator.)

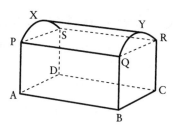

a Find the area of one end of the barn, for example BQYRC.

b Find the volume of the barn.

c Find the angle of depression of A from Y, giving your answer in degrees correct to the nearest 0.1 of a degree.

(SEG S90)

9 ABCDEFGH is a conservatory with face CDHG attached to a vertical wall. The horizontal base ABCD is a rectangle with AB 10 m, and BC 6 m. AE is 3.1 m, and DH is 3.8 m.
Find the angle which the plane EFGH makes with the horizontal.

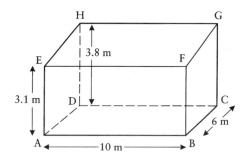

10 PQRSTUVW is a cuboid with PQ = 15 cm, QR = 11 cm and RV = 7.1 cm.
Find the angle which plane WRQ makes with plane PQRS.

11 An exhibition centre is in the shape of a pyramid with a hexagonal base of side 15 m. Each slant height of the pyramid is 18 m. Find:

a the vertical height of the pyramid,

b the angle which each face makes with the horizontal.

12 Part of a hillside slopes uniformly at an angle of 34° with the horizontal. A straight path makes an angle of 49° with the line of greatest slope. Find the angle which the path makes with the horizontal.

*46 Further Trigonometry

In all the trigonometry you have met so far, you have always used triangles which contain a right angle. This is obviously not typical of the real world. Hence we need to find a method of finding the lengths of sides and sizes of angles when the triangle does not contain a right angle.

46.1   Graphs of trigonometric functions

Using your calculator, you can obtain the value of $\sin x°$ for values of x other than those between $0°$ and $90°$.
If you plot the values obtained you will obtain the graph below.

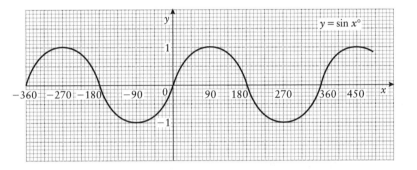

This is called a *sine curve* and such a curve is frequently found in the real world. Examples are the waves at sea (not near a beach!) and soundwaves. The wavelength is $360°$, being the change in x necessary to repeat the curve. You will notice that $\sin x$ is positive for values of x between $0°$ and $180°$ and that $\sin x$ is negative for values of x between $180°$ and $360°$ and between $0°$ and $-180°$.

In a similar manner you can obtain the graph of $y = \cos x$:

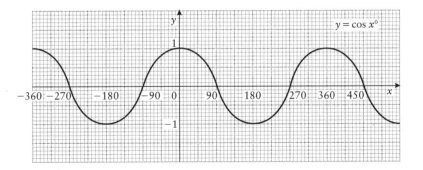

This curve is the same shape as the sine curve but the initial starting point (at 0) is moved along a quarter of a wavelength (i.e. $90°$), or the cosine curve is obtained by translating the sine curve by $\begin{pmatrix} -90 \\ 0 \end{pmatrix}$ (see page 395).

The tangent function is significantly different. Its graph is shown here:

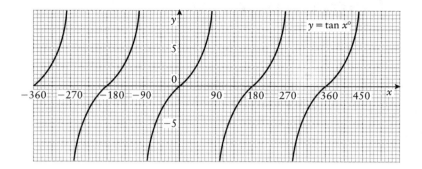

You will see that tan x is positive for values of x from 0 to 90, and from 180° to 270°, etc. Tan x is negative for the remaining angles.

Example 1

Find sin 280°.

Using your calculator you can find:
sin 280° = −0.9848

Example 2

Find three angles whose tangent is 2.21.

Inverse tan on your calculator gives 65.7° as one angle. Studying the tangent graph, shown above, you will note that it is repeated every 180°. Hence 180° + 65.7° = 245.7° is also an angle whose tangent is 2.21°. Similarly for 360° + 65.7° = 425.7°

∴   Three angles whose tangent is 2.21 are 65.7°, 245.7° and 425.7°.

Exercise 46A

1   Find, to three decimal places:

   a   sin 354°   c   tan 395°

   b   cos (− 156°)   d   sin 231.7°.

2   Find two angles whose sine is 0.876.

3   Find two angles satisfying tan x = −1.25.

4   Find two angles satisfying cos x = 0.352.

46.2   The area of a triangle

The formula for the area of a triangle is:

$$\text{area of a triangle} = \tfrac{1}{2}bh$$

where b is the length of the base of the triangle and h is the height of the triangle.

Suppose we need to find the area of triangle ABC when AC = 5 cm, BC = 8 cm and angle C = 55°.

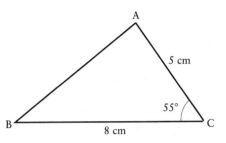

First drop a perpendicular line, AD, from A to BC. This is the height of the triangle. We now have two right-angled triangles.

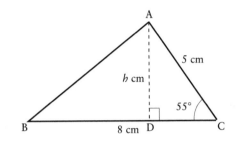

Let AD = h cm

$$\frac{h}{5} = \sin 55°$$

$$h = 5 \sin 55°$$

Area of triangle ABC = $\frac{1}{2} \times$ Base $\times$ Height

$$= \frac{1}{2} \times BC \times h$$

$$= \frac{1}{2} \times 8 \times 5 \sin 55°$$

$$= 16.4 \, \text{cm}^2$$

We can apply this method to a general triangle to obtain a formula for the area. In triangle ABC the lengths of AC and AB and the size of angle C are given. A perpendicular line, AD, is dropped from A to BC and its length is called h.

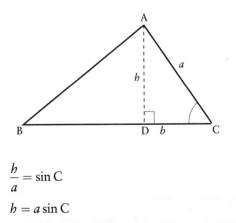

$$\frac{h}{a} = \sin C$$

$$h = a \sin C$$

Area of triangle ABC $= \frac{1}{2} \times$ Base $\times$ Height

$$= \frac{1}{2} \times b \times a \times \sin C$$

$$= \frac{1}{2} ab \sin C$$

The area can also be written as $\frac{1}{2} bc \sin A$ or $\frac{1}{2} ac \sin B$.

Note. The angle used is always the angle between the two given sides.

─ Example ─

Find the area of triangle PQR if PQ = 13 cm, PR = 16 cm and angle Q = 120°.

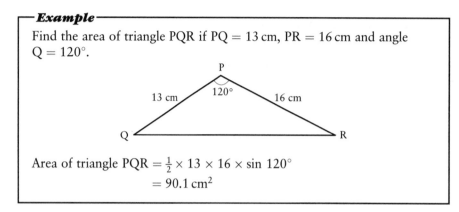

Area of triangle PQR $= \frac{1}{2} \times 13 \times 16 \times \sin 120°$

$$= 90.1 \text{ cm}^2$$

Exercise 46B

1 Find the areas of the triangles given below:

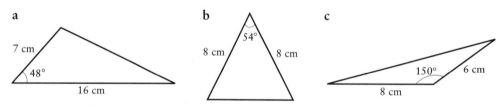

a

b

c

2 In triangle PQR, PQ = 23.7 m, QR = 18.2 m and angle PQR = 74°. What is the area of triangle PQR?

3 A kite is made with dimensions as shown in the diagram. What is the area of the paper covering the kite?

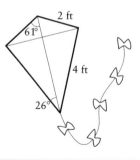

46.3   The sine rule

Consider triangle ABD:

From triangle ABD:

$$\frac{h}{c} = \sin B, \text{ i.e. } h = c \sin B$$

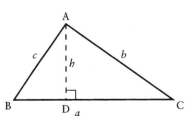
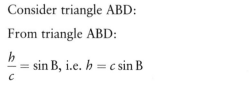

From triangle ACD:

$\dfrac{h}{b} = \sin C$, i.e. $h = b \sin C$

$\therefore \quad h = c \sin B = b \sin C$

or $\dfrac{c}{\sin C} = \dfrac{b}{\sin B}$

By dropping a perpendicular from C to AB it can be shown similarly that

$$\frac{a}{\sin A} = \frac{b}{\sin B}$$

The sine rule states that $\dfrac{a}{\textbf{sin A}} = \dfrac{b}{\textbf{sin B}} = \dfrac{c}{\textbf{sin C}}$

Example

In the triangle A = 62°, angle B = 70° and AC = 3.8 cm.
Find angle C and the sides AB and BC.

All the angles of a triangle add up to 180°.
$\therefore$ Angle C = $180° - (62° + 70°) = 48°$
Use the sine rule:

$\dfrac{a}{\sin 62°} = \dfrac{3.8}{\sin 70°} = \dfrac{c}{\sin 48°}$

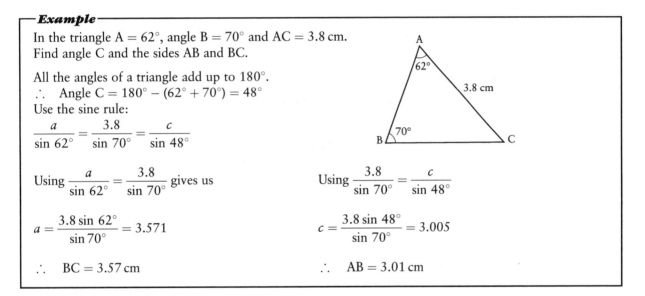

Using $\dfrac{a}{\sin 62°} = \dfrac{3.8}{\sin 70°}$ gives us

$a = \dfrac{3.8 \sin 62°}{\sin 70°} = 3.571$

$\therefore$ BC = 3.57 cm

Using $\dfrac{3.8}{\sin 70°} = \dfrac{c}{\sin 48°}$

$c = \dfrac{3.8 \sin 48°}{\sin 70°} = 3.005$

$\therefore$ AB = 3.01 cm

If an angle is to be found, the sine rule may be rearranged to give

$$\frac{\sin A}{a} = \frac{\sin B}{b} = \frac{\sin C}{c}$$

When choosing the appropriate formula to use, the first item to write down is
the unknown side or angle. For example, if side c is to be found and side a is
known, the appropriate version of the sine rule to use is:

$$\frac{c}{\sin C} = \frac{a}{\sin A}$$

Exercise 46C

1 Find:

   a angle C

   b AB

   c AC

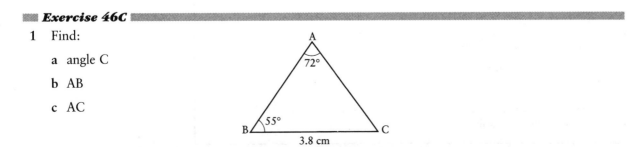

2 Find:

   a angle R

   b angle Q

   c PR.

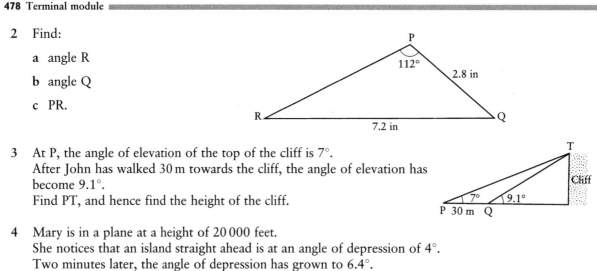

3 At P, the angle of elevation of the top of the cliff is 7°.
  After John has walked 30 m towards the cliff, the angle of elevation has
  become 9.1°.
  Find PT, and hence find the height of the cliff.

4 Mary is in a plane at a height of 20 000 feet.
  She notices that an island straight ahead is at an angle of depression of 4°.
  Two minutes later, the angle of depression has grown to 6.4°.
  Find the speed of the plane.

Ambiguous case

When you are given two sides, and you are finding a second angle, note that
there could be **two** solutions.

Figure 1 ⎯⎯⎯ **Figure 2**

In the diagrams shown, AB = 4 cm, AC = 3.7 cm and angle B is 63°.
However, ABC_1 and ABC_2 are two different triangles, both satisfying these
conditions.

Example

In triangle ABC above, AB = 4 cm, AC = 3.7 cm and angle B = 63°. Find
angle C.

Using the sine rule gives us

$$\frac{3.7}{\sin 63°} = \frac{4}{\sin C}$$

$$\sin C = \frac{4 \sin 63°}{3.7}$$

$$\therefore \quad C = 74.4°$$

This satisfies Figure 1 above.

For Figure 2 the angle is
$180° - 74.4° = 105.6°$

Unless you are given further information, you must assume that both
triangles are possible solutions.

Exercise 46D

1 In triangle LMN, LM = 7.3 cm, angle L is 41.1°, and MN = 5.1 cm. Find:

   a angle LNM

   b angle LMN.

   Hence find the two possible lengths for LN.

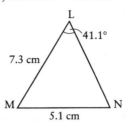

2 A ship sails on a bearing of 158° from a port. A lamp on an island is 7.8 km south of the port. After 30 minutes the ship is 3.5 km away from the lamp. What are the possible values of the ship's speed?

3 A ladder of length 4.6 m is supported by a leg of length 2.7 m. The ladder makes an angle of 25° with the ground.

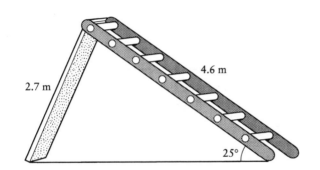

   What is the distance between the end of the ladder and the end of the leg?

46.4 The cosine rule

Consider the figure opposite. Let AD = x. Then CD = $b - x$.

Using Pythagoras' theorem on triangle ABD gives us

$$c^2 = h^2 + x^2$$

and from triangle BCD,

$$a^2 = (b - x)^2 + h^2.$$

Subtracting the two equations, we get

$$
\begin{aligned}
c^2 - a^2 &= h^2 + x^2 - (b - x)^2 - h^2 \\
&= x^2 - (b - x)^2 \\
&= x^2 - (b^2 - 2bx + x^2) \\
&= x^2 - b^2 + 2bx - x^2 \\
&= 2bx - b^2 \\
b^2 + c^2 - 2bx &= a^2
\end{aligned}
$$

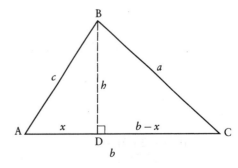

From triangle ABD,

$$\frac{x}{c} = \cos A$$

Hence $x = c \cos A$

$$\therefore \quad a^2 = b^2 + c^2 - 2bc \cos A, \text{ which is the cosine rule.}$$

This is used if you need to find one side of a triangle when you know the other two sides and the angle in between them.

If you want to find an angle of a triangle and you know all three sides the cosine rule is rearranged to:

$$\cos A = \frac{b^2 + c^2 - a^2}{2bc}$$

If the triangle is not lettered ABC, note that the side *opposite* the angle required is the a in the formula above.

── Example 1 ──

In triangle PQR, angle P is $57°$, PQ $= 3.2$ cm, and PR $= 5.1$ cm. Find QR.

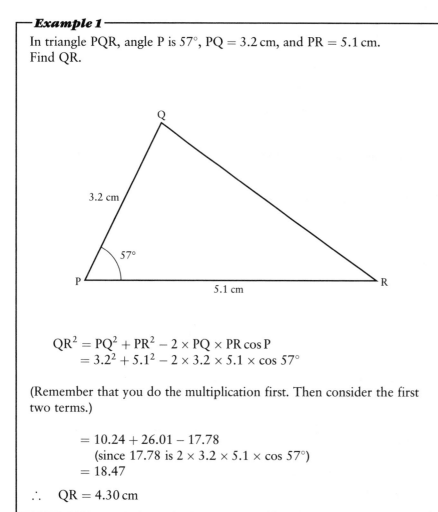

$$QR^2 = PQ^2 + PR^2 - 2 \times PQ \times PR \cos P$$
$$= 3.2^2 + 5.1^2 - 2 \times 3.2 \times 5.1 \times \cos 57°$$

(Remember that you do the multiplication first. Then consider the first two terms.)

$$= 10.24 + 26.01 - 17.78$$
$$\text{(since 17.78 is } 2 \times 3.2 \times 5.1 \times \cos 57°)$$
$$= 18.47$$

$$\therefore \quad QR = 4.30 \text{ cm}$$

Example 2

A sailor at a quay intends to sail directly towards a restaurant, R, 11.1 miles away. He knows that the lighthouse, L, is 7.8 miles due east of the quay and 14.8 miles from the restaurant. On what bearing should he sail?

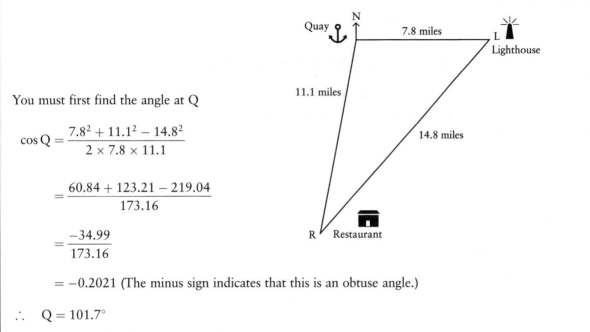

You must first find the angle at Q

$$\cos Q = \frac{7.8^2 + 11.1^2 - 14.8^2}{2 \times 7.8 \times 11.1}$$

$$= \frac{60.84 + 123.21 - 219.04}{173.16}$$

$$= \frac{-34.99}{173.16}$$

$$= -0.2021 \text{ (The minus sign indicates that this is an obtuse angle.)}$$

$$\therefore \quad Q = 101.7°$$

The sailor should sail on a bearing of $101.7° + 90° = 191.7°$.

Exercise 46E

1 In triangle XYZ, XY = 3.4 cm, YZ = 5.2 cm and XZ = 4.9 cm.
  Find the three angles of the triangle.

2 In quadrilateral ABCD, AB = 4.7 in,
  AC = 8.1 in, BD = 9.1 in and CD = 7.5 in. Angle CAB is 72°.
  Find:

  a BC

  b angle BDC.

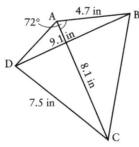

3 A yacht sails two miles from port on a bearing of 141°. A windsurfer is
  1.3 miles from the port on a bearing of 107°.
  What is the distance between the yacht and the windsurfer?

4 An orienteer runs 3 km due east and then 4.1 km south west.
  How far is she from the starting point?

Choosing the rule to solve a problem

Draw a diagram and mark on it all the given sides and angles.

If you are given three sides and wish to obtain an angle, you must use the cosine rule:

$$\cos A = \frac{b^2 + c^2 - a^2}{2bc}$$

If you want the third side of a triangle, once again the cosine rule should be used:

$$a^2 = b^2 + c^2 - 2bc \cos A$$

In cases when you are not given the three sides, or want to find the third side, you should use the sine rule. Use

$$\frac{a}{\sin A} = \frac{b}{\sin B} = \frac{c}{\sin C}$$

when you want to find a side,

and

$$\frac{\sin A}{a} = \frac{\sin B}{b} = \frac{\sin C}{c}$$

when you want to find an angle.

Exercise 46F

1   A kite is being held by two children, John and Elizabeth. John's string is 224 m long, and Elizabeth's is 128 m long. The two children are 174 m apart on horizontal ground.

What is the angle which John's string
makes with the horizontal?
Hence find the height of the kite.

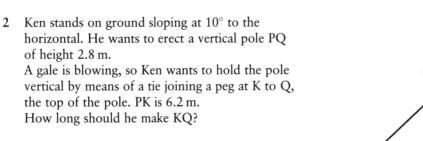

2   Ken stands on ground sloping at 10° to the horizontal. He wants to erect a vertical pole PQ of height 2.8 m.
A gale is blowing, so Ken wants to hold the pole vertical by means of a tie joining a peg at K to Q, the top of the pole. PK is 6.2 m.
How long should he make KQ?

3 A plane is 75 miles south west of Land's End when the pilot hears an SOS from a fishing boat 32 miles north 24° west from Land's End. On what bearing should the plane fly to reach the boat?
Flying at 320 mph, how long will the journey take?

4 London is 832 miles due north of Valencia. From Valencia, Nice is 480 miles on a bearing of 053°. From London, Munich is 563 miles on a bearing of 114°. How far is it from Nice to Munich?

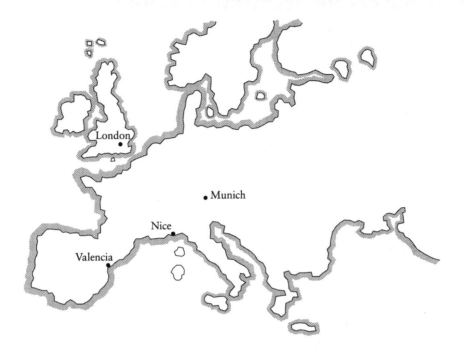

5 Three towns, Doncaster (*D*), Lincoln (*L*) and Sheffield (*S*), lie at the vertices of a triangle, as shown. The direct distance from Doncaster to Lincoln is 36 miles and from Sheffield to Lincoln is 42 miles. The angle *DLS* is 25°.

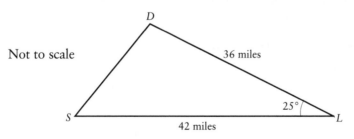

a Calculate the direct distance from Doncaster to Sheffield.

b Calculate the angle *DSL*. (SEG S94)

47 Further Mensuration

47.1   Arcs, sectors and segments of a circle

If an arc is drawn which is one quarter of the circumference, it will **subtend** an angle of 90° at the centre of the circle.
(i.e. one quarter of 360°.)

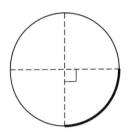

If an arc subtends an angle of 60° at the centre of the circle, the length of the arc must be one sixth of the length of the circumference.
(60° is one sixth of 360°.)

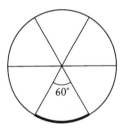

Similarly, it can be deduced that the length of an arc is proportional to the angle it subtends at the centre of the circle.

In general $\dfrac{\textbf{Length of arc}}{\textbf{Length of circumference}} = \dfrac{\textbf{Angle at centre}}{360°}$

Example

Find the length of the arc which subtends an angle of 30° at the centre of a circle of radius 5 cm.

$$\frac{\text{Length of arc}}{\text{Circumference}} = \frac{30°}{360°}$$

$$\text{Length of arc} = \frac{30}{360} \times \text{Circumference}$$

$$= \frac{1}{12} \times 2 \times \pi \times 5$$

$$= 2.617$$

∴   The length of the arc is 2.62 cm.

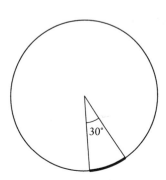

Exercise 47A

1   Find the lengths of the arcs marked in the following diagrams:

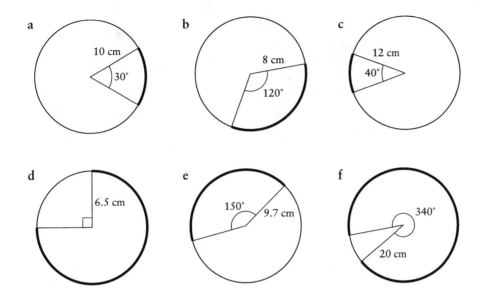

a   10 cm   30°

b   8 cm   120°

c   12 cm   40°

d   6.5 cm

e   150°   9.7 cm

f   340°   20 cm

2   In the following diagram, PQ is an arc of length 7.2 cm.
    Calculate:

    a   the circumference of the circle

    b   the radius of the circle.

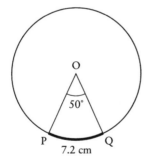

O   50°   P   Q   7.2 cm

3   The radius of the circle shown is 4.5 cm and the arc PQ has a length of
    14.3 cm.
    Calculate the size of angle POQ.

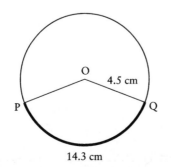

O   4.5 cm   P   Q   14.3 cm

4 Three lengths of wire are bent into the shape of an arc of a circle, and they are then used to make a child's mobile.

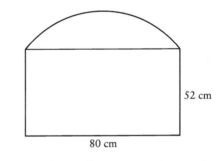

The circle has a radius of 11.8 cm and each piece of wire subtends an angle of 170° at the centre of the circle.
Find the total length of wire used.

5 The handle of a cup is an arc of a circle which subtends an angle of 176° at the centre. The length of the handle is 9 cm.
Find the radius of the circle of which the handle is an arc.

6 The diagram shows the cross-section of a cabin trunk.
The lower part of the trunk is rectangular in shape with height 52 cm and width 80 cm.
The lid is an arc of a circle of radius 62.5 cm and subtends an arc of 80° at the centre of the circle.
A metal strap is fastened tightly around the trunk and the ends are welded together.
What is the length of the metal strap?

52 cm

80 cm

The area of a sector

The area of sector OPQ is proportional to the angle at O. If the angle is 30° ($\frac{1}{12}$ of the whole angle at O which is 360°), the area of the sector is $\frac{1}{12}$ of the whole area of the circle.

In the general case:

$$\frac{\text{Area of sector}}{\text{Area of circle}} = \frac{\text{Angle of sector at O}}{360°}$$

Example

Find the area of a sector of a circle radius 8 cm subtended by an angle of 50° at the centre of the circle shown below.

$$\frac{\text{Area of sector}}{\text{Area of circle}} = \frac{50°}{360°}$$

$$\text{Area of sector} = \frac{50}{360} \times \text{Area of circle}$$

$$= \frac{50}{360} \times \pi \times 8^2$$

$$= 27.93 \text{ cm}^2$$

∴ The area is 27.9 cm².

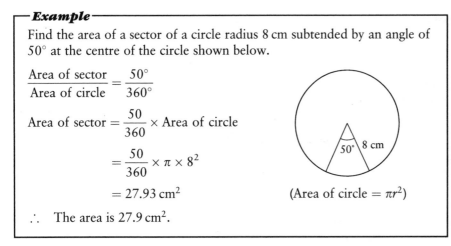

(Area of circle = πr^2)

Exercise 47B

In questions 1–3, find the areas of the shaded sectors:

1

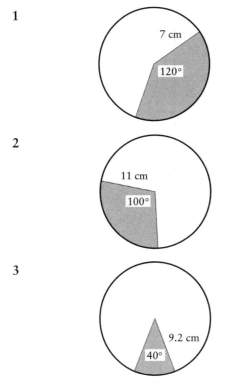

7 cm

120°

2

11 cm

100°

3

9.2 cm

40°

4 The area of the sector is 70 cm². Find the radius of the circle.

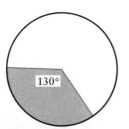

130°

5 The area of the sector is 32 in². Find the radius of the circle.

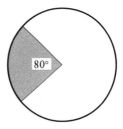

80°

6 Find the angle x if the area of the sector is 30 cm².

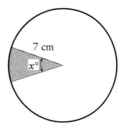

7 cm

$x°$

7 Find the angle y if the area of the sector is 87 cm².

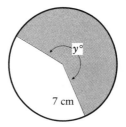

$y°$

7 cm

The area of a segment

The segment PRQ is the shaded area
in the diagram opposite.

The area of this segment is found by
subtracting the area of the triangle OPQ
from the area of the sector OPRQ.

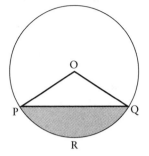

Example

The radius of a circle is 3 cm and the angle at the centre, $\angle$ ACB, is 40°.
Find the area of the shaded segment.

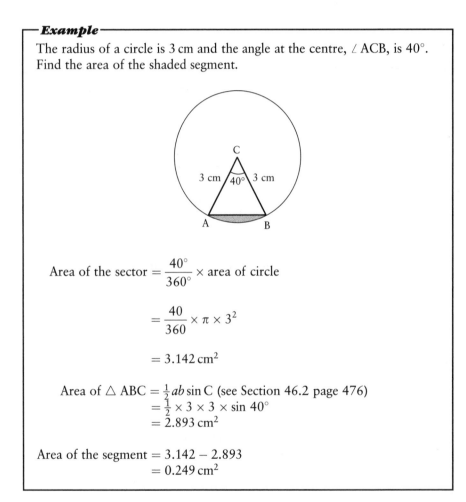

$$\text{Area of the sector} = \frac{40°}{360°} \times \text{area of circle}$$

$$= \frac{40}{360} \times \pi \times 3^2$$

$$= 3.142 \text{ cm}^2$$

$$\text{Area of } \triangle \text{ ABC} = \tfrac{1}{2}ab \sin \text{C} \quad \text{(see Section 46.2 page 476)}$$
$$= \tfrac{1}{2} \times 3 \times 3 \times \sin 40°$$
$$= 2.893 \text{ cm}^2$$

$$\text{Area of the segment} = 3.142 - 2.893$$
$$= 0.249 \text{ cm}^2$$

Exercise 47C

1   Find the area of the segment shown shaded in the following questions:

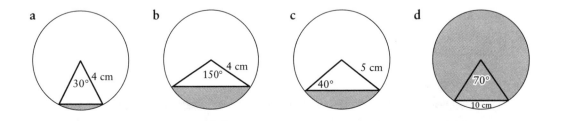

a   b   c   d

2   Find the area of the shaded segment:

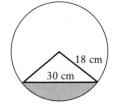

3   A circle has a diameter of 16 cm. A chord is drawn so that it is perpendicular to the diameter and divides the diameter in the ratio of 3 to 1.
Find the area of the minor segment (i.e. the smaller part of the circle) cut off by the chord.

4   A log of wood, 15 ft long, has a diameter of 2.5 ft.
The log is floating in a river with the top of the log 1.75 ft above the surface of the water.
What volume of the log is under the water?

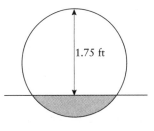

5   The base of a wooden toy has dimensions 15 cm by 6 cm, as shown in the diagram below:

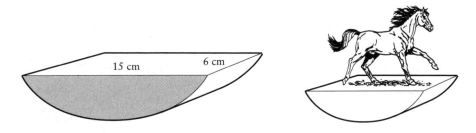

Two bases can be cut from a cylinder of wood which is 6 cm long and has a diameter of 20 cm.
What volume of wood is wasted?

47.2   The areas of other polygons

The areas of other polygons may be obtained by dividing each polygon into a number of simpler shapes (triangles, rectangles, etc.). Then the areas of each of these shapes can be worked out and the total found.

Example

Find the area of a regular hexagon with sides 20 cm.

Draw the hexagon and add in extra lines as shown:

Exterior angle of a regular polygon is $\dfrac{360°}{6} = 60°$.

$\therefore$ Interior angle (e.g. AFE) = 120°.

Hence $\angle FAE = \angle AEF = 30°$ (angles of $\triangle AFE$ add up to $180°$).

$\therefore \quad \angle BAE = 90°$

From $\triangle AFP$, $AP = AF \cos FAP = 20 \cos 30° = 17.32$ cm.

Area of rectangle $ABDE = AB \times AE$

$$= 20 \times (17.32 \times 2) = 693 \text{ cm}^2$$

From $\triangle AFP$, $FP = AF \cos FAP = 20 \sin 30° = 10$ cm.

Area of $\triangle AEF = \frac{1}{2} \times AE \times FP$

$$= \frac{1}{2} \times 34.64 \times 10$$

$$= 173 \text{ cm}^2$$

$\therefore \quad$ Area of polygon $ABCDEF = $ Area $ABDE + $ Area $\triangle AEF + $ Area $\triangle BCD$

$$= 693 + 173 + 173$$

$$= 1039$$

The area of polygon $ABCDEF$ is 1040 cm^2.

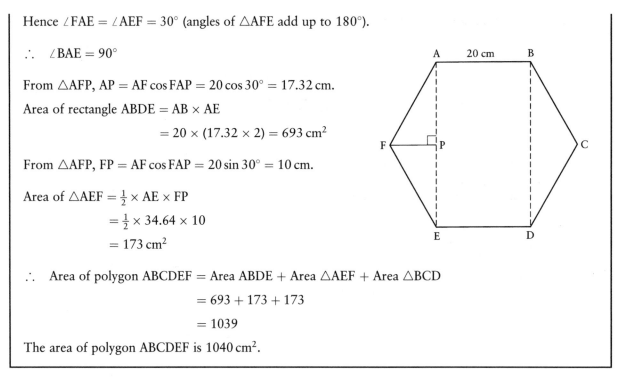

Exercise 47D

1 Find area ABCDE in the shape shown.

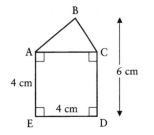

2 Find the area of a regular octagon of side 8 cm.

3 Find the area of the hexagon ABCDEF.

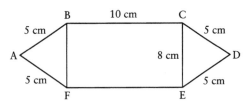

4 Find the area of ABCD.

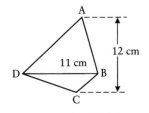

5 Find the area of PQRS.

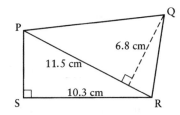

6 This shape is a regular hexagon of side length 8 centimetres.

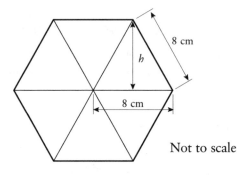

Not to scale

a Calculate the length marked h.

'Super Chocs' are packed in regular hexagonal
prisms with dimensions as shown.

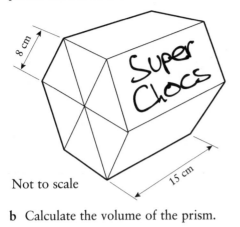

Not to scale

b Calculate the volume of the prism.

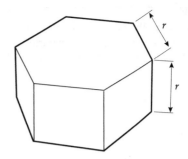

c By considering the dimensions, choose the
correct formula that represents the total
surface area of this hexagonal prism.

$11.2r^3$, $2\pi r^2 + 6r^2$, $11.2r$, $11.2r^2$.

(SEG W95)

47.3 Areas and volumes of similar shapes

Areas of similar shapes

If a plane figure is **enlarged** to form a new figure, all the corresponding
lengths of the two figures will be in the same ratio.

The two figures are said to be **similar**.

The larger square above has sides which are 3 times the length of the sides of
the smaller square.

The area of the larger square is 9 times the area of the smaller square:

$$\text{Larger area} = 3^2 \times \text{Smaller area}$$

The smaller circle has a radius r; the larger circle has a radius R.

Area of smaller circle : area of larger circle $= \pi r^2 : \pi R^2$
$$= r^2 : R^2$$

For all similar shapes:

$$\textbf{Larger area} : \textbf{Smaller area} = L^2 : l^2$$

where L and l are the lengths of corresponding dimensions.

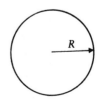

The areas of two similar shapes are in the ratio of the squares of the
corresponding dimensions.

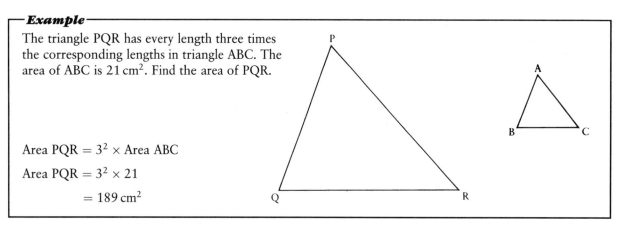

Example

The triangle PQR has every length three times the corresponding lengths in triangle ABC. The area of ABC is 21 cm². Find the area of PQR.

Area PQR = 3² × Area ABC

Area PQR = 3² × 21

$\qquad$ = 189 cm²

Exercise 47E

1 The quadrilaterals ABCD and PQRS are similar. The area of quadrilateral ABCD is 7 cm². PQ = 2AB. Calculate the area of quadrilateral PQRS.

2 The area of a pentagon is 38 cm². The pentagon is enlarged by scale factor 5. Find the area of the new shape.

3 The area of quadrilateral PQRS is 31 cm². The quadrilateral is enlarged by scale factor $\frac{1}{2}$. Find the area of the new shape.

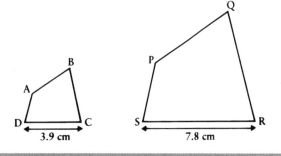

4 The radii of two circles are in the ratio 1:6. The area of the larger circle is 306 cm². Calculate the area of the smaller circle.

5 The area of a circle is 74 cm². After an enlargement, the area is 296 cm². Find the scale factor of the enlargement.

6 The area of a triangle is 24 in². After an enlargement, the area is 384 in². Find the scale factor of the enlargement.

7 The area of a triangle is 38 cm². After an enlargement, the area is 9.5 cm². Find the scale factor of the enlargement.

8 A semicircle has an area of 108 m². It is enlarged to a semicircle of area 192 m². In what ratio are the diameters of the two semicircles?

Volumes of similar solids

When a Russian doll, like the ones shown, is lifted, inside is another doll looking just the same but smaller, and inside that doll is another, and so on.

All the dolls in the set are similar to each other, but each one is a scale model of another.

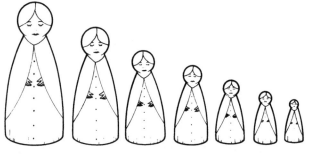

Just as areas can be enlarged, so can volumes.

Two solids are **similar** if the ratios between their corresponding dimensions are equal.

The larger cube below has sides which are three times the length of the sides of the smaller cube.

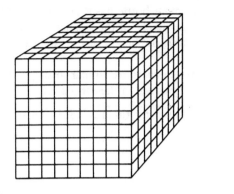

The volume of the larger cube is 27 times the volume of the smaller cube.

$$\text{Larger volume} = 3^3 \times \text{Smaller volume}$$

For all similar shapes:

$$\text{Larger volume} : \text{Smaller volume} = L^3 : l^3$$

where L and l are the lengths of corresponding dimensions.

Example

A small tin of cocoa has a height of 8 cm. A similar tin is an enlargement of scale factor $1\frac{1}{4}$.
The volume of the smaller tin is $320\,\text{cm}^3$.

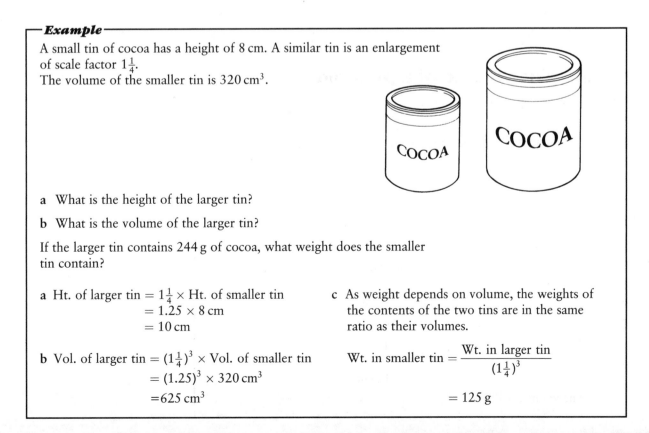

a What is the height of the larger tin?

b What is the volume of the larger tin?

If the larger tin contains 244 g of cocoa, what weight does the smaller tin contain?

a Ht. of larger tin $= 1\frac{1}{4} \times$ Ht. of smaller tin
$= 1.25 \times 8\,\text{cm}$
$= 10\,\text{cm}$

b Vol. of larger tin $= (1\frac{1}{4})^3 \times$ Vol. of smaller tin
$= (1.25)^3 \times 320\,\text{cm}^3$
$= 625\,\text{cm}^3$

c As weight depends on volume, the weights of the contents of the two tins are in the same ratio as their volumes.

$$\text{Wt. in smaller tin} = \frac{\text{Wt. in larger tin}}{(1\frac{1}{4})^3}$$

$$= 125\,\text{g}$$

1 A child's set of 64 building blocks are packed into a box which is similar in shape to the blocks.

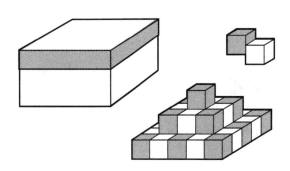

a The volume of a block is 3 in³.
What is the volume of the box?

b The height of a block is 2 in.
What is the height of the box?

2 Two cylinders are similar in shape. Their dimensions are in the ratio of 1:5.

a If the volume of the larger cylinder is 875 cm³, what is the volume of the smaller cylinder?

b If the area of the base of the smaller cylinder is 1.4 cm², what is the area of the base of the larger cylinder?

3 A small bottle of medicine has a capacity of 20 ml.
A similar bottle has a capacity of 160 ml. If the height of the smaller bottle is 7.5 cm, what is the height of the larger bottle?

4 Two buckets are similar in shape. Their volumes are in the ratio 27:8.

a In what ratio are their heights?

b In what ratio are the areas of their circular bases?

c If the height of the smaller bucket is 28 cm, what is the height of the larger bucket?

5 The dimensions of a giant ice cream cone are $1\frac{1}{2}$ times those of a small cone.

a The volume of ice cream in the small cone is 112 cm³. How much ice cream does the giant cone hold?

b If the price of a small cone is 60p, what should the ice cream vendor charge for a giant cone?

*47.4 The volume of a pyramid

The volume of a pyramid is given by:

$$\text{Volume} = \frac{1}{3} \times \text{Base area} \times \text{Height}$$

$$= \frac{1}{3}Ah$$

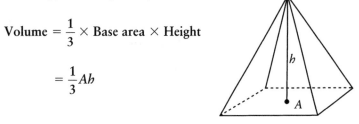

Example

A pyramid has a square base of side 8 in. The height of the pyramid is 6 in. What is its volume?

$$\text{Base area} = 8 \times 8$$
$$= 64 \text{ in}^2$$

$$\therefore \quad \text{Volume} = \frac{1}{3} \times 64 \times 6$$
$$= 128 \text{ in}^3$$

∴ The volume is 128 in³.

---

*47.5   The volume of a cone

The volume of a cone is given by:

$$\text{Volume} = \frac{1}{3}\pi r^2 h$$

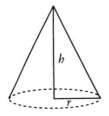

Example 1

Find the volume of a cone of height 12 cm and base radius 8 cm.

$$\text{Volume} = \frac{1}{3}\pi r^2 h$$

$$= \frac{1}{3} \times \pi \times 8^2 \times 12$$

$$= \frac{1}{3} \times \pi \times 64 \times 12$$

$$= 804.2 \text{ cm}^3$$

$\therefore$ The volume is 804 cm^3.

Example 2

Find the volume of a cone of slant height 20 cm, and radius 12 cm.

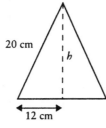

Since you are given the slant height, you must
first find the vertical height h.
Pythagoras' theorem is used to do this.

$$h^2 = 20^2 - 12^2$$

$$= 400 - 144$$

$$= 256$$

$$h = 16$$

$$\therefore \quad \text{Height} = 16 \text{ cm}$$

$$\text{Hence} \quad \text{Volume} = \frac{1}{3} \times \pi \times 12^2 \times 16$$

$$= 2413 \text{ cm}^3.$$

$\therefore$ The volume is 2410 cm^3.

*47.6   The volume of a sphere

The volume of a sphere of radius r is given by:

$$\text{Volume} = \frac{4}{3}\pi r^3$$

Example 1

Find the volume of a sphere of radius 6 cm.

$$\text{Volume} = \frac{4}{3}\pi r^3$$

$$= \frac{4}{3} \times \pi \times 6^3$$

$$= \frac{4}{3} \times \pi \times 216$$

$$= 904.8 \text{ cm}^3.$$

∴   The volume is 905 cm³.

Example 2

The volume of a sphere is 280 cm³. What is its radius?

$$\frac{4}{3}\pi r^3 = 280$$

$$\therefore \quad 4\pi r^3 = 3 \times 280$$

$$\therefore \quad \pi r^3 = \frac{3 \times 280}{4}$$

$$= 210$$

$$\therefore \quad r^3 = \frac{210}{\pi}$$

$$\therefore \quad r = \sqrt[3]{\frac{210}{\pi}}$$

$$= 4.058 \text{ cm}$$

∴   The radius is 4.06 cm.

Exercise 47G

1   Find the volume of a sphere of radius:

   a 6 cm   **b** 3.8 cm   **c** 5.2 in

2   Find the volume of a pyramid of height 12 cm on a square base of side 4 cm.

3   Find the volume of a pyramid of height 6 cm on a rectangular base 8 cm by 10 cm.

4   Find the volume of a pyramid of height 9 cm on a rectangular base 6 cm by 8 cm.

5   Find the volume of a sphere of diameter:

   a 8 cm   **b** 9.4 cm   **c** 3.4 in

6   Find the volume of a cone:

   a height 9 cm, base radius 8 cm

b height 7 cm, base diameter 11 cm

c height 2.8 in, base radius 6.4 in.

7 The volume of a sphere is 252 cm³.
Find its radius.

8 The volume of a sphere is 412 cm³.
Find its diameter.

9 The volume of a cone is 280 in³. Its height is 21 in.
Find its base radius.

10 The slant height of a cone is 25 cm. The base radius is 7 cm.
Find its volume.

47.7 The surface areas of bodies

When a body has a polygon for each face (e.g. a cube or cuboid) the surface area is obtained by adding together all the areas of the faces. The faces are all shapes identified on pages 428 and 431.

Example 1

Find the surface area of a cuboid of length 12 cm, width 8 cm and height 6 cm.

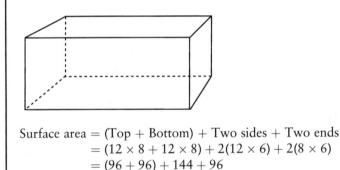

Surface area = (Top + Bottom) + Two sides + Two ends
= $(12 \times 8 + 12 \times 8) + 2(12 \times 6) + 2(8 \times 6)$
= $(96 + 96) + 144 + 96$
= 432 cm²

∴ The area is 432 cm².

Example 2

Find the surface area of a pyramid which has a square base of side 6 cm and for which each sloping edge is 10 cm long.

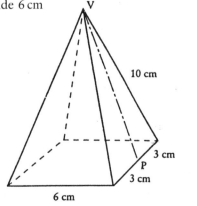

The diagram shows a sloping face.

To find the area, you first need to find the length VP.

By Pythagoras' theorem

$$VP^2 = 10^2 - 3^2$$

$$\therefore \quad VP = \sqrt{91}$$

$$= 9.539 \text{ cm}$$

∴ The area of a sloping face $= \dfrac{1}{2} \times 6 \times 9.539$ cm².

$$= 28.62 \text{ cm}^2$$

The area of the base is $6 \times 6 = 36 \, \text{cm}^2$.

There are four sloping faces.

$\therefore$ Total surface area = Area of base + 4 × 28.62
$$= 36 + (4 \times 28.62)$$
$$= 150.48 \, \text{cm}^2$$

$\therefore$ The total area is 150 cm² (to 3 s.f.).

Exercise 47H

1   Find the surface area of a cube with edges:

    a   5 cm   **b**   6 cm   **c**   8.1 cm

2   Find the surface area of a cuboid with dimensions:

    a   6 cm   3 cm   8 cm

    b   4.1 cm   5.2 cm   3.8 cm

    c   8 in   6 in   3.5 in

3   The volume of a cube is 64 cm³. Find the length of each edge, and hence the surface area of the cube.

4   The volume of a cube is 28 cm³. Find the length of each edge, and hence the surface area of the cube.

5   The volume of a cuboid is 32 cm³. Its length and width are 8 cm and 5 cm. Find the height and surface area of the cuboid.

6   A pyramid with slant edge 10 cm is on a square of side 8 cm. Find the surface area of the pyramid.

7   A pyramid on a triangular base (i.e. a tetrahedron) has every edge 12 cm. Find its surface area.

8   A tetrahedron (see question 7) has an equilateral base of side 20 cm. The slant edges are of length 26 cm. Find the surface area of the tetrahedron.

47.8   The surface area of a cylinder

The curved area of a cylinder of height h and radius r is $2\pi rh$.

Closed cylinders also have two circular ends, each of which is of radius r.

Example 1

A tin of baked beans has height h cm, and has a radius of r cm.
A label is wrapped around the tin as shown.

What area of paper is used for the label?

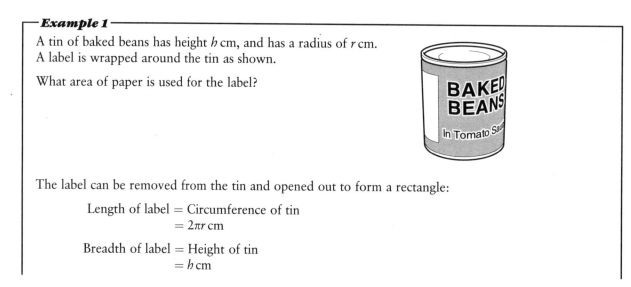

The label can be removed from the tin and opened out to form a rectangle:

       Length of label = Circumference of tin
$$= 2\pi r \, \text{cm}$$

      Breadth of label = Height of tin
$$= h \, \text{cm}$$

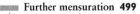

Area of label = Area of rectangle
= Length × Breadth
= 2πr × h
= 2πrh cm²

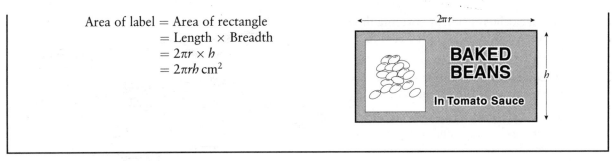

This example verifies the formula for the curved surface of a cylinder:

Curved surface area = 2πrh

Example 2

Find the curved surface area of a cylinder, radius 12 cm and height 10 cm.

Curved surface area = 2πrh

= 2 × π × 12 × 10

= 754 cm²

∴ The curved surface area is 754 cm².

Example 3

A drinking glass is in the shape of a cylinder, radius 4 cm and height 8 cm. Find its outer surface area.

Curved surface area = 2πrh

= 2 × π × 4 × 8

= 201.1 cm²

Area of one end (the other is open!) = πr²

= π × 4²

= 50.26 cm²

∴ Total surface area = 201.1 + 50.26

= 251.4 cm²

∴ The total surface area is 251 cm².

Exercise 47I

1 Kitchen roll is wrapped around a tube of cardboard of length 23 cm and diameter 4.2 cm. What area of cardboard is used in making the tube?

2 A cylindrical block of wood is 25 cm high and has a diameter of 14 cm. Calculate the area of the curved surface.

3 A garden cloche is made by covering three wires, each bent into a semicircle, with polythene.

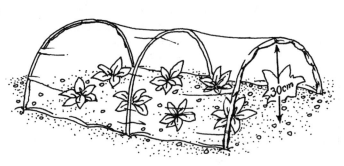

a If the height of the cloche is 30 cm, what length of wire is required?

b The length of each cloche is 60 cm. What area of polythene is required to cover the wires? (The ends of the cloche are left open.)

*47.9 The surface area of a sphere

Surface area of a sphere $= 4\pi r^2$

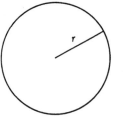

Example 1

Find the surface area of a sphere of radius 5 cm.

$$\text{Surface area} = 4\pi r^2$$
$$= 4 \times \pi \times 5^2$$
$$= 4 \times \pi \times 25$$
$$= 314 \text{ cm}^2$$

∴ **The surface area is 314 cm².**

Example 2

A sphere has curved surface area 220 cm². Find its radius.

$$4\pi r^2 = 220$$
$$r^2 = \frac{220}{4\pi}$$
$$= 17.51$$
$$r = \sqrt{17.51}$$
$$= 4.184$$

∴ **The radius is 4.18 cm.**

Exercise 47J

1 Find the surface area of a sphere of radius:

   a 5 cm   **b** 8 cm   **c** 6 in

2 Find the surface area of a sphere of diameter:

   a 8 cm   **b** 20 in   **c** 12.1 cm

3 Find the radius of a sphere with surface area:

   a 28 cm²,   **b** 28.4 in²

4 Find the volume of a sphere with surface area:

   a 20 cm²   **b** 36 cm²

Appendix on Non-calculator Methods

In mathematics examinations, on or after Winter 1999, a significant proportion of a candidate's performance will be assessed by means of examination papers in which a calculator is not allowed.

In the Winter 1999 examination (and in future sessions) of the SEG Modular Mathematics GCSE, half the marks on the written papers will be awarded on papers on which no calculator is allowed.

In the Statistics and Probability module, and in the Terminal module, the majority of the questions in the non-calculator paper will be of the type in which a calculator is not helpful. However students will be expected to carry out simple additions, subtractions, multiplications and divisions in these modules.

The majority of the non-calculator questions which require students to use non-calculator arithmetic will appear in the Money Management and Number question papers. This appendix (pages 501–509) contains a selection of non-calculator techniques which are needed in the examinations. It supplements the material in the main text.

Students should be aware of the four rules of arithmetic, which are **addition**, **subtraction**, **multiplication** and **division**.

When numbers are   **added**   the result is called the **sum**.
When numbers are **subtracted** the result is called the **difference**.
When numbers are **multiplied** the result is called the **product**.
When numbers are   **divided**   the result is called the **quotient**.

Place value is very important and, in all calculations, the digits *must* be written in the correct column.

A1 Calculators involving whole numbers (i.e. integers)

Addition and subtraction

Example 1

Find the sum of 1009, 124 and 2188

The numbers are written with the digits in the correct units, tens, hundreds or thousands column:

$$
\begin{array}{cccc}
1 & 0 & 0 & 9 \\
  & 1 & 2 & 4 \\
2 & 1_1 & 8_2 & 8 \\
\hline
3 & 3 & 2 & 1 \\
\hline
\end{array}
$$

$9 + 4 + 8 = 21$, the 1 is written in the unit column of the answer and the 20 (2 tenths) is added to the tens column.

$0 + 2 + 8 + 2 = 12$ (tens), 2 is written in the tens column and 1 (hundred) is carried to the hundreds column etc.

Example 2

Subtract 234 from 2013

The numbers are written in their correct columns:

2 0 1 ^{1}3   4 cannot be subtracted from 3, so ten must be 'borrowed'
  2 3$_1$ 4   from the tens column and added to the 3 giving 13.
     9   Then $13 - 4 = 9$, which is written in the answer. The ten
————   which was 'borrowed' is then 'paid back'.

2 0 ^{1}1 ^{1}3 $3 + 1 = 4$ which cannot be subtracted from 1, so 10
  2$_1$ 3$_1$ 4   tens are 'borrowed' from the hundreds column and
   7 9   added to the 1 in the tens column giving 11.
————   Then $11 - 4 = 7$, which is written in the answer. The
    hundred which was 'borrowed' is then 'paid back'.

2 ^{1}0 ^{1}1 ^{1}3 $2 + 1 = 3$ which cannot be subtracted from 0, so 10
$_1$ 2$_1$ 3$_1$ 4   hundreds are 'borrowed' from the thousands column
  7 7 9   and added to the 0 in the hundreds column giving 10.
————   Then $10 - 3 = 7$, which is written in the answer. The
    thousand which was 'borrowed' is then 'paid back'.

2 ^{1}0 ^{1}1 ^{1}3 $2 - 1 = 1$ which is written in the answer.
$_1$ 2$_1$ 3$_1$ 4
1 7 7 9

Exercise A1

1 In the following questions, find the sum of the numbers listed

  a 345, 47, 1029, 76   **d** 840, 56, 102, 35, 43

  b 1205, 736, 54, 108   **e** 519, 23, 7, 1900, 820

  c 6203, 1007, 8042

2 In the following questions, find the difference between the given numbers:

  a 589 and 135   **c** 1307 and 844   **e** 2004 and 807

  b 724 and 206   **d** 5920 and 765

Multiplication

'Short' multiplication is generally used when multiplying by a single-digit number and 'long' multiplication when multiplying by a number with two or more digits. (If you know the multiplication tables, then 'short' multiplication may be used when multiplying by 11 or 12.)

Example 1

Multiply 693 by 7

  6 9 3 $3 \times 7 = 21$, the 1 is written in the answer and the 2 (tens)
$_4$ $_6$ $_2$ 7   carried on to the tens column.
4 8 5 1 $9 \times 7 + 2 = 63 + 2 = 65$, the 5 is written in the answer
    and the 6 (hundred) carried on to the hundreds column.
    $6 \times 7 + 6 = 42 + 6 = 48$, the 8 is written in the answer and
    the 4 carried on to the thousands column. As there are no
    thousands to multiply by 7, the 4 is written in the answer.

Example 2

Multiply 693 by 47

```
      6 9 3
      4 7
  ---------
  4 8 5 1
```
First, 693 is multiplied by 7 (as for 'short' multiplication in the example above).
The result (4851) is written down.

```
      6 9 3
      4 7
  ---------
  4 8 5 1
2 7 7 2 0
```
Next, 693 is multiplied by 4 tens.
The result (2772 tens) is written down, beginning in the **tens** column.
(A zero is written in the unit column.)

```
      6 9 3
      4 7
  ---------
  4 8 5 1
2 7 7 2 0
  ---------
3 2 5 7 1
```
Finally, the two products (693 × 7 and 693 × 40) are added together, to give 32 571
which is 693 × 47

Exercise A2

Multiply:

1 37 by 8

2 136 by 9

3 215 by 5

4 924 by 7

5 548 by 12

6 107 by 23

7 59 by 34

8 651 by 59

9 839 by 18

10 738 by 82

Division

As with multiplication, 'short' division is generally used when dividing by a single digit number and 'long' when dividing by a number with two or more digits.

Example 1

Divide 994 by 7

```
    1
  --------
7)9 ²9 4
```
First, 7 is divided into 9 (hundred).
The result is 1 (hundred) with a remainder of 2 (tens).
The 1 (hundred) is written in the answer and the 2 (tens) are carried into the tens column.

```
    1 4
  --------
7)9 ²9 ¹4
```
Next, 7 is divided into 29, giving 4 (tens) and a remainder of 1 (unit).
The 4 is written in the answer and the 1 is carried into the units column.

```
    1 4 2
  --------
7)9 ²9 ¹4
```
Finally, 7 is divided into 14 and result, 2, is written in the answer.

'Long' division follows the same procedure as short division, but, because the remainders are larger (often two or more digits), the setting out is different.

Example 2

Divide 782 by 23

23)7 8 2   As 23 will not divide into 7, the 7 (hundred) is immediately 'carried' into the tens column and nothing is written in the answer.

```
      3
23)7 8 2
   6 9
     9
```
23 is now divided into 78. It divides in 3 times, which is written in the answer.
$23 \times 3 = 69$ and, in order to find the remainder, 69 is written under 78 and subtracted.

```
      3
23)7 8 2
   6 9
     9 2
```
Instead of the remainder, 9 (tens), being carried to the 2 units, the 2 is carried down to the 9.

```
      3 4
23)7 8 2
   6 9
     9 2
     9 2
     0 0
```
The process is then repeated by dividing 23 into 92. The result is 4 times (which is written in the answer) and there is no remainder.

Exercise A2

Divide:

1 98 by 7   5 684 by 12   8 945 by 35

2 345 by 5   6 420 by 15   9 858 by 26

3 612 by 9   7 304 by 19   10 527 by 31

4 768 by 8

A2 Calculations involving decimals

Addition and subtraction of two numbers

Example

a Add 34.5, 9.7 and 56.12

b Subtract 91.72 from 164.6

First, write the numbers in a column so that the decimal points are in line, then each digit is in its correct place.

```
a    3 4.5              b  1 6 4. 6 0
       9.7                  9 1. 7 2
     5 6.12                 7 2. 8 8
   1 0 0.32
```

Calculate, without the aid of a calculator:

1	$36.2 + 5.7$	**3**	$6.7 + 51.09 + 76.18$	**5**	$89.13 - 72.42$
2	$104.9 + 75.4$	**4**	$72.9 - 56.2$	**6**	$121.6 - 69.85$

Multiplication and division of two numbers

Example

a Multiply 271.3 by 9

b Divide 384.3 by 7

$$\begin{array}{r} \textbf{a} \quad 271.3 \\ 9 \\ \hline 2441.7 \end{array}$$

$$\textbf{b} \quad \begin{array}{r} 54.9 \\ 7\overline{)384.3} \end{array}$$

Multiplication of two decimal numbers

$$0.2 \times 0.03 = \frac{2}{10} \times \frac{3}{100}$$
$$= \frac{6}{1000}$$
$$= 0.006$$

You will see that to multiply two decimals:

(i) you multiply the numbers, 2×3

(ii) count up the total number of digits **after** the decimal point in the original question (1 digit in 0.2 plus 2 digits in 0.03 gives a total of 3 digits)

(iii) position the decimal point so that the number of digits after the decimal point is the same as the total you have just found.

Example 1

Find 0.03^2

$$0.03^2 = 0.03 \times 0.03$$
$$3 \times 3 = 9$$

There are four digits after the decimal points (03 and 03).

Thus there must be four digits after the decimal point in the answer (0009), the answer must be .0009

So $0.03^2 = 0.0009$

Example 2

Find 0.002×0.04

$$0.002 \times 0.04 = 0.000\,08$$

(There must be five digits after the decimal point.)

Division of decimals

When you divide two decimals, the denominator must be a whole number.

So move the position of the decimal point in **both** numbers by the same amount until the denominator (i.e. the one you are dividing by) becomes a whole number.

Example 1

Find $\dfrac{0.02}{0.004}$

Move the decimal point three places to make the denominator become a whole number:

$$\frac{0.02}{0.004} = \frac{20}{4}$$

$$= 5$$

Example 2

Find $\dfrac{0.12}{0.0008}$

Move the decimal point 4 places:

$$\frac{0.12}{0.0008} = \frac{1200}{8}$$

$$= 150$$

Example 3

Find $\dfrac{0.015}{0.24}$

$$\frac{0.015}{0.24} = \frac{1.5}{24} \quad \text{(moving the decimal point two places)}$$

$$= 0.0425$$

To divide 1.5 by 24, move the decimal point in 1.5 up to the answer and keep adding zeros on to 1.5 as necessary.

$$
\begin{array}{r}
0.0625 \\
24\overline{)1.5000} \\
\underline{1\,44} \\
60 \\
\underline{48} \\
120 \\
\underline{120} \\
0
\end{array}
$$

Evaluate the following.

1	24.2×7	9	2.4×0.003
2	78.04×9	10	0.3×0.005
3	147.5×12	11	0.06×0.0035
4	$54.6 \div 7$	12	$0.27 \div 0.03$
5	$650.79 \div 9$	13	$0.8 \div 0.002$
6	$526.46 \div 11$	14	$2.4 \div 0.3$
7	$200 \times 30\,000$	15	$2 \div 0.004$
8	0.7×0.04	16	$0.003 \div 0.12$

A3 Calculations involving fractions

Addition and subtraction

Before fractions can be added or subtracted, they must have the same denominator.

This denominator could be the lowest common multiple (LCM, see page 11) or you can simply multiply the denominators together.

Example 1

Add $\frac{3}{4}$ and $\frac{2}{5}$

$4 \times 5 = 20$. (In this case, multiplying gives the common denominator.)

$\frac{3}{4} = \frac{15}{20}$ (using equivalent fractions, see page 20)

$\frac{2}{5} = \frac{8}{20}$

So $\quad \frac{3}{4} + \frac{2}{5} = \frac{15}{20} + \frac{8}{20} = \frac{23}{20} = 1\frac{3}{20}$

Example 2

$7\frac{3}{4} - 2\frac{3}{10}$

In this case we could multiply the denominators ($4 \times 10 = 40$) and work the answer out in $\frac{1}{40}$'s but since the LCM of 4 and 10 is 20 we will work in $\frac{1}{20}$'s:

$$7\frac{3}{4} - 2\frac{3}{10}$$

$$= 7\frac{15}{20} - 2\frac{6}{10}$$

$$= 5\frac{9}{20}$$

A4 Approximations

To find an approximate answer it is normal to approximate the numbers in the question to either whole numbers (integers) or to one significant figure.

Example 1

Find an approximate value of:

a $\dfrac{9.23 \times 8.11}{6.04}$

b $\dfrac{9.23 \times 8.11}{6.04} \times 0.03^2$

a $\dfrac{9.23 \times 8.11}{6.04}$

9.23 is approximately equal to 9, i.e. $9.23 \approx 9$.

Similarly $8.11 \approx 8$ and $6.04 \approx 6$.

$$\therefore \quad \frac{9.23 \times 8.11}{6.04} \approx \frac{9 \times 8}{6}$$

$$= \frac{72}{6}$$

$$= 12$$

$\therefore$ An approximate value is 12.

b $0.03^2 = 0.03 \times 0.03$

$$= 0.0009$$

$$\approx 0.001$$

$$\frac{9.23 \times 8.11}{6.04} \times 0.03^2 \approx 12 \times 0.001$$

$$\approx 0.012$$

$\therefore$ An approximate value is 0.012.

Example 2

By rounding each number to one significant figure find an approximate value of

$$\frac{394 \times 215}{4172}$$

$$\frac{394 \times 215}{4172} \approx \frac{400 \times 200}{4000} \quad \text{(see page 17 for rounding to 1 significant figure)}$$

$$= \frac{80\,000}{4000}$$

$$= 20$$

Exercise A6

Write each of the numbers in questions 1–5 to the nearest whole number.
Hence calculate an estimate of each calculation.

1 $\dfrac{3.81 \times 5.22}{1.98}$

3 $\dfrac{42.8 \times 8.89}{5.92}$

5 $\dfrac{11.9 \times 4.01}{5.2 + 2.9}$

2 $\dfrac{6.14 \times 11.87}{8.97}$

4 $\dfrac{4.97 \times 63.1}{3.14}$

Write each of the numbers in questions 6–10 to 1 significant figure. Hence
find an estimated value of each calculation.

6 $\dfrac{42.8 \times 17.1}{39.2}$

9 $11.24 \times 12.1 + (8.1 - 2.2)^2$

7 $\dfrac{41.8 + 21.7}{2.81 + 2.556}$

10 $\dfrac{19.2 \times 24.1}{52.3 - 13.8}$

8 $(29.3 - 17.1)(34.1 + 13.2)$

See page 19 for further examples.

Other techniques you will need to know on the non-calculator papers include:

A5   Substitution in formulae (see pages 236–7)

A6   Percentages (see pages 48–53)

A7   Standard form (see pages 20–23)

Solutions to *Exercises*

Note. Solutions are given for exercises with a 'closed' numerical answer. They are not provided for 'open-ended' questions and for some questions leading purely to an illustration (e.g. a pie-chart or a graph).

Exercise 1A

1 a 24 b 4, 49 c 17 or 41 d 4 e 41
2 16
3 a 601 300
  b six hundred and one thousand three hundred
4 17 5 a 24 b 1
6 three thousand one hundred and eighty three
7 a −4 b 5 c 4 d 3 e 4
8 a $\sqrt{19}, \sqrt{7}$ b $\sqrt{16}, \sqrt{81}$ c $\sqrt{16}$ d $\sqrt{49}, \sqrt{81}$
  e $\sqrt{49}$
9 62 10 17 11 417
12 a −8, 4, 19 b 4, 19 c $\sqrt{5}, \pi$ d 19 e −8, 4
13 18043 14 20 106 15 one hundred and thirty thousand four hundred and two
16 1, 2, 3, 4, 6, 8, 12, 24
17 a 2, 19 b Any two of $\sqrt{16}$, 16, 9 c $\sqrt{101}$
  d 15 or 2 e 45 or 9
18 9.34, 3^2, 2.9^2, $\sqrt{65}$, $\sqrt{63}$
19 a 10.5 b 0.105 c 0.03 d 0.006 e 67030 f 0
  g 4570 h ∞ i 120

Exercise 1B

1 a 2 b −4 c −10 d 4 e 6 f 40 g 0 h −49
2 a −1 b 17 c −13 d 20 e −3 f 0 g 50 h 36
3 a 12 b 637 c 5295 d −1186 e −3462 f 1766

Exercise 1C

1 £25.68 2 29°
3 −£33.75 or £33.75 OD 4 £191

Exercise 1D

1 a −18 b −3 c 28 d $1\frac{3}{4}$ e −15 f −27 g 8
  h 36 i −1 j −10
2 a $-5\frac{1}{4}$ b 3431 c −390 d 168 e 1.2 f $-\frac{1}{2}$
3 a $-\frac{1}{4}$ b $-\frac{3}{4}$ c $-\frac{3}{4}$

Exercise 1E

1 a 13 b −16 c 44 d −14
2 a 21.1°C b −2.2°C c −20°C d −23.3°C
3 a 36 b −125 c −9.96

Exercise 1F

1 1, 4, 9, 16, 25, 36, 49, 64, 81, 100
2 1, 8, 27, 64, 125, 216
3 a 32 b 729 c 25 d $\frac{1}{8}$ e −64
4 a 32 b 2187 c 4 d 9
5 a 5 b 11 c 25 d 15 e 14
6 a (i) 3 (ii) −4 (iii) 9 b 2 c (i) 3 (ii) 5

7 a 6 b $\frac{4}{3}$ c 8 d $\frac{4}{15}$ e $\frac{2}{19}$
8 a $\frac{3}{4}$ b $1\frac{1}{3}$ 9 $1 + 3 + 5 + 7 + 9$
10 a $\frac{1}{121}$ b $\frac{1}{12}$ c $\frac{9}{8}$ d −27 e 1
11 8 12 27 13 81 14 125
15 a 32 b 81 c $\frac{1}{16}$ d −8 e 0.25
16 a 2 b 3 c 6 d $\frac{1}{2}$ e −4
17 3 18 a $\frac{1}{4}$ b 9 c 2 d $\frac{2}{7}$ e $\frac{2}{17}$ 19 100

Exercise 1G

1 $3\sqrt{2}$ 2 $6\sqrt{6}$ 3 $2\sqrt{5}$ 4 $7\sqrt{3}$ 5 $9\sqrt{3}$
6 $9\sqrt{5}$ 7 $3\sqrt{2}$ 8 $5\sqrt{3}$ 9 60 10 60
11 2 12 $\frac{1}{\sqrt{5}}$ or $\frac{\sqrt{5}}{5}$ 13 $\frac{1}{\sqrt{10}}$ or $\frac{\sqrt{10}}{10}$
14 $\sqrt{\frac{28}{27}}$ or $\frac{2\sqrt{7}}{3\sqrt{3}}$ 15 $\sqrt{3}$

Exercise 1H

1 $(\sqrt{3})^2$
2 b (i) irrational (ii) cannot be simplified to a fraction or an exact decimal.
3 a e.g. $a + b$, where a and b are rational.
  b (i) e.g. $\sqrt{a} + \sqrt{b}$, where a and b are integers.
    (ii) e.g. $(a + \sqrt{b}) + (a - \sqrt{b})$, where a and b are integers.
4 a $\sqrt{5}, \pi, (\sqrt{3})^3$
  b e.g. $(a + \sqrt{b})$ and $(a - \sqrt{b})$, where a and b are rational.

Exercise 1I

1 a 1×15 b 1×27 c 1×36 d 1×64
     3×5 3×9 2×18 2×32
                                      3×12 4×16
                                      4×9 8×8
                                      6×6
  e 1×100
     2×50
     4×25
     5×20
     10×10
2 a (i) 1, 2, 3, 4, 6, 8, 12, 16, 24, 48
    (ii) 1, 2, 3, 4, 6, 8, 9, 12, 18, 24, 36, 72
  b 1, 2, 3, 4, 6, 8, 12, 24
3 64, 144, 116, 2620
4 a 5, 10, 20, 25 b 5, 10, 25 c 5, 10 d 5, 25
  e 5, 10, 20
5 a 240, 87, 96, 255 b 240, 96 c 240, 255

Exercise 1J

1 a $2^2 \times 3 \times 5$ b $3^2 \times 5 \times 7$ c $3^2 \times 5^2$
  d $2^2 \times 3^2 \times 5 \times 7^2$ e $2 \times 7 \times 11^2$
2 a (i) $2^2 \times 3^2$; $2 \times 3^2 \times 5$ (ii) 2×3^2
  b (i) $2^3 \times 3 \times 7$; $2^3 \times 3^2$ (ii) $2^3 \times 3$
  c (i) $2^3 \times 3 \times 13$; $2^2 \times 5 \times 13$ (ii) $2^2 \times 13$
  d (i) $2 \times 7 \times 11$; $2^2 \times 5 \times 11$ (ii) 2×11
  e (i) $2 \times 3^3 \times 17$, $2^2 \times 3^2 \times 13$ (ii) 2×3^2
  f (i) $3 \times 5 \times 7^2$, $2 \times 3 \times 5^2 \times 7$ (ii) $3 \times 5 \times 7$

3 a (i) 14, 28, 42   (ii) 30, 60, 90   (iii) 24, 48, 72
   (iv) 30, 60, 90
 b (i) multiples of 14   (ii) multiples of 30
   (iii) **multiples of 24**   (iv) **multiples of 30**

Exercise 1K

1 a 32 b 15 c 28 d 14 e 115 f 22
2 a 150 b 144 c 156 d 12 600 e 252 f 144
3 a 12 b $\frac{13}{18}$ 4 a 225 b $\frac{4}{13}$ 5 72
6 a 5 b 4 c 40 metres 7 1.24 am
8 a 252 b Artemis, 14; Khons, 9; Soma, 7
9 a 250 b 15, 28 **10** 18

Exercise 2A

1 9.74 2 0.36 3 147.5 4 29 5 0.53 6 4.20
7 1250 8 0.004 9 270 10 460.0

Exercise 2B

1 a 3500 b 240 c 1200 d 300 e 250 f 10 g 40
 h 5
2 a 9 b 0.06 c 24
3 Incorrect calculations are **c, d, e, f, h, j**
4 a 4000 and 250 **b** Yes
5 a 20 and 15 (14) **b** £60 (£56)

Exercise 2C

1 a 7.9×10^5 b 4.6×10^{-3} c 3.13×10^4
 d 9.41×10^{-5} e 1×10^5 f 2.82×10^{-4}
 g 1.57×10^4 h 4.7×10^7 i 3.4×10^{-2}
 j 2.7×10^{-6} k 5×10^{-1} l 1.4×10^{-7}
2 a 7530 b 240 c 0.0019 d 0.837 e 0.0451
 f 40420 g 3 192 000 h 0.000 974 i 0.000 006 8
3 a 6×10^4 b 5.18×10^0 c 4×10^1
 d 1.35×10^4 e 2.174×10^3 f 4×10^{-2}
 g 4.32×10^{-5} h 2.84×10^4 i -5×10^{-4}
 j 6×10^1
4 a 2.3×10^{-5} kg b 8.70 m
5 3.5×10^4 cm; 266.6 m
6 W$ 756.25 7 4.096×10^7
8 1.378×10^5 square miles 9 4.68×10^9 lire
10 £6.206×10^9 11 1.0457×10^9
12 a 5.8×10^7 b £3.207×10^3
13 a 4.17×10^{-5} kg (4.17×10^{-2} g)
 b 0.0417 g
14 a 2.4×10^8 b 64.8 15 1.28 s

Exercise 2D

1 11.8 2 73 700 3 0.137 4 29.0 5 21.4
6 0.579 7 2.70 8 8.14 9 7.472 10 24.6
11 −0.6 *12 d 5.828 *13 c 2.236
14 (i) 1.732 (ii) 3.162
15 2.236; 1.732; 3.162. The method used in questions
 13 and 14 calculates the square root of a given
 number.

Exercise 3A

1 a 8 b 8 c 18 d 9 e 7 f 9 g 5 h 16
2 a $\frac{2}{3}$ b $\frac{3}{4}$ c $\frac{2}{9}$ d $\frac{1}{3}$ e $\frac{3}{4}$ f $\frac{2}{3}$ g $\frac{6}{13}$ h $\frac{2}{3}$
3 a $\frac{20}{9}$ b $\frac{31}{6}$ c $\frac{59}{12}$ d $\frac{111}{10}$ e $\frac{35}{4}$ f $\frac{38}{3}$ g $\frac{143}{20}$ h $\frac{143}{7}$
4 a $1\frac{2}{7}$ b $7\frac{1}{5}$ c $3\frac{1}{3}$ d $7\frac{5}{8}$ e $13\frac{10}{11}$ f $10\frac{7}{12}$
 g $8\frac{4}{5}$ h $23\frac{1}{2}$

Exercise 3B

1 $1\frac{3}{20}$ 2 $\frac{1}{24}$ 3 $\frac{2}{3}$ 4 $\frac{3}{4}$ 5 $\frac{1}{2}$ 6 $5\frac{1}{20}$ 7 2 8 $1\frac{11}{18}$
9 $1\frac{5}{9}$ 10 $\frac{1}{20}$ 11 $\frac{7}{8}$ 12 $1\frac{1}{4}$

Exercise 3C

1 $\frac{7}{10}$ 2 $1\frac{1}{3}$ 3 12 4 $1\frac{1}{9}$ 5 $7\frac{3}{20}$ 6 $1\frac{5}{7}$ 7 8
8 $\frac{7}{8}$ 9 $1\frac{3}{17}$ 10 5 11 3

Exercise 3D

1 a $\frac{8}{12} = \frac{2}{3}$, $0.\dot{6}$ b $\frac{1}{4}$, 0.25 c $\frac{5}{8}$, 0.625
2 a 0.1 b 0.5 c 0.75 d 1.45 e 4.84 f $2.8\dot{3}$
 g $7.\dot{4}$ h $2.\dot{8}$ i $0.8\dot{1}$ j $3.\dot{1}4285\dot{7}$

Exercise 4A

1 5:6 2 4:5:10 3 20:9 4 a 3:2 b 10:1
 c 12:1 d 1:4 e 4:3 f 3:2 g 3:8 h 3:2 i 2:3
5 a 8 b 2 c 1.2 d 0.6 e 3.2 f 2.6
6 42 cm 7 a 13.5 cm b 8 inches
8 a 3.5 kg b 5.25 kg 9 410 and 1025
10 a 5:4 b £10 c £15

Exercise 4B

1 £260, £390 2 £1500, £4500, £6000
3 £48, £32, £24 4 £1750, £1250, £1000
5 £375, £625 6 a £75 b £50
7 a (i) 5:3 (ii) £7.50, £4.50 b £7.33, £4.67
8 £26154, £34872, £17436, £26154
9 1.36 kg; 0.24 kg (or 1360 g and 240 g)
10 £61; £76.25; £45.75

Exercise 4C

1 £3.18 2 £2.66 3 £1.26 4 £1.84 5 £1.79
6 4 hours 7 88p
8 a 516.5 g b 34p c no (≈ 17% more)

Exercise 4D

1 £200 2 3 hours 3 a 9.6 days b 12 people
4 36 minutes 5 10 extra 6 £10 625

Exercise 4E

1 81p 2 £374.50
3 a 90 miles b 7 gallons c approximately 8 miles/litre
4 £201.57 5 6 hours 43 minutes 6 9.7 kg
7 a 8p b £42.24 8 1.17 g/cm^3

Exercise 4F

1 a 54 mph b 8.6 m/s (31.0 km/h) c 88 km/h
2 a 204 km b 2760 miles c 98.125 km
3 a 3 hours b 7 hours 39 minutes
 c 3 hours 48 minutes
4 $7\frac{1}{2}$ miles 5 800 m
6 a 3 hours 52 minutes b 6 km/h
7 48.75 mph
8 a 5 hours 50 minutes b 21.9 mph
9 a 63.3 mph b 85.5 mph

Exercise 4G

1 24 km/h 2 a 45 mph b 12 minutes c 49.6 mph
3 a 10 miles b $3\frac{1}{2}$ mph c 3 mph

4 a 85 km b 135 km c 1 hour 40 minutes
  d 75.4 km/h

Exercise 5A

1 a 1.232 km b 32 mm c 0.626 kg d 73.1 cl
  e 16.2 ml f 12 700 m g 0.0591 litres h 3400 kg
  i 13.9 mg j 85 kg
2 230 g 3 200 bags
4 25 000 mg 5 a 2.93 m b 293 cm
6 a 28.4 g b 28 400 mg c 2.84×10^4 7 1.2 litres
8 27.7 litres

Exercise 5B

1 a 3.25 ft b 2 lb 8 oz c 5.5 yards d 27 inches
  e 55 oz f 2.5 gal g 28 pints h 14 fl oz
2 40 bottles 3 $\frac{3}{4}$ pint 4 1.2 tons
5 $8\frac{3}{4}$ oz 6 a 46.5 ft b 15.5 yards
7 $6\frac{1}{8}$ (6.125) oz 8 5 bottles

Exercise 5C

1 a 15 oz b 15.5 oz c 14.6 oz
2 a 454 g b 5.3 oz c 2.3 kg d 336 g e 0.5 oz
3 a 2 lb 3 oz b 14 oz c 3 lb 5 oz d 4 lb 10 oz
4 a 0.22 gal b 4.4 gal 5 8.8 miles per litre
6 a 75 miles b 202 miles c 268 miles d 35 miles
  e 231 miles
7 Yes ($\approx$ 15 cm wide)
8 Omar, by 0.72 pints or 0.41 litres.

Exercise 6A

1 a 20% b $12\frac{1}{2}$% c 70% d 65% e $66\frac{2}{3}$% f 36%
  g 175% h 250%
2 a $\frac{3}{5}$ b $\frac{1}{4}$ c $\frac{1}{10}$ d $\frac{17}{20}$ e $\frac{3}{20}$ f $1\frac{3}{10}$ g $\frac{3}{8}$ h $\frac{1}{3}$
3 a   0.75   75%
  b $\frac{1}{2}$ 50%
  c   0.125 $12\frac{1}{2}$%
  d $\frac{1}{3}$ $0.\dot{3}$
  e $\frac{3}{8}$ $37\frac{1}{2}$%
  f   0.7   70%
  g $\frac{7}{20}$ 0.35
  h $\frac{2}{3}$ $66\frac{2}{3}$%
  i   0.6   60%
  j $\frac{5}{8}$ 0.625

Exercise 6B

1 a £1.38 b £318.62 c £22.12 d £13.91 e £5.17
  f £8.73 g £97.20 h £44.20

Exercise 6C

1 a £10.12 b £4.07 c £5.50 d £10.79 e £10.04
2 a £108.18 b £1.05 c £792.00 d £233.75
  e £156.00 f £21.69

Exercise 6D

1 a £24.00 b £8.40 c £14.08 d 36p e £10.39
2 a £49.77 b 59p c £51.20 d £23.76 e £59.43
  f £1.09

Exercise 6E

1 a 80.0% b 3.45% c 175% d 50%
  e 54.5% f 130%
2 a +23.1% b +20.0% c +375% d −20.0%
  e −45.5% f +464%

Exercise 6F

1 £2.86 2 29.6%
3 10% if earning greater than £100, £10 if earnings less
  than £100
4 6195 5 9.18% 6 a 406 b 146
7 a 200 b 75 c $\frac{3}{10}$ 8 £946.20 9 12%
10 a £13.60 b $1759.12 or £1196.68 c £1728
  d £1808 e 51.1%

Exercise 7A

1 38.46 2 450 3 52p 4 7p 5 5
6 10 7 £10.91 8 £3.86

Exercise 7B

1 $0.7\dot{7}$ 2 $0.27\dot{2}\dot{7}$ 3 $0.63\dot{6}\dot{3}$ 4 $0.\dot{1}4285\dot{7}$
5 $0.\dot{3}5714\dot{2}\dot{8}$ 6 2 7 $0.083\dot{3}$ 8 $0.266\dot{}$ 9 $0.05\dot{5}$
10 0.25 11 One less than the value of the divisor.

Exercise 7C

1 $\frac{2}{3}$ 2 $\frac{5}{9}$ 3 $\frac{2}{45}$ 4 $\frac{1}{300}$ 5 $\frac{7}{30}$ 6 $\frac{67}{90}$ 7 $\frac{8}{33}$
8 $\frac{13}{33}$ 9 $\frac{263}{450}$ 10 $\frac{18}{55}$ 11 $1\frac{142}{495}$ 12 $\frac{245}{666}$ 13 $\frac{86}{495}$

Exercise 7D

1 11.5 m, 12.5 m
2 4.65 m, 4.75 m
3 37.5 ton, 38.5 ton
4 £285, £294.99
5 16 years, 16 years 364 days
6 347.5 g, 352.5 g

Exercise 7E

1 58 m, 54 m 2 268.75 ft², 235.75 ft²
3 12.28 m, 12.24 m
4 433.0375 m², 428.8075 m²
5 £52.98, £51 6 29.78 m, 27.32 m
7 1.117 in, 0.9687 in
8 69 802.5 m³, 66202.5 m³
9 £3.91, £3.45
10 a 531.25 miles b 477.25 to 531.25 miles
11 a (ii) 25 g (ii) 15 g b 85 p
12 a 497.5 g
  b (i) 755 g (ii) a 37p b 35p
13 Yes: Max. length is 459 ft
14 a 516 mph b 46 mph
15 a 303.875 ft³ b 47.9 days to 28.6 days

Exercise 8A

1 £172.20 2 £479.67 3 £6.43 4 £11582.40
5 a £259.70 b £155.75 c £441

Exercise 8B

1 £249.90 2 £344.10
3 Andrews, £163.18; Collins, £214.92; Hammond,
  £210.94; Jali, £254.72; Longman, £246.76

4 £259.89
5 a £3.50 b 4 hours 6 £47.03 7 3 hours

Exercise 8C

1 £1560 2 £24.72 3 £304 4 £1456
5 a Firm B b Firm A, £250

Exercise 8D

1 £59.04 2 £125.44 3 £52.38 4 £152.22
5 a Mrs English, £520; Mrs Beckett, £432
  b Mrs English, £530; Mrs Beckett, £424
  c Increase for Mrs English, decrease for Mrs Beckett;
    the new scheme favours the more productive
    worker.

Exercise 8E

1 a (i) £4335 (ii) £165 (iii) £16.50
  b (i) £4335 (ii) £615 (iii) £61.50
  c (i) £4335 (ii) Nil (iii) Nil
2 £137 3 £9.72 (3)

Exercise 8F

1 a £1579.45 b £984.67 c £2554
2 £242.12 3 £258.12
4 a £3083 b 3083
5 a £62.15 b £36.29 c £98

Exercise 8G

1 a (i) £8111 (ii) £675.92
  b (i) £8631.50 (ii) £719.25
  c (i) £7469 (ii) £622.42
2 a £8511 b £783.50
3 £17 262 4 a £6741 b Yes, by £728

Exercise 8H

1 a £4.87 b £30.94 c £56.91 d £29.90 e £5.64
2 a Nil b £2704
3 a £184 per week b £12 c £140.75

Exercise 9A

1 £26.60 2 £40.95 3 £1.66 4 25p 5 £3.36
6 £105.16
7 A, £14.81; B, £14.30; C, £14.97; B offers the best deal
8 £97.49 9 £7.53, £95.88
10 a 59p, £5.54 b £27.20, £367.20 c £2.60, £31.50
   d £46.92, £322.92 e £1.92, £15.66

Exercise 9B

1 £1.22 2 £9114.89 3 £17.98 4 £1042.70
5 £106.55 6 a £134.03 b £765.87
7 a £3.11, £25.88 b £1.93, £16.06 c £5.89, £49.10
  d £21.43, £178.56 e £35.36, £294.63
  f £12.32, £102.67 g £64.28, £535.71
  h £1.28, £10.71
8 $4.73 9 £172 10 £15.30

Exercise 10A

1 a (i) £1 (ii) 25%
  b (i) £7 (ii) 35%
  c (i) 9p (ii) 12%

d (i) £18 (ii) 20%
e (i) £100 (ii) 19%
f (i) £3.99 (ii) 11.1%
2 33.3% 3 a £4.32 b 12% 4 £9719.80
5 20% 6 22.4% 7 a 95.1% b 1.46%

Exercise 10B

1 a £35 b 17.9% 2 a £1174.50 b £225 c 23.7%
3 a £28.20 b 24.5% 4 a £3.25 b 18.1%
5 a £1200 b £1518.30 c £219.70 d £1738
  e £538; 44.8%
6 a £1080 b £6.75 c £1404 d 71.22%

Exercise 11A

1 a £252 b £82.50 c 2 years d 5 years
2 £2295, £9045 3 £231.25 4 $7\frac{1}{2}$%

Exercise 11B

1 £7986 2 £2315.25 3 £7504.67, £984.67
4 £4655.45; £1551.82 5 a £69.16 b 6.92%
6 The second paying 8.75% per annum, ease of access to
  the account/services available/extras, e.g. cheque book,
  credit card.
7 £826.69

Exercise 11C

2 a Tax is deducted from the interest earned at the
    current rate.
  b (i) e.g. National Savings bonds, NSB Investment
       Account, special building society accounts
    (ii) non-taxpayers
5 a Tax exempt special savings account
  b (i) maximum of £9000 to be invested over 5 years
       (£1800 per year or £500 per month or £3000
       in first year and £600 in year 5).
    (ii) only interest after tax may be withdrawn during
       the 5 years, otherwise income tax must be paid.
6 (i) the type of account
  (ii) the amount of money invested.

Exercise 12A

1 (i) Year is incorrect – should be 1992.
  (ii) Alteration should be initialled.
  (iii) Amount in figures should read £43.
  (iv) It is usual to add 'only' to an amount in whole
      pounds and/or draw a line through the remaining
      space to prevent additions to the amount.
2 (i) Insufficient funds in the account to cover the
      cheque.
  (ii) The cheque is unsigned.
  (iii) The cheque was completed incorrectly.
  (iv) The cheque is out of date.
3 The cheque has 'bounced' and the payee must go back
  to the drawer (writer of the cheque) for payment.
4 The cheque must be paid into the bank account of the
  payee.
5 The card guarantees that the bank will honour cheques
  up to the amount on the card.
6 In the case of theft, one cannot be used by the thief
  without the other.

Exercise 12B

1 **a** Southford **b** 0123456 **c** 30/3/89 **d** £444.01
e £29.65 **f** £215.00 **g** Direct debit **h** £225
i The account is in credit. **j** Cheque numbers

Exercise 13A

1 **a** £43.00 **b** £38.70 **2** £26.25
3 **a** £223.44 **b** £33.45 **c** 8.8% per annum **d** 17.6%
4 **a** £5309.74 **b** 19.7%
5 **a** £1849.75 **b** £5549.25 **c** £8809.75 **d** 25.4%
e 25.4%
6 **a** £11 138.70, £12 111.90 **b** 17.6%, 15.6%
7 A because APR values are 28% and 34.9%.
8 **a** (i) £1680 (ii) £9240 **b** £1890
c (i) $33\frac{1}{3}$% (ii) APR ≈ 22.2%

Exercise 13B

1 **a** (i) £1111.32   (ii) £111.32   (iii) 11.1%
b (i) £1220.88   (ii) £220.88   (iii) 11.0%
c (i) £1337.04   (ii) £337.04   (iii) 11.2%
d (i) £372   (ii) £72   (iii) 12%
e (i) £3209.40   (ii) £809.40   (iii) 11.2%
f (i) £2075.76   (ii) £375.76   (iii) 11.1%
2 **a** Something of value which the borrower possesses
(e.g. a house) which can become the property of the
lender if the borrower is unable to repay the loan.
b The loan is insured. The insurance company would
repay the loan if the borrower died.
3 **a** (i) £7429.80 **b** (i) £8871 **c** (i) £7953.60
4 **a** £1 **b** 96p, 89p
5 **a** (i) £67.78 (ii) £1.48
b (i) £79.74 (ii) £11.96 (iii) 35.3%

Exercise 13C

7 **a** £12.50 **b** £237.50 **c** £4.75
8 **a** £237.50 **b** £4.28 **c** Mrs Magee should change.
9 **a** (i) £203.50 (ii) 33.9% **b** (i) £104.41 (ii) 26%
c The catalogue gives the better deal.

Exercise 13D

1 **a** (i) £1.52; £1.34 (ii) 500 g size **b** 250 g size
2 **a** (i) 4.49 g (ii) 4.65 g **b** 600 g jar
c possible wastage with larger sizes/sell by date/taste
3 **a** E15: £5.49; E10: £5.68½; E3: £6.10
b (i) E15: more economical for a family
(ii) E3: less to pay out of pension, not heavy to
carry
(iii) E10: good compromise, but an argument could
be made for any size
(iv) E3: small size for limited storage space.
5 **a** (i) A, £14.51 (ii) £14.01
6 **a** 575 g **b** 27.0p
c (i) 25.65p (ii) 18p

Exercise 14A

1 **a** 235.8 Sch **b** 24920 Pts **c** $161.16
2 **a** £19.90 **b** £5.03 **c** £2.44
3 £16 **4 a** £57.09 **b** 28.5p
5 **a** 1678.80 DM **b** 6023.78 Sch
6 **a** 68 250 000 **b** 15 193 000 **c** £70.67
d £7.42
7 **a** 945 Fr **b** £94.03 **c** £5.97

Exercise 14B

1 **a** £3 **b** £3 **c** £11.97 **2 a** £12 **b** £6
3 **a** 299.39 DM; 238.43 SFr; 2299.05 Sch
b (i) £570 (ii) £5.70 **c** £226.43

Exercise 14C

1 6.15 am; 0615
2 Ten past eleven in the evening, 2310
3 Half past nine in the morning, 9.30 am
4 Twenty to two in the morning; 0140
5 Ten to two in the afternoon, 1.50 pm
6 Twenty past ten in the evening; 10.20 pm
7 9.50 pm; 2150
8 Quarter to eleven in the morning; 1045
9 12.25 am; 0025
10 Four minutes to five in the afternoon; 4.56 pm

Exercise 14D

1 14 minutes **2** 0655 **3** 9.30 am **4** 0942
5 **a** 1623 **b** 28 minutes **6** 1 hour 45 minutes
7 **a** 1945 **b** 2030

Exercise 14E

1 **a** 20 minutes **b** 4 **c** 39 minutes
d 1451, 1543 (or 1551)
2 **a** 5 **b** 0030 Belgian time **c** 25 minutes **d** 0930
e 1 hour **f** 1335

Exercise 15A

1 £67 500, £11 000 **2** £95 400, £14 600
3 £59 375, £1625 **4** £113 600, £33 400
5 £160 000, £100 000

Exercise 15B

1 £51 000 **2** £78 750 **3** £56 250
4 less than £16 667 **5** £83 250

Exercise 15C

1 **a** (i) £138.90   (ii) £33 336
b (i) £103.68   (ii) £37 324.80
c (i) £260.78   (ii) £78 234
d (i) £304.20   (ii) £54 756
2 **a** £240.45 **b** £72 134(5)
3 **a** £28 025 **b** £242.14 **c** £87 170.40 **d** £2975
e £90 145.40
4 **a** (i) £524.70   (ii) £157 410
b (i) £433.44   (ii) £104 025.60
c (i) £419.58   (ii) £75 524.40
d (i) £583.95   (ii) £210 222
5 £508.32

Exercise 15D

1 £6240 **2** £4440 **3** £36 **4** £175 p.c.m.
5 **a** £5640 **b** (i) £6100 (ii) £5797.44
c Yes – they will own their own house eventually at
not too much extra cost. Also because rent would
rise in due course.
6 **a** Repairs and maintenance are usually carried out by
the landlord.
Freedom to change jobs and move around the
country or abroad.

b You own the property which, over a period of time, will appreciate in value. Freedom to do what you wish with the house and garden.

Exercise 15E

1 a £550   b £1071.67   c £1260
2 £20.62 (or £20.63)   3 £25

Exercise 15F

1 £417.08   2 £127.27
3 £92.42, £9.24   4 36.7p
5 a £119.70   b £66.76   c £94.79   d £106.84
6 £257.81
7 Yes, saving £39.77 per year

Exercise 15G

1 a 66 865   b 42 735   c 68 138   d 94 238
2 a £94.65   b £136.24   c £99.27   d £104.82
  e £33.34
3 a £21.38   4 a 902   b 65 210

Exercise 15H

1 a £168.30   b £118.15   c £164.72   d £190.19
2 388.5   3 £41.49
4 a 148   b 155.4   c £66.82   d £76.22
5 a (i) 5839   (ii) 211
  b (i) 221.55   (ii) £88.18   (iii) £96.88

Exercise 15I

1 £39.29   2 £37.19   3 a £31.77   b £37.33
4 a £24.96   b 169   c £7.10   d £32.06   e £5.61
  f £37.67

Exercise 16A

1 £56   2 £84   3 a £36   b £114
4 a £375   b £31.25   5 a £179.28   b £14.94

Exercise 16B

1 £67.64   2 £137.92   3 £14.78   4 £36; 23.1%
5 £290.75; £30.28
6 Not correct (estimate £360)

Exercise 16C

1 a £556   b £702   c £1059   d £1129
2 a £631.80   b £568.62   c £284.31
3 £189.04   4 a £800   b £720   c £504
5 £221.47   6 £228.60

Exercise 17A

1 a £25   b £54   c £1477.33   d £222   e £45
2 a £60   b £34   c £1.01   d £1200   e £85
3 £180   4 £215.20   5 £7180   6 £2.66
7 £9500   8 £49.93

Exercise 17B

1 50%   2 21%   3 $10\frac{1}{2}$%   4 9.6%   5 6.5%   6 4.4%

Exercise 17C

1 £2304   2 £630   3 £328.05   4 £2414.45

Exercise 18A

1 4.1   2 5.2   3 3.2   4 4.69   5 3.33   6 3.8
7 4.36   8 3.141   9 5.385

Exercise 19A

1
2	4	8	9.5
10	40	80	95

$y = 10x$

2 9   3 a 4   b 3
4 a 15   b 9.8   5 a 21   b 50
6 a $s = 45t$ b 135 miles
7 a 66 mm   b 5 N

Exercise 19B

1 108   2 b (i) 12   (ii) $\frac{3}{4}$
3 £320   4 a 272 ft   b 109 mph
5 a 0.44 hours (26.5 min)   b 1800 mph
6 a 0.032 N   b 5060 miles
7 a £2330   b 4 years 8 months

Exercise 20A

(D = discrete, C = continuous, QL = qualitative, QN = quantitative)
1 QN,D   2 QN,C   3 QN,C   4 QN,D   5 QL
6 QN,C   7 QN,D   8 QN,D   9 QN,D   10 QL
11 QN,C   12 QL

Exercise 21A

1 a total population of the town
  b people asked at a particular time in a particular shopping street
  c Specify the intended catchment area of the new shopping centre. Use the electoral role for this area to select a random sample.
2 a all the students in the College
  b She only asks people in the common room, and she only asks girls.
  c Find students' enrolment numbers, and select 50 random numbers in the range of numbers given.
3 a all cars in the UK
  b The police only check cars between 5pm and 6pm.
  c Sample at different times of day on different days of the week on different roads.
4 a the soil in his garden
  b only one sample. Soil beyond the range of his throw cannot be sampled.
  c Draw a plan of the garden and divide into numbered squares. Use random sampling. Choose several squares and take samples from these areas.
5 a All teenagers
  b Carol only asked people entering a tobacconist's. The sample is biased towards smokers. Carol should only ask teenagers.
  c Obtain a list of pupils in the school; select a random number from the list.
6 a any homes in the telephone area
  b The salesgirl only asks those with a telephone. The sample includes only those who are at home when the salesgirl rings.
  c The salesgirl needs to know the number of homes in the area. She needs to use the rating register (used in water rates) to identify homes; use their reference numbers to obtain a random sample.

7 **a** all cars owned by people in an area
  b Peter only asked people at home during one afternoon. The responses came from a small part of the town.
  c Use the DVLC in Swansea to produce a list of cars registered as being owned by people in the required area. Use random numbers to obtain a sample of the required size.

8 **a** all people who travel to work in John's area
  b People at the station would be biased towards those using a train to go to work.
  c Use the electoral role and ask a random sample from this how they travel to work. If they do not work, delete them from the sample. Continue with a random selection until you have a suitable number who do work.

9 **a** No guarantee that all work sections will be represented or that there will be the correct proportion of men to women.
  b (i) 16   (ii) 10

Exercise 21B

10 **a** (i) Change the question
      e.g. How many days were you absent from work last year because of stress related illness?
    (ii) Increase the lower categories
      e.g. 1 day, 2–3 days, 4–10 days
  b Take a random sample of 10 from each of the 10 factories.

11 **a** Change the question to:
    How often did you come into the bank last month?
  b Change the response:
    split the over 30 category into smaller classes, delete the under 18 category.
    Make sure 25 is not in two categories.

Exercise 21C

1 **a** Analyse rainfall for the given period over several years.
  b Survey conducted over a typical college week, or a questionnaire to include both staff and students.
  c Devise a questionnaire to find the viewing requirements and compare with a survey of what TV is actually providing.
  d Devise a questionnaire, survey, or experiment.

Exercise 22A

1 1: 5 times. 2: 4 times. 3: 4 times. 4: 5 times. 5: 6 times. 6: 4 times
2 0–10: 9 times. 11-20: 7 times. 21–30: 10 times. 31–40: 6 times. 41–50: 5 times. 51–60: 6 times.
3 0–4: 9 times. 5–9: 11 times. 10–14: 8 times. 15–19: 11 times. 20–24: 5 times. 25–29: 8 times. 30–34: 12 times. 35–39: 11 times. 40–44: 5 times.
4 A: 19. E: 29. I: 12. O: 26. U: 4.
5 **b** 1, 4, 4, 7, 6, 2, 5, 1

Exercise 23F

2 **b** A low absence rate is more likely on Mondays and a high one on Fridays.
3 **b** The experimental results are very close to the expected results.

4 **b** Book 1: Science fiction.
    Book 2: Child's story.
    Book 1 has a greater number of long words and Book 2 has a greater number of short words.
6 There are more women than men over the age of 60.

Exercise 23G

1 **b** Bimodal
  c Both female and male handspans have been recorded.
2 Symmetrical.
3 **a** Symmetrical and rectangular.
  b The die is biased towards a score of 5.
5 **a, d** and **g**.
6 **a** (vi)   **b** (iv)   **c** (iii)   **d** (v)   **e** (i)   **f** (ii)

Exercise 24A

1 **a** 160   **b** 120   **c** 220   **d** 140
2 **a** 325   **b** 290   **c** 210, 87.8%
3 **a** 240   **b** 210   **c** 210   **d** 120
4 Conservative by 960 votes
5 40 476, 20 238, 85 714
6 **a** 298   **b** 46   **c** 306
7 **a** (i) 35%   (ii) 21   **b** Cycling
  c Over $2\frac{1}{2}$ times as many people travel by car in 1994 (as a percentage). Nearly twice as many people walked in 1964 (as a percentage). The proportion travelling by bus has not changed significantly.
    Cycling was the most popular mode of transport in 1994, followed by bus travel, whereas in 1994, bus travel was the most popular followed by car, etc.
  d 49
8 **a** 72.2%   **b** 458 300 acres
9 **a** 75   **b** 65   **c** 40

10

Cost of lunch (£)	Frequency
0–1.99	2
2.00–3.99	4
4.00–5.99	12
6.00–7.99	16
8.00–9.99	16
10.00–11.99	8
12.00–13.99	6
14.00–15.99	10

11

Distance (m)	Number of lobsters
0–5	30
5–10	10
10–15	20
15–20	30
20–25	30
25–30	30
30–35	40
35–40	10

12 Number of peas	Frequency
100–199	15
200–249	20
250–299	30
300–349	30
350–399	22
400–499	20

13 81   14 140, 56%   15 56

Exercise 24B

1 b For example:
The second author generally uses much longer sentences than the first.
$\frac{4}{5}$ of Fleming's sentences contain less than 15 words, whereass less than half the second author's do.
14% of the second author's sentences are longer than any sentence of Fleming's.
i.e. the sentences in the second novel are longer on average and have a larger variation in length.
2 a Normal distribution.
b Variety B is generally taller than variety A and has a greater variation in height.
3 a 1, 115, 380, 4   b 65.87 inches
d The mean height of females is lower than that of males and the standard deviation is less, i.e. females are generally shorter than men and there is less variation in their heights.

Exercise 24C

1 Missed: 1988, 1989. No scale on sales.
2 Different widths of bars. Non-linear vertical scale.
3 Vertical scale does not start at origin. No time interval on horizontal scale.
4 Different width bars. No vertical scale. No indication of meaning of 'goodness'. Food not identified.
5 Other drinks not identified. Different width bars. No vertical scale. Probably picked one constituent of milk not in other drinks.
6 No vertical scale. Different width bars. 'Typical other drink' meaningless.
7 No vertical scale. 'Pollution' not defined.

Exercise 25A

1 £130   2 69.06 mph   3 68.27 mph   4 38.58
5 The ages are given to the last completed year. A person aged 17 is between 17 years 0 days and 17 years 364 days old. Therefore, the true meaning is probably above $28\frac{3}{4}$, hence John is probably correct.
6 a £1710   b £1820   c £182   7 1.848 m
8 46.5 people.

Exercise 25B

1 4   2 44p   3 8–10 m   4 110–114   5 red
6 a 2, 5, 7, 10, 5, 3   c 20.00 and less than 20.10

Exercise 25C

1 34   2 12.5   3 4   4 50p   5 6.35 kg
6 a Mean b Mode   c Median   d Mode   e Mean
f Mean or median   g Median or mode

Exercise 25D

1 1   2 6   3 26.60 min (= 26 min 36 s)
4 7.79 words   5 £167.64
6 a 16 ppm
b Taking mid-points assumes an even distribution of the data in every class, which may not be true.

Exercise 26A

1 a 32   b 28   c 35   d 7   e 8
2 a 73 cm   b 66 cm   c 79 cm   d 13 cm
e (i) 4   (ii) 55
3 a £12 700   b £11 200   c £13 800   d £2600
e £14 100
4 a 58   b 44   c 71   d 27
e (i) 84   (ii) 83   (iii) 80
5 b (i) 190 g   (ii) 74   c (i) 34 kg   (ii) 21 kg
6 a 30, 90, 130, 150, 160
c 1 min 34 s (1.57 min)
d 20

Exercise 27A

1 34   2 11   3 86   4 8.9
5 a 4.71 cm, 3.9 cm, 4.73 cm, 0.8 cm
b Similar means, but plant I has a greater range of diameters (i.e. more variable in size). Plant 2 greater consistency. Other factors influence choice such as disease or weather resistance.   c 4.8 cm 4.7 cm
d Mean or median
6 a £16.10, £2.55, £16.82, £2.60
Similar ranges, but club 2 has a higher mean.
b £15.80, £16.95
c Club 1, distribution is possibly skewed since median < mean; club 2, distribution is negatively skewed since median > mean.

Exercise 27B

1 31, 15.7   2 15, 1.14
3 459, 5.44. Minimum wt 1 lb   4 16, 2.30.
5 a Town 23 min, 4.9 min; by-pass 23 min, 2.14 min
b The by-pass. There is less variation in the times, i.e. it is more reliable.
6 69.8; 11.9. 58.8; 8.62
Saturday's cars are travelling more slowly, and at a more uniform speed.

Exercise 27C

1 0.986   2 1.22   3 11.6
4 a 75, 10.23   b smaller, the marks are more consistent
c 10.9, 4.33, 10.2; the first and third sets have an almost equally wide spread of ability, while the students in the second set are all of a similar ability.
5 a 20, 9.64   b Increase, more goods to consider/buy
c If the increased amount of time is the same for all shoppers.

Exercise 28B

1 a Weak positive   b High negative   c None
2 Weak positive   3 Quite high negative
4 Quite high positive
5 a Positive   b None   c Negative   d Positive

6 **b** No

  c Over a period of time both variables have changed, but this is a coincidence. The smaller cod catches are probably due to depleted fish stocks. The percentage vaccinated refers to an independent event.

Exercise 28C

See p. 267. The equations of the lines are given to enable students to check the position of their lines. Answers given are by calculation, answers from graphs should be correct to two s.f.

1 **a** $\bar{x} = 25$, $\bar{y} = 2.09$
  c $y = -0.113x + 4.91$
  d (i) 25.8 min (ii) 16.9 min
2 **a** (i) 64 kg (ii) 73.8 kg
  b (ii) $y = -48.8 + 0.662x$
  (iii) 60.4 kg
3 **a** $y = 13.1 + 2.28x$
  b 79 marks
4 **a** $y = 52.9 - 0.781x$
  b 39.6 kcal/hour/m²
  c 33.3 kcal/hour/m²
  d The relationship is not linear outside the given range.
5 **b** 14 500
6 **b** (i) About 5.4 years (ii) About 31 days
7 **b** The greater the number of storks, the greater the number of babies born (or vice versa).
8 **a** Reported cases of illness increase as the population increases (i.e. a positive correlation).
  c (i) 300 (ii) 350
  d actual figures are considerably lower than the predictions.

Exercise 29A

1 $\frac{5}{8}$ 2 0 3 $\frac{1}{3}$ 4 0
5 **a** Statistical evidence **b** Subjective estimate
  c Statistical evidence **d** Statistical evidence
  e Subjective estimate
6 **a** $\frac{3}{5}$ or 0.6 **b** $\frac{57}{100}$ or 0.57
  c greater amount of data used.

Exercise 29B

1 $\frac{3}{10}$ 2 $\frac{1}{6}$ 3 $\frac{1}{2}$ 4 **a** $\frac{1}{2}$ **b** $\frac{1}{13}$ **c** $\frac{1}{52}$
5 $\frac{6}{11}$ 6 **a** $\frac{7}{20}$ **b** $\frac{1}{10}$ 7 $\frac{17}{20}$

Exercise 29C

1 **a** $\frac{1}{9}$ **b** $\frac{1}{18}$ **c** $\frac{1}{18}$ 2 **a** $\frac{1}{4}$ **b** $\frac{1}{2}$ 3 **a** $\frac{1}{4}$ **b** $\frac{3}{16}$
4 **a** $\frac{1}{6}$ **b** $\frac{1}{12}$ 5 **a** $\frac{1}{3}$ **b** $\frac{1}{8}$

6 **a**

4	6	8	10	12
2	3	4	5	6
2	3	4	5	6
0	0	0	0	0

  b $\frac{1}{3}$

Exercise 29D

1 $\frac{2}{13}$ 2 **a** $\frac{3}{10}$ **b** $\frac{3}{20}$ 3 $\frac{4}{9}$ 4 $\frac{16}{25}$ 5 $\frac{1}{2}$ 6 $\frac{4}{11}$ 7 $\frac{7}{10}$

Exercise 29E

1 **a** $\frac{9}{49}$ **b** $\frac{16}{49}$ **c** $\frac{24}{49}$ 2 $\frac{17}{30}$ 3 $\frac{53}{100}$
4 **a** $\frac{1}{36}$ **b** $\frac{5}{18}$ 5 **a** $\frac{1}{4}$ **b** $\frac{31}{50}$
6 0.9999
7 **a** (i) 20 (ii) Probability is an estimate.
  b

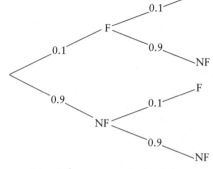

1st switch   2nd switch

  c (i) 0.81 (ii) 0.18
8 **a** $\frac{4}{13}$ **b** $\frac{48}{91}$ 9 **b** $\frac{2}{9}$ **c** $\frac{5}{9}$ **d** $\frac{1}{12}$
10

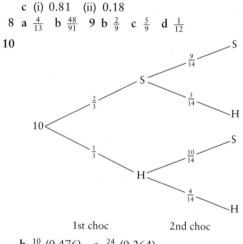

1st choc   2nd choc

  b $\frac{10}{21}$ (0.476) **c** $\frac{24}{91}$ (0.264)

Exercise 30A

1 $3a + 4b + 2c$ 2 $3a + 5c$ 3 $6f + 3s$
4 $7x + 5y$ 5 $2p - q$ or $q - 2p$ 6 $\frac{2a}{b}$
7 $5x - 8$ 8 $\frac{y}{4} + 6$ 9 $\frac{2n + 3}{8}$
10 $\frac{4n - 1}{2n}$

Exercise 30B

1 $5x$ 2 $7n$ 3 $8a$ 4 $3b$ 5 $2y$ 6 $5x + 5y$
7 $x + 5y$ 8 $3a + b$ 9 $x + 5y + 9z$
10 $7a + 6b + 4c$ 11 $9p + 2q + r$ 12 $4x + 2y - 3z$
13 $5x - 3y + z$ 14 $-a - b + 4c$ 15 $\frac{1}{4}p + \frac{5}{4}q$
16 $3x + 5y$

Exercise 30C

1 **a** 11 **b** -1 **c** 13 **d** 6 **e** $\frac{1}{2}$ **f** 2 **g** -4 **h** 17
2 **a** 1 **b** -9 **c** -5 **d** -6 **e** 0.3 **f** -14 **g** 8
  h -5
3 **a** 9 **b** -24 **c** 8 **d** -2 **e** -5 **f** 38 **g** 7
  h 28 **i** 0

Exercise 30D

1 $5x$ 2 $6a$ 3 $3mn$ 4 $8xy$ 5 abc 6 $6pqr$
7 $\dfrac{2y}{z}$ 8 $\dfrac{4bc}{d}$ 9 $6xy$ 10 $\dfrac{-2p}{q}$ 11 $\dfrac{9xy}{z}$ 12 $-2a$

Exercise 30E

1 x^3 2 x^2y 3 x^3 4 x^5 5 x^8 6 x^6 7 x^6 8 x
9 x^3 10 x^4 11 $8x^6$ 12 $4x^6$ 13 $3x^3y^2$
14 $-6a^2b^2c$ 15 $6a^2b^3c^2$ 16 $9a^2b$ 17 $-18ab^3$
18 $2x$ 19 x^3y^3 20 $\dfrac{-9a^2b}{c}$

Exercise 30F

1 $3x + 3y$ 2 $6x + 2y$ 3 $4x - 4y$ 4 $7a + 14y$
5 $6p - 12q$ 6 $20x + 8y$ 7 $6a - 12b$ 8 $-6p + 12q$
9 $5x - 15y + 10z$ 10 $-a + b + c$
11 $3x^2 - xy + 2xz$ 12 $-a^2 - ab + ac$
13 $2x^3 + 3x^2 + 2x$ 14 $4y^3 - 12y^2 + 4y$
15 $5x + 2y$ 16 $y^2 + y - 12$ 17 $9a + 17b$
18 $8p - 15q$ 19 $x^2 - 11xy$ 20 $-2x + 4y - 4z$
21 $3x + 5y$

Exercise 30G

1 $4(x + 3y)$ 2 $3(p - 2q)$ 3 $5(a + 2)$ 4 $5(2b - 1)$
5 $7(2m - 3n)$ 6 $4(3s + 5t)$ 7 $x(y + z)$ 8 $x(y + x)$
9 $y(y - 2)$ 10 $2y(y - 2)$ 11 $3x(2y + 1)$
12 $5x(1 - 2x)$ 13 $2(a + 4b - 2c)$ 14 $3(3x - y - 2z)$
15 $5(3x - y + 2)$ 16 $7(p - 2q + r)$ 17 $2a(a + 2b - 4)$
18 $3p(p + 2q - 3)$ 19 $xy(x + z + y)$ 20 $4pq(r - 3p)$

Exercise 30H

1 $\dfrac{3x}{5}$ 2 a 3 $\dfrac{x}{4}$ 4 $\dfrac{7x}{4}$ 5 $\dfrac{7x}{12}$ 6 $\dfrac{x}{15}$ 7 $\dfrac{13x}{12}$ 8 $\dfrac{a}{4}$

9 $\dfrac{y}{3}$ 10 $\dfrac{y}{15}$ 11 $\dfrac{3}{x}$ 12 $\dfrac{1}{10x}$ 13 $\dfrac{3}{2y}$ 14 $\dfrac{1}{2y}$

15 $\dfrac{3}{20a}$ 16 $\dfrac{2x + 7}{2x}$ 17 $\dfrac{x + 8}{12}$ 18 $\dfrac{12x + 1}{10}$

19 $\dfrac{2x + 1}{2}$ 20 $\dfrac{3(x - 5)}{10}$

Exercise 30I

1 $n + 5 = 12$ 2 $n - 7 = 13$ 3 $2n + 4 = 10$
4 $7n = 21$ 5 $5n - 6 = 29$ 6 $\frac{1}{2}n + 10 = 22$
7 $2n - 3 = 3$ 8 $\frac{1}{4}n - 3 = 3$ 9 $2(w + 2w) = 18$
10 $2(w + w + 4) = 28$

Exercise 30J

1 7 2 20 3 3 4 3 5 7 6 24 7 3 8 24
9 width = 3 10 length = 9 11 4 12 3 13 6
14 $1\frac{1}{2}$ 15 8 16 3 17 6 18 10 19 9 20 8
21 9 22 20 23 1.4 24 $1\frac{1}{2}$

Exercise 30K

1 4 2 3 3 2 4 $\frac{1}{3}$ 5 9 6 2 7 0 8 10 9 4
10 1 11 3 12 -5.2

Exercise 30L

1 a (i) $(x + 25)$ pence (ii) $12x$ pence (iii) $10(x + 25)$
  b (i) $12x + 10(x + 25) = 690$ (ii) $x = 20$p (iii) 45p
2 a £$(2x + 2.50)$ b £1.25
3 a £$9x$ b £$5(20 - x)$ c £$(4x + 100)$
  d $4x + 100 = 148$ e 12 matches
4 a £$(x - 1.8)$ b $(18x - 18)$ c (i) £8.60 (ii) £6.80
5 55p 6 50, 47, 42 7 68 kg 8 £1.50 9 15 min

Exercise 30M

1 a 105 cm^2 b 3 cm c 32 ft
  d (i) 20 cm (ii) 28.5 ft e 28 mph
2 a No, 170°F b 70°F c 35°C
3 a (i) 30 m/s (ii) 36 m/s
  b (i) 67.5 mph (ii) 81 mph
4 5 5 a 10 b 25 c -10 6 754 cm^2
7 £1918.44 8 a 24 cm^2 b 41.44 cm^2

Exercise 30N

1 a $B = \dfrac{P}{2} - L = \dfrac{P - 2L}{2}$ b $h = \dfrac{S}{2\pi r}$

  c $r = \sqrt{\dfrac{A}{\pi}}$ d $t = \dfrac{2s}{u + y}$ c $a = \dfrac{v - u}{t}$

2 a $x = \dfrac{y - c}{m}$ b (i) 2 (ii) 7

3 a 12 b $a = 3M - b - c$ c 4

4 a $h = \dfrac{2A}{a + b}$, 5 b $a = \dfrac{2A}{h} - b = \dfrac{2A - bh}{h}$, 8

5 a £113 b $C = 20\left(\dfrac{D - A}{N}\right)$ c £340

Exercise 30P

1 a $5x + 3y$ b $9p - 2q - 2r$ 2 a 4 b -4 c 8
3 a $12ab$ b $8a^2b$ c x^2 d $10a^6$ e 1 f $12p^2q^3r^3$

4 a x^3 b $\dfrac{1}{x^3}$ c 1 d $\dfrac{3a^3}{2}$ e $\dfrac{6}{x}$

5 a $9x - 4$ b $9x + 3$ c $-2x$
6 a $3a(2b + 3c + 1)$ b $xy(z - y)$ c $2abc(c - 3a + 2b)$

7 a $\dfrac{x}{12}$ b $\dfrac{1}{2x}$ c $\dfrac{5x + 1}{12}$

8 a 8 b 2 c 2 d 30 9 a 210 b 1275

10 a 2.5 b $x = \dfrac{2}{y - 3}$ c (i) -4 (ii) $\frac{1}{2}$

Exercise 31A

1 a 1040 b 1927, 1951 and 1974 c 1940 d 1960
2 a £75 b £200
  c There are more points plotted around 1984 which
    fit the straight line
  d 1976
3 a Sarah b 0.6 m and 1.3 m c 6.2 m d 3.1 m
4 a December, October b June, June c Feb
  d South of France (Nice)

Exercise 31B

1 a 10 st 4 lb ($\pm$ 1 lb) b 10 st 8 lb
   c February 1996 and August 1997 d Yes
2 a 60 cm^2 b 4.1 cm c 90 cm^2
3 a October and February b 14.8 and 15.8 hours
   c December to January and May to June
4 a 25 mm b 0.7 minutes c 235 mm
5 a 2.6 s b −10 cm c 0.45 s d 1.75 s

Exercise 31C

1 a 22 miles
   b Orleans, 66 miles; Tours, 153 miles; Poitiers,
     219 miles; Bordeaux, 350 miles c 96 km
2 a 32°F b 77°F c 28°C d −10°C
3 Milk, 6 fl oz; sugar, 9 fl oz; flour, 4 fl oz;
   apples, 53 fl oz; butter, 1 fl oz
4 a 3 m b 5 m c 11 m

Exercise 32A

1 All horizontal 2 All vertical
3 All pass through the origin 4 All parallel
5 All parallel
12 a (i) The same coefficient of x (ii) the same gradient
   b The coefficient of x is the gradient of the line.
13 a All the lines pass through the origin.
   b A graph of the form $y = ax$ passes through the
     origin.
14 The point on the y-axis through which the line passes.
15 The line slopes downwards, i.e. the gradient of the line
   is negative.

Exercise 32B

1 a 3 b 2 c −2 d $\frac{1}{2}$ e $\frac{2}{3}$ f −2 g −1 h −2
   i $-\frac{1}{3}$
2 a 2 b −5 c 4 d −1 e 2 f $\frac{1}{2}$ g −2 h −4
   i 2
3 a 2 b 3 c $\frac{1}{2}$ d −1 e −1 f $\frac{3}{2}$ g −4 h $-\frac{1}{2}$
4 a Any graph of the form $y - 2x = b$
   b Any graph of the form $y - bx = 7$
   c $y = \frac{2}{3}x - 3$
5 a $y = 2x + 1$ b $y = 3x - 1$ c $y = -x - 1$
   d $y = -\frac{1}{2}x + 2$
6 a (v) b (iii) c (vi) d (iv) e (i) f (ii)
8 b $L = 8W + 20$ or $y = 8x + 20$ c 42.4 cm d 20 cm
9 b $P = -0.6A + 132$ c (i) 99 (ii) 121

Exercise 32C

1 a

x	−3	−2	−1	0	1	2	3
y	11	6	3	2	3	6	11

   c 4.2(3)
2 b Graph 1 crosses the y axis at $y = 2$, and graph 2
     crosses the y axis at $y = 3$.
   c (i) 8.8 (ii) ±2.2
3 b 1.4(5) c ±2.4(5)
5 b (i) 115 m (ii) 15.8 s(16 s) (iii) 52 m (iv) 5.2 m/s

6 a

x	−5	−4	−3	−2	−1
y	−0.20	−0.25	−0.33	−0.50	−1

x	1	2	3	4	5
y	1	0.50	0.33	0.25	0.20

   c (i) 0.67 (ii) −0.29 (iii) −2.5 (iv) 1.2
7 a 36 b

N	4	6	8	10	12
T	9	6	4.5	3.6	3.0

   d 5.1 hours
8 2.4 9 1.8 10 0.8 (0.76)
12 a (vi) b (iii) c (i) d (vii) e (ii) f (iv) g (v)

Exercise 32D

1 a $y = -20, -7, 0, 1, -4, -15$
   b It has a maximum value for y instead of a minimum
     value.
     The coefficient of x^2 is a negative number.
   c −0.75 d 2.6, −1.3
2 a $y = 15, 8, 3, 0, -1, 0, 3, 8, 15$
   c (i) 2.06 (ii) 6.2, −0.2 (iii) −1 (iv) $x = 3$
3 a $y = 0, 1.99, 3.96, 5.91, 7.84, 9.75$
   c (i) no (ii) yes d 2 m and 2.5 m
4 a

v	0	20	40	60	80	100
d	0	9	28	57	96	145

   c (i) 12 m (ii) 64 m d yes
5 a $A = 0, 144, 336, 576, 864, 1200$
   c (i) 426 cm^2 (ii) 7.6 cm d 584 cm^2 e 1.4%

Exercise 32E

1 a (i) ±2.1 (ii) ±3.3 b −1.2, 1.7 c −1.8, 0.8
2 −1.3 3 −5.2, 0.2 4 2.6 5 −1.8, −0.4, 1.2
6 a

x	−3	−2	−1	0	1	2	3
y	−22	−3	4	5	6	13	32

   b (ii) −2.4, 0.4, 2
   c Any value greater than 9 or less than 1.

Exercise 33A

1 9, 11, Add 2 to previous term.
2 13, 15, Add 2 to previous term.
3 81, 243, Multiply previous term by 3.
4 5, 3, −1, Subtract 2 from previous term.
5 0.001, 0.00001, Divide previous term by 10.
6 13, 21, 34, Add together the two previous terms.
7 $\frac{1}{16}, \frac{1}{64}$, Divide previous term by 2.
8 $\frac{5}{8}, \frac{7}{16}$, Add 2 to previous numerator, multiply previous
   denominator by 2.
9 25, 36, 49, Square the number of the next term.
10 10, 21, Add to the previous term one more each time.

Exercise 33B

1 $T_n = T_{n-1} + 2$ 2 $T_n = T_{n-1} + 2$ 3 $T_n = 3T_{n-1}$
4 $T_n = T_{n-1} - 2$ 5 $T_n = 0.1T_{n-1}$
6 $T_n = T_{n-1} + T_{n-2}$ 7 $T_n = \frac{1}{2}T_{n-1}$
8 4, 7, 10, 13   9 2, 8, 32, 128   10 2, -2, -6, -10
11 4, 2, 1, $\frac{1}{2}$
12 -2, 2, 4, 5   13 $\frac{1}{2}$, $-\frac{1}{2}$, $-\frac{3}{2}$, $-\frac{5}{2}$

Exercise 33C

1 a 5, 7, 9, 11, 13   b 3, 2, 1, 0, -1
  c $-1\frac{1}{2}$, -1, $-\frac{1}{2}$, 0, $\frac{1}{2}$
  d 2, -1, -4, -7, -10
2 a $T_n = 3n$ b $T_n = 4n - 3$ c $T_n = 7n + 2$
  d $T_n = 60 - 10n$ e $T_n = 2n - 8$
3 a $T_n = 3n + 1$ or $y = 3x + 1$
  b $T_n = 7 - 4n$ or $y = 7 - 4x$
  c $T_n = \frac{1}{2}n + \frac{1}{2}$ or $2y = x + 1$
  d $T_n = \frac{1}{2}n - 2$ or $2y = x - 4$
  e $T_n = 10 - 6n$ or $y = 10 - 6x$
4 a

n	2	3	4	5	6
T_n	31	36	41	46	51

  b $T_n = 5n + 21$ c (i) 191
5 a

a	4	5	6	7
$T_{n,a}$	$5n + 12$	$5n + 15$	$5n + 18$	$5n + 21$

  b 3   c $T_{n,a} = 5n + 3a$
7 a 1, 5, 9, 13 $T_n = 4n - 3$
  b 1, 9, 17, 25 $T_n = 8n - 7$
  c 1, 7, 13, 19 $T_n = 6n - 5$
8 a 1, $\frac{2}{3}$, $\frac{3}{5}$, $\frac{4}{7}$, $\frac{5}{9}$ b $\dfrac{n}{3n - 1}$

Exercise 33D

1 a (i) 25, 36   (ii) n^2
  b (i) 5 is added to each term   (ii) $n^2 + 5$
2 a $\frac{2}{2}$, $\frac{3}{5}$, $\frac{4}{10}$, $\frac{5}{17}$, $\frac{6}{26}$ or 1, $\frac{3}{5}$, $\frac{2}{5}$, $\frac{5}{17}$, $\frac{3}{13}$
  b $\dfrac{n}{n^2 + 3}$
3 a 162   b $2n^2$
4 a 15, 21   b (i) $(10 \times 11) = 110$ (ii) $n(n + 1)$
  c $\frac{1}{2}n(n + 1)$ (ii) 210
5 a 28, 65   b 6   6 2, 4, 8, 16, 32: $U(x) = 2^x + 5$

Exercise 33E

1 a -3.2, 1.2   b -1.54, 4.54   c 0.32   d 3.34
2 0.3, 7.7   3 a (i) -3 (ii) 22   b The change of sign
  means the graph has crossed the x-axis ($y = 0$) between
  $x = 4$ and $x = 5$, therefore there is a solution between
  these values.   c 4.17
4 a 3.2 (and -2.2)   b (i) $(\frac{1}{2}, -7\frac{1}{4})$ (ii) -2.2, 3.2

Exercise 33F

1 $f(-3) = 14$; $f(-1) = 6$; $f(\frac{1}{2}) = 0$; $f(1\frac{1}{2}) = -4$;
  $f(2) = -6$
2 a -9, -8, -7, -6, -5, -4 b 28, -11
  c 2, $\frac{1}{2}$, 0, $\frac{1}{2}$, 2, 8   d 5, $3\frac{1}{2}$, 3, $3\frac{1}{2}$, 5, 11
  e $\frac{1}{2}$, $-\frac{1}{5}$, 4, -5 3 $y = \frac{1}{2}x + 2$

Exercise 33G

1 a $x \rightarrow$ [Multiply by 7] $\rightarrow$ [Add 5] $\rightarrow y$

  b $x \rightarrow$ [Multiply by 3] $\rightarrow$ [Subtract 11] $\rightarrow y$

  c $x \rightarrow$ [Square] $\rightarrow$ [Multiply by 3] $\rightarrow$ [Add 5] $\rightarrow y$

  d $x \rightarrow$ [Multiply by 2] $\rightarrow$ [Add 1] $\rightarrow$ [Square] $\rightarrow y$

2 a $y = 3x + 3$ b $y = \left(\dfrac{x}{5} + 4\right)^2$ c $y = \{3(x + 6)\}^2$

Exercise 33H

1 a x divide by 4 subtract 2 $f^{-1}(x)$
  b x add 3 divide by 5 $g^{-1}(x)$
  c x subtract 7 multiply by 2 $h^{-1}(x)$

2 a $f^{-1}(x) = \dfrac{x + 5}{3}$ b $g^{-1}(x) = \dfrac{x - 7}{2}$

  c $h^{-1}(x) = \dfrac{x + 11}{7}$

Exercise 33I

1 $f(x) = x^2 + 3$: graph has moved vertically upwards by 3
  $f(x) = x^2 - 1$: graph has moved vertically downwards by 1
  $f(x) = x^2 - 2$: graph has moved vertically downwards by 2
2 Graph moves a units along y-axis (i.e. a translation of a units along y-axis.
8 $f(x + a)$: translation of $-a$ units along x-axis
  $af(x)$: stretch of a units vertically

  $f(ax)$: stretch of $\dfrac{1}{x}$ units horizontally

  $-f(x)$: reflection in x-axis
  $f(-x)$: reflection in y-axis
9 a $f(x) = x$: straight line of gradient 1, passing through origin

  $f(x) = \dfrac{1}{x}$: two separate curves which approach the

  axes but never touch or cross them
10 (ii) d   (iii) a   (iv) c   (v) f   (vi) e   (vii) b   (viii) g

Exercise 33K

1 a
  b
  c
  d
  e
  f
  g
  h

2 a $-2 < x < 2$ **b** $-1 \leqslant x \leqslant 3.5$ **c** $x < 0$
d $-2 < x < 3$ **e** $x \geqslant 1.5$ **f** $x < -1, x > 1$
3 a $8, 9, 10, 11, 12$ **b** $-3, -2, -1$
c $-5, -4, -3, -2, -1$ **d** $-1, 0, 1$ **e** 0 **f** $4, 5, 6$
g $-8, -7, -6, -5, -4, -3, -2, -1, 0, 1$

Exercise 33L

1 a $x < 3$ **b** $x > 7$ **c** $x \leqslant 3$ **d** $x \geqslant 17$ **e** $x \geqslant 8\frac{1}{2}$
f $x > 15$ **g** $x > 8$ **h** $x < -5, x > 5$
i $-1.5 \leqslant x \leqslant 1.5$ **j** $x > 2$
2 $-3 < x < 3$

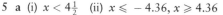

3 a Cost of chips + cost of apple $\leqslant 95$p **b** 60p
4 a $4x - 7$ **b** $4x - 7 < 20$ **c** 6

Exercise 33M

1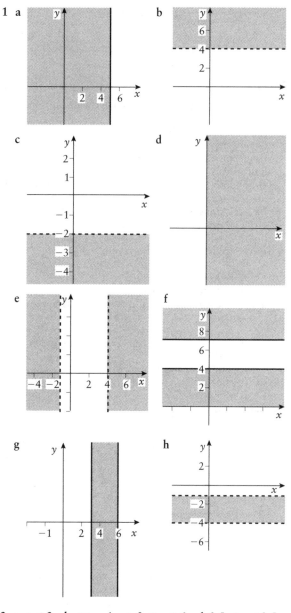

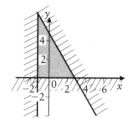

2 a $y \leqslant 2$ **b** $x > -1$ **c** $2 \leqslant x \leqslant 6$ **d** $0.5 < y < 2.5$
e $-6 \leqslant x < 4$ **f** $y \geqslant 3, y \leqslant 6$

3 a LHS = 8, RHS = 6: $\therefore$ (1,8) does not satisfy
inequality
b LHS = 8, RHS = 5: $\therefore$ (1,8) does satisfy inequality.

4 a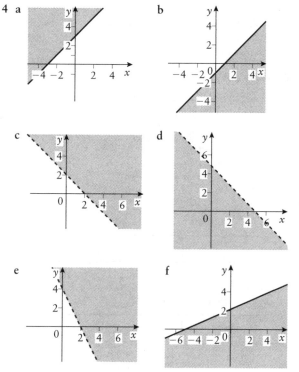

5 a (i) $x < 4\frac{1}{2}$ (ii) $x \leqslant -4.36, x \geqslant 4.36$

b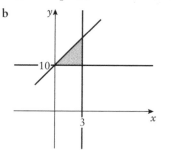

Exercise 33N

1 a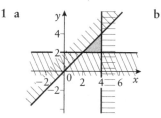

b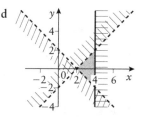

c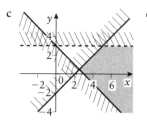

e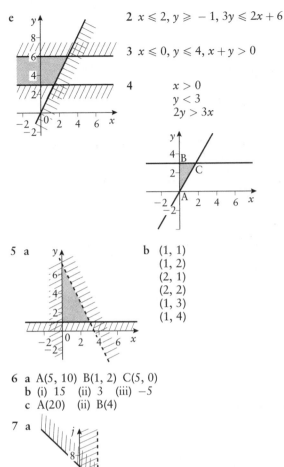

2 $x \leqslant 2, y \geqslant -1, 3y \leqslant 2x + 6$

3 $x \leqslant 0, y \leqslant 4, x + y > 0$

4 $\quad x > 0$
$\quad\quad y < 3$
$\quad\quad 2y > 3x$

5 a   **b** (1, 1)
(1, 2)
(2, 1)
(2, 2)
(1, 3)
(1, 4)

6 a A(5, 10) B(1, 2) C(5, 0)
  b (i) 15   (ii) 3   (iii) −5
  c A(20)   (ii) B(4)

7 a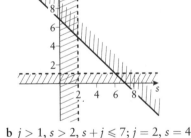

  b $j > 1, s > 2, s + j \leqslant 7; j = 2, s = 4$

Exercise 33P

1 a $5x + 3y \leqslant 75, 0 \leqslant x \leqslant 12, 0 \leqslant y \leqslant 9$
  c 10 CDs and 8 cassettes or 9 CDs and 9 cassettes.
  d For the lesser cost choose 9 CDs and 9 cassettes. (Other answers are possible, provided the reason is valid.)
2 b $0 \leqslant x \leqslant 12, 0 \leqslant y \leqslant 10$
  d 18 vehicles: 10 minibuses and 8 cabs or 11 minibuses and 7 cabs.
3 a AC: $2p + 3c = 60$
     HE: $p + 2c = 20$
  b Quadrilateral BCFG   **c** G
  d 8 plain and 6 chocolate   **e** 68p
4 a $2b + 3h \leqslant 18, b + 2h \geqslant 8, b > h$ **b** 8
  c 4 bungalows and 3 houses or 6 bungalows and 2 houses earn £40.
5 a $c + f \leqslant 14, 5c + 3f \leqslant 60, f \geqslant 3, c \geqslant 5$.
  c (choc., fruit): (9, 5), (8, 6), (7, 7), (6, 8), (5, 9).
  d 5 bars of chocolate and 9 pieces of fruit cost the least at £2.60.
6 b 30 scones

Exercise 34A

1 a 18 minutes   **b** 60 mph   **c** 45 minutes   **d** 8.45 am
  e 35 miles   **f** 20 mph
2 a 8 miles   **b** 11 am   **c** 12 noon
3 a Ali, 9.15 am: Barry, 9.25 am   **b** 8 mph
4 a 9.45 am   **b** Ali, 3 miles; Barry, 2 miles
5 a 88 km/h   **b** 58 km/h   **c** 80 km/h

Exercise 34B

1 b 0   **c** 9 m   **d** 6 m/s   **3 b** (i) −7   (ii) 8
3 b 21 m/s   **4 b** (i) −13   (ii) 20
5 a $A = 12, 20, 32, 48, 68, 92$ **c** $5.5\,\text{cm}^2$ **d** $8\,\text{cm/s}^2$
6 b (i) 20 m   (ii) 0.7 s and 3.3 s   **c** (i) 10 m/s
    (ii) −5 m/s. The negative gradient indicates that the stone is falling back to earth.

Exercise 34C

1 a $1.3\,\text{m/s}^2$ **b** 91.3 m   **c** 18.3 m
2 a 3.99 s   **b** 56 m   **c** $7\,\text{m/s}^2$
3 a (i) $0.6\,\text{m/s}^2$ (ii) 140 m;   **b** 36.1 m
4 a 6 min   **b** 3.9 km
5 a $20\,\text{m}^2$ **b** 3.2 years
6 a

Time	1	2	3	4	5	6	7
Mass	200	100	50	25	12.5	6.25	3.125

  b $1\frac{1}{2}$ days
7 b £2860   **c** (i) £890 per year   (ii) £370 per year

Exercise 34D

1 a $4.8\,\text{m/s}^2$ **b** 26 m   **c** 19 m
2 a $-6\,\text{m/s}^2$ **b** 127 m   **c** 27.6 m/s
3 a $4.8\,\text{m/s}^2$ **b** 49 m   **c** 23 m
4 a $6.7\,\text{m/s}^2$ **b** 235 m
5 a 59 m   **b** The distance travelled in 6 s.

Exercise 34E

1 a (i) 25.65 m   (ii) 25.61 m   **b** 19.40 m
2 a 6.5 s   **b** 126.75 m   **3** (i) 3.05 m   (ii) 3.01 m

Exercise 35A

1 $x = 2$ **2** $x = 9$ **3** $x = 7$ **4** $x = 6$
  $y = 4$ $y = 1$ $y = 5$ $y = 4\frac{1}{2}$
5 $x = 3$ **6** $x = 2$ **7** $x = 2$ **8** $x = 5$
  $y = 1$ $y = 1$ $y = 6$ $y = 0$
9 $x = 5$ **10** $x = 3$ **11** $x = 1\frac{1}{2}$ **12** $x = -2$
  $y = 6$ $y = -12$ $y = 2\frac{1}{2}$ $y = -1$

Exercise 35B

1 64p, 44p   **2** 7, 3   **3** $88\,\text{m}^2$ **4** £616   **5** 10, 3
6 24 years, 3 years   **7** 12 sweets

Exercise 35C

1 $x = 2$ **2** $x = 3$ **3** $x = 5$ **4** $x = 2$ **5** $x = 2\frac{1}{2}$
  $y = 3$ $y = 1$ $y = 4$ $y = 3$ $y = 2$
6 $x = 2$ **7** $x = -1$ **8** $x = -1$ **9** $x = 4$
  $y = 3$ $y = 2$ $y = -7$ $y = -2$

Exercise 35D

1 $x^2 + 7x + 12$
2 $x^2 + x - 2$
3 $x^2 - 8x + 15$
4 $x^2 + 5xy + 4y^2$
5 $x^2 - xy - 2y^2$
6 $x^2 - 10x + 21$
7 $x^2 - x - 2$
8 $x^2 - 8x + 12$
9 $x^2 - 9$
10 $x^2 - y^2$

11 $4x^2 - 25$
12 $9x^2 - 16$
13 $x^2 + 4x + 4$
14 $x^2 - 2x + 1$
15 $9x^2 + 6x + 1$
16 $4x^2 - 12xy + 9y^2$
17 $2x^2 + 9x + 4$
18 $3x^2 - 4x - 4$
19 $6x^2 + 7xy + 2y^2$
20 $12x^2 + xy - 6y^2$

Exercise 35E

1 $(x + 1)(x + 7)$
2 $(x + 2)(x + 5)$
3 $(x - 2)(x + 3)$
4 $(x - 2)(x - 3)$
5 $(x + 4)(x - 2)$
6 $(x + 3)(x - 5)$
7 $(x - 3)(x - 3)$
8 $(x - 6)(x + 2)$
9 $(x - 3)(x - 4)$

10 $(x + 12)(x - 1)$
11 $(x - 6)(x + 1)$
12 $(x - 8)(x + 2)$
13 $x(x - 7)$
14 $2x(x + 3)$
15 $3x(2x - 1)$
16 $x(1 - 2x)$
17 $(x + 6)(x + 6)$
18 $(x - 4)(x - 4)$

Exercise 35F

1 $(2x + 1)(x + 2)$
2 $(2x + 3)(x + 2)$
3 $(2x - 1)(x + 5)$
4 $(2x + 1)(x - 7)$
5 $(3x + 5)(x + 1)$
6 $3(x + 1)(x - 3)$

7 $(3x - 2)(x - 2)$
8 $(3x + 2)(x - 5)$
9 $(2x - 3)(x - 4)$
10 $(3x + 2)(x + 6)$
11 $(3x + 2)(x - 8)$
12 $(2x - 7)(x + 2)$

Exercise 35G

1 $(x - 1)(x + 1)$
2 $(x - 4)(x + 4)$
3 $(x - 5)(x + 5)$
4 $(2x - 3)(2x + 3)$
5 $(3x - 1)(3x + 1)$
6 $(4x - 5)(4x + 5)$

7 $(7 - x)(7 + x)$
8 $(6 - 5x)(6 + 5x)$
9 $(5 - 7x)(5 + 7x)$
10 $2(x - 2)(x + 2)$
11 $9(x - 2)(x + 2)$
12 $3(2x - 5)(2x + 5)$

Exercise 35H

1 $-2, -1$ 2 $1, 2$ 3 $-2, 1$ 4 $-1, 2$ 5 ± 2 6 ± 3
7 ± 4 8 0 or 2   9 -3 or 0   10 0 or 6   11 1 or 4
12 -3 or -3 13 -4 or 3   14 2 or 6   15 2 or 7
16 -7 or -4 17 $x = -3$ or $-\frac{1}{2}$ 18 $x = -\frac{4}{3}$ or 2

19 $x = 0$ or $\frac{2}{5}$ 20 $x = \frac{2}{3}$ or 6   21 $x = -\frac{7}{2}$ or 2
22 $x = 5$ or -4 23 $x = \frac{1}{2}$ or -1

Exercise 35I

1 $-1.82, 0.82$
2 $-0.29, -1.71$
3 $0.73, -2.73$
4 $0.77, -0.43$
5 $1.36, -7.36$
6 $5.30, 1.70$

7 $3.24, -1.24$
8 $1.58, 0.42$
9 $0.89, -3.39$
10 $2.72, 0.61$
11 $1.45, -3.45$
12 $2.62, 0.38$

Exercise 35J

1 a $(x + 3)$ by $(x - 1)$ b $(x + 3)(x - 1) = 21$
   c 4 m by 4 m
2 a 14   b 8   3 a $\frac{1}{2}x(x - 1)$ b $x^2 - x - 56 = 0$ c 8
4 a $(20 - x)$ metres   b $x(20 - x)\,\text{m}^2$ c 8 m by 12 m
   d 14.4 m by 5.6 m   c 10 m by 10 m
5 $(7 - x)(4 - x) = 14$, $x = 1.47$
6 $(5 + x)(5 + 2x) = 80$, $x = 2.7$, 10.4 cm $\times$ 7.7 cm
7 a 24   b 54   c $S = 6(n - 2)^2$ d (i) 4   (ii) 6

Exercise 35K

1 a $v = \dfrac{2s - ut}{t}$ b 18   c 0; the vehicle has stopped

2 a 8 cm   b $u = \dfrac{fv}{v - f}$ c 4.3 cm

3 a $l = 10\left(\dfrac{T}{2\pi}\right)^2$ b 25.3 cm

4 a $r = 100\left(\sqrt{\dfrac{A}{P}} - 1\right)$ b 22.9%   c $A = PR^2$

   d $R = \sqrt{\dfrac{A}{P}}$ e (i) 1.051   (ii) 5.12%

5 a $S = \dfrac{100 - v^2}{20}$

   b (i) 3.75 m; above its starting position
      (ii) 5 m; above its starting position, i.e. at the highest point
      (iii) 0 m; level with its starting position
      (iv) -6.25 m; 6.25 m below its starting position

6 a $P = 2x + \pi x + 4h$ b $x = \dfrac{P - 4h}{2 + \pi}$ c 1.5 m

7 a $P = 2(x + y)$ b $A = xy$ c $xy = 6(x + y)$

   d $x = \dfrac{6y}{y - 6}$ c 15 cm

Exercise 35L

1 a (i) $x = 3$ (ii) $x = 2\frac{1}{2}$ b $x = 3$ 2 $x = 50$
      $y = -2$ $y = 1$ $y = 5$ $y = 24$
3 a $x^2 + 7x + 10$ b $x^2 - 49$ c $2x^2 - 17x + 8$
   d $2x^2 - xy - 3y^2$
4 a $(x - 9)(x + 1)$ b $3x(x + 4)$ c $3(x - 2)(x + 2)$
   d $(x + 3)(x - 2)$ e $3(x + 3)(x + 1)$
   f $(2x - 1)(x + 6)$
5 a $5, -1$ b $0, 3$ c ± 5 d $-4, -3$ e 2, 0
   f $\frac{4}{3}, -\frac{4}{3}$ g $-2, -\frac{3}{2}$
6 a $1.83, -3.83$ b $2.28, 0.22$
7 $1.8, -3.8$ 8 a $x \geqslant 6$ b $x > 3$ c $-3 \leqslant x \leqslant 3$
9 $U_n = 4n - 2$
10 a $(x^2 + 2x)\,\text{cm}^2$ b $(4x + 4)\,\text{cm}$ c $x^2 - 2x - 4 = 0$
    d 3.24 cm

11 a Nx pence   b $\dfrac{Ny}{10}$ pence   c $P = \dfrac{Ny}{10} - Nx$ d £60

    e $N = \dfrac{10P}{y - 10x}$ f 5000

Exercise 36A

1 7.745 97   2 a 7.742   b 7.742, 0.258
3 a (i) and (ii) 0.258   b (i) 6   (ii) 4
4 a $x_2 = -3.8125$, $x_3 = -3.7936$, $x_4 = -3.7915$,
      $x_5 = -3.7913$, $x_6 = -3.7912$, $x_7 = -3.7912$
   b $x = -3.791$ c 4
5 a $x_2 = 5.125$ $x_3 = 4.987\,804\,8(78)$
      $x_4 = 5.001\,222\,4(94)$ b 5
   d (i) $(2x + 1)(x - 5)$ (ii) $x = 5$ or -0.5

6 a $x_{n+1} = \dfrac{1}{3}\left(11 + \dfrac{4}{x_n}\right)$ and $x_{n+1} = \dfrac{3x_n^2 - 4}{11}$

   b $x_2 = 4.111\,111\,1$ $x_2 = 2.090\,909\,1$
      $x_3 = 3.990\,990\,9$ $x_3 = 0.828\,700\,2$
      $x_4 = 4.000\,752\,4$ $x_4 = -0.176\,342\,5$

c The formula $x_{n+1} = \frac{1}{3}\left(11 + \frac{4}{x_n}\right)$ is approaching

a positive root of value 4.

The formula $x_{n+1} = \frac{3x_n^2 - 4}{11}$ is approaching a

negative root.

d $x_2 = 8.111\,111\,1$ $x_2 = -0.339\,090\,9$
   $x_3 = 3.831\,050\,2$ $x_3 = -0.332\,277\,4$
   $x_4 = 4.014\,700\,0$ $x_4 = -0.333\,525\,0$
   $x_5 = 3.998\,779\,5$ $x_5 = -0.333\,298\,4$
   $x_6 = 4.000\,101\,7$ $x_6 = -0.333\,339\,6$

e The roots are 4 and $-0.\dot{3}$ $(-\frac{1}{3})$

g $x_{n+1} = 2 - \dfrac{1}{2x_n}$

Exercise 36B

1 a Does not converge
   b Limit is 2
2 a Does not converge
   b Converges to -2.095
3 Yes, converges to -1.236 when $x_1 = 1$, other values do not converge.

Exercise 37A

1 a $44°$ b $48°$ c $124°$ d $54°$
2 a acute   b reflex   c obtuse   d obtuse   e reflex
   f acute
3 a $145°, 35°$ b $125°, 55°, 95°$ c $60°, 90°, 30°$

Exercise 37B

1 $a = 52°, b = 128°, c = 128°, d = 52°$
2 $p = 65°, q = 115°, r = 65°, s = 115°, t = 65°$
3 $r = 32°, s = 148°, t = 32°, u = 148°, v = 32°,$
   $w = 148°$
4 $l = 74°, m = 74°, n = 74°, p = 106°, q = 74°$
5 $a = 42°, b = 42°, c = 48°, d = 48°$
6 $a = 120°, b = 120°, c = 30°, d = 30°$
7 a $\angle PLQ = \angle LQS$ (alternate angles)
   b $116°$

Exercise 37C

1 a 1   b 5   c 3   d 2
2

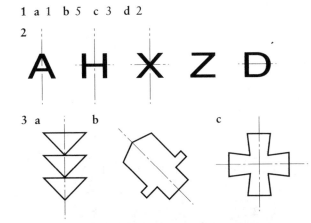

3 a   b   c

Exercise 37D

1 H, 2; I, 2; N, 2; O, ∞; S, 2; X, 4; Z, 2.
2 a Shapes b, c and d   b 5; 3; 2.
3 a (i) 0   (ii) 5   b (i) 0   (ii) 3   c (i) 4   (ii) 4
   d (i) 2   (ii) 2   e (i) 0   (ii) 6   f (i) 1   (ii) 1
4 a (i) B (ii) A (iii) 3
   b (i)

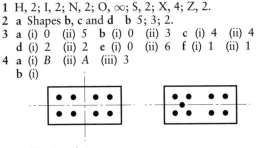

   (ii) 1 and 0

Exercise 37E

1 4   2 4   3 1   4 3   5 5   6 9   7 6   8 ∞ 9 7
10 ∞

Exercise 37F

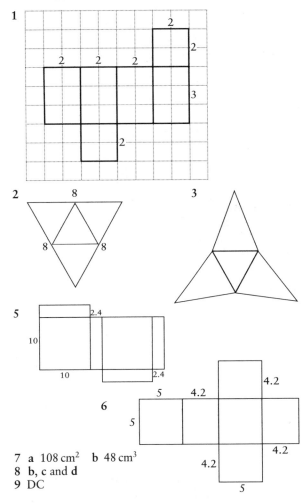

7 a $108\,\text{cm}^2$ b $48\,\text{cm}^3$
8 b, c and d
9 DC

Exercise 37G

1 $20°, 80°$ 2 $A = B = 62\frac{1}{2}°$
3 $72°, 36°, 108°, 36°$
4 $70°, 40°, 140°$ 5 $60°, 50°$

Exercise 37H

1 Two pairs of adjacent sides are equal.
  The diagonals bisect each other at right angles.
  The diagonals bisect the angles between equal pairs of sides.
2 $\angle P = 136° = \angle Q$; $\angle R = 44°$ 3 $45°$
4 a $32°$ b $58°$ 5 a $57°$ b $81°$ c $28°$
6 a (i) $37°$ (ii) the angles of a triangle $= 180°$
   b (i) $115°$
     (ii) $\angle PSR = 115°$ (angles on a straight line)
         $b = \angle PSR$ (opposite angles of a parallelogram)

Exercise 37I

1 a $45°$ b $135°$ 2 a $36°$ b $144°$
3 18   4 9   5 $540°$ 6 $1440°$
7 a 11 b 15

Exercise 37J

1 $C = 90°$, $B = 55°$ 2 $R = 90°$, $P = Q = 45°$
3 a $65°$, b $50°$ 4 a $48°$, b $96°$ 5 $56°$

Exercise 37K

1 $55°$ 2 $70°$
3 a $53°$ b $118°$ 4 $B = 28°$, $D = 52°$, $E = 100°$
5 a $71°$ b $48°$ 6 a $32°$ b $106°$ c $42°$
7 a $41°$ b $41°$ 8 a $65°$ b $25°$ c $70°$
9 a $50°$ b $130°$ c $65°$
10 $\angle OBC = 21°$ 11 a $37°$ b $37°$ c $90°$
12 a $79°$ b $25°$

Exercise 38A

1 $\triangle ABC$ and $\triangle RPQ$ 2 $\triangle ABD$ and $\triangle CBD$ or
  $\triangle ABD$ and $\triangle CDB$
3 $\triangle ABD$ and $\triangle ACD$
4 Not necessarily congruent   5 Not congruent
6 $\triangle LMN$ and $\triangle QOP$ 7 A and D

Exercise 38B

1 $PQ = 3.2$ cm, $PR = 10.4$ cm
2 $OP = 4.5$ cm, $OQ = 5.4$ cm, $PQ = 6.3$ cm
3 a 3 b 8.1 cm   4 a 4.5 cm b 5.0 cm c 6.25 cm
5 a 3.6 b $153°$ 6 No

Exercise 38C

1 a 7.6 cm by 16 cm b 6.4 cm by 4.0 cm c 1.6 cm
2 a (i) 7 cm (ii) 14 cm × 12 cm (iii) 8 mm
   b (i) 93 cm (ii) $13\frac{1}{2}$ cm (iii) 69 cm × 60 cm
3 a 9 ft 2 in b 39.7 cm
4 a 20.4 m b 68 feet
5 a 17 mm b (i) 5 ft 9 in (ii) 172.5 cm

Exercise 38D

1 a 1 cm b 4.8 cm c 1.1 m d 28.8 cm
2 a 20 m b 124 m c 168 m d 9000 in (250 yd)
   e 9000 in
3 a 1 : 10 000 b 1 : 50 000 c 1 : 25 000
   d 1 : 12 500 e 1 : 1 000 000
4 a 250 m b (i) 1850 m (ii) 4.4 km (iii) 500 m
   (iv) 1550 m c (i) 20 cm (ii) 7.2 cm (iii) 19 cm
   (iv) 42.4 cm
5 a 2 cm b (i) 2 km (ii) 24.4 km (iii) 8.15 km
   (iv) 18.45 km c (i) 54 cm (ii) 21.2 cm
   (iii) 1.16 mm (iv) 1.188 m
6 a 2.5 km b 6.25 km² c 3.75 km² d 4 cm²
7 a (ii) 1050 m b (i) 700 m (ii) 950 m

Exercise 39A

1 a $090°$ b $135°$ c $315°$ d $330°$ e $195°$
2 a $113°$ b $074°$ c $312°$ d $237°$ 3 $012°$
4 $241°$ 5 $144°$ 6 $216°$
7 a $126°$ b $306°$ 8 2.30 miles
9 a (i) $258°$, 4 cm (ii) $335°$, 4.5 cm

Exercise 40A

1 6.5 cm   2 3.6 cm   3 $88.4°$ 4 $61.3°$ 5 4.9 cm

Exercise 40B

2 a $52°$ b 4.1 cm   5 148 m   6 375 m   7 26.8 m

Exercise 40C

1 Perpendicular bisector of the line joining the buoys.
2 a a pair of lines parallel to the fixed line
   b the circumference of a circle with AB as diameter

3 A: ——— B: ∿∿ C: ⌒⌒
                              Ground Level

4 A, a vertical line; B, a quarter circle radius $\frac{1}{2}$ AC;
  C, a horizontal line

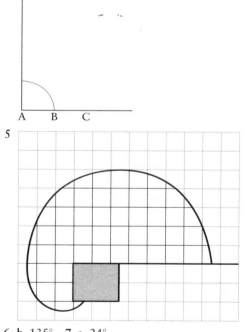

6 b $135°$ 7 c $24°$

Exercise 41A

1 a (i) $\begin{pmatrix} 6 \\ 3 \end{pmatrix}$ (ii) $6\mathbf{i} + 3\mathbf{j}$ b (i) $\begin{pmatrix} 6 \\ -4 \end{pmatrix}$ (ii) $6\mathbf{i} - 4\mathbf{j}$

  c (i) $\begin{pmatrix} 6 \\ 3 \end{pmatrix}$ (ii) $6\mathbf{i} + 3\mathbf{j}$ d (i) $\begin{pmatrix} -7 \\ 2 \end{pmatrix}$ (ii) $-7\mathbf{i} + 2\mathbf{j}$

2 a $\begin{pmatrix} 0.5 \\ 2 \end{pmatrix}$ b $\begin{pmatrix} 7 \\ 0 \end{pmatrix}$ c $\begin{pmatrix} -4.5 \\ -2 \end{pmatrix}$ d $\begin{pmatrix} 5 \\ 4 \end{pmatrix}$

  e $\begin{pmatrix} -6.5 \\ 2 \end{pmatrix}$

3 a (i) $\begin{pmatrix} 4 \\ 1 \end{pmatrix}$ (ii) $4i + j$ b (i) $\begin{pmatrix} 4 \\ -4 \end{pmatrix}$ (ii) $4i - 4j$

c (i) $\begin{pmatrix} 9 \\ -1 \end{pmatrix}$ (ii) $9i - j$ d (i) $\begin{pmatrix} -10 \\ -5 \end{pmatrix}$

(ii) $-10i - 5j$ e (i) $\begin{pmatrix} -7 \\ 7 \end{pmatrix}$ (ii) $-7i + 7j$

f (i) $\begin{pmatrix} -3 \\ 7 \end{pmatrix}$ (ii) $-3i + 7j$

Exercise 41B

1 a 10   b 8.60   2 a 13   b 17   c 5
3 a 4.47   b 10

Exercise 41C

1 (i) $\begin{pmatrix} 20 \\ -48 \end{pmatrix}$ (ii) $\begin{pmatrix} 2\frac{1}{2} \\ -6 \end{pmatrix}$ (iii) $\begin{pmatrix} -10 \\ 24 \end{pmatrix}$ (iv) $\begin{pmatrix} -5 \\ 12 \end{pmatrix}$

(v) 13   (vi) 39
2 b, c and e
3 a parallel   b equal magnitude   c both   d parallel
   e equal

Exercise 41D

1 a $\begin{pmatrix} 6 \\ 4 \end{pmatrix}$ (ii) $\begin{pmatrix} -2 \\ -6 \end{pmatrix}$ (iii) $\begin{pmatrix} 8 \\ 9\frac{1}{2} \end{pmatrix}$ (iv) $\begin{pmatrix} 8 \\ 17 \end{pmatrix}$

(v) $\begin{pmatrix} 1 \\ 0 \end{pmatrix}$ (vi) $\begin{pmatrix} 1 \\ 3 \end{pmatrix}$ (vii) $\begin{pmatrix} -\frac{1}{2} \\ -2 \end{pmatrix}$ (viii) $\begin{pmatrix} 0 \\ 6 \end{pmatrix}$

2 (i) $-5i + 3j$ (ii) 5.83 (iii) $11i$ (iv) 11
   (v) $4\frac{1}{2}i - \frac{1}{2}j$ (vi) 4.53

3 (i) $p = \begin{pmatrix} 17 \\ 19 \end{pmatrix}$ (ii) $q = \begin{pmatrix} 17 \\ 19 \end{pmatrix}$

(iii) p and q are parallel and of equal lengths.

Exercise 41E

1 (i) $\overrightarrow{PQ} = q - p$, $\overrightarrow{OR} = \frac{3}{4}p$, $\overrightarrow{OS} = \frac{3}{4}q$, $\overrightarrow{RS} = \frac{3}{4}(q - p)$
   (ii) RS and PQ are parallel and $RS = \frac{3}{4}PQ$.
2 (i) $\overrightarrow{AP} = \frac{1}{2}b$, $\overrightarrow{CD} = (a - b - c)$, $\overrightarrow{CR} = \frac{1}{2}(a - b - c)$,
     $\overrightarrow{PS} = \frac{1}{2}(a - b)$, $\overrightarrow{QR} = \frac{1}{2}(a - b)$,
     $\overrightarrow{PQ} = \frac{1}{2}(b + c)$, $\overrightarrow{SR} = \frac{1}{2}(b + c)$
   (ii) PQRS is a parallelogram.
3 $\overrightarrow{RS} = -p$, $\overrightarrow{ST} = -q$, $\overrightarrow{QT} = -2p$, $\overrightarrow{QR} = q - p$,
   $\overrightarrow{PR} = 2q - p$, $\overrightarrow{OR} = 2q$
4 (i) $c + \frac{1}{2}a$, $\frac{1}{2}c + \frac{1}{4}a$, $-a$, $c - a$, $\frac{1}{2}(c - a)$, $\frac{1}{2}(a + c)$, $\frac{1}{4}a$
   (ii) $XY = \frac{1}{4}OA$ and $XY| |OA$
5 (i) $\overrightarrow{AB} = b - a$, $\overrightarrow{AD} = \frac{2}{3}(b - a)$,
     $\overrightarrow{CD} = \frac{1}{3}a + \frac{2}{3}b = \frac{1}{3}(a + 2b)$
     $\overrightarrow{BE} = \frac{1}{2}a$, $\overrightarrow{DB} = \frac{1}{3}(b - a)$,
     $\overrightarrow{DE} = \frac{1}{6}a + \frac{1}{3}b = \frac{1}{6}(a + 2b)$
   (ii) CD and DE are in the same straight line, CDE.
       D divides CE in the ratio $2:1$.
6 a 3a   b 2b   c $3a - 2b$

Exercise 42A

1 a $\begin{pmatrix} 0 \\ -6 \end{pmatrix}$ b $\begin{pmatrix} 9 \\ -5 \end{pmatrix}$ c $\begin{pmatrix} 9 \\ 0 \end{pmatrix}$ d $\begin{pmatrix} 5 \\ 3 \end{pmatrix}$ e $\begin{pmatrix} -5 \\ 4 \end{pmatrix}$

2 A'(6, 4); B'(7, 8); C'(3, 1)

Exercise 42B

2 a A'(2, −1); B'(3, 2); C'(5, 3)
   b A'(−2, 1); B'(−3, −2); C'(−5, −3)
   c A'(−1, −2); B'(2, −3); C'(3, −5)
   d A'(1, 2); B'(−2, 3); C'(−3, 5)
3 b $y = x - 2$

Exercise 42C

2 a rotation about (0, 0) through 78° anticlockwise
   b rotation about (3, 0) through 45° clockwise
   c rotation about (−1, −2) through 90° anticlockwise
   d rotation about (−1, −1) through 233° clockwise
3 a (3, −13)   b 37° anticlockwise
4 A'(1, −2); B'(−2, −2); C'(−2, 1); D'(1, 1)
5 rotation about $(1\frac{1}{2}, -1\frac{1}{2})$ through 90° anticlockwise.

Exercise 42D

2 a (i) 3 (ii) (−5, −1)   b (i) 2 (ii) (7, −2)
   c (i) $2\frac{1}{2}$ (ii) (5, −3)   d (i) $-\frac{1}{2}$ (ii) (−2, 3)
3 a 2   b A'(4, 7)
4 enlargement scale factor 3 with centre $(9\frac{1}{2}, \frac{1}{2})$.
5 45 cm^2 6 750 in^2.

Exercise 42E

1 b (i) translation of $\begin{pmatrix} 0 \\ -10 \end{pmatrix}$

    (ii) translation of $\begin{pmatrix} -10 \\ 0 \end{pmatrix}$

    (iii) reflection in $y = -x$
    (iv) reflection in $y = x$
2 a rotation of 120° clockwise about (0, 0)
   b rotation of 180° about (0, 0)
   c no change
   d rotation of 180° about (0, 0)
3 a reflection in y-axis

   b translation of $\begin{pmatrix} 3 \\ 5 \end{pmatrix}$

   c rotation through 180° about $(-1\frac{1}{2}, 0)$ or
     enlargement S.F. −1, C of E $(-1\frac{1}{2}, 0)$
   d reflection in line $x = -2\frac{1}{2}$
   e rotation of 180° about $(0, -2\frac{1}{2})$ or enlargement
     S.F. −1, C of E $(0, -2\frac{1}{2})$

Exercise 42F

1 a $\begin{pmatrix} -2 \\ 3 \end{pmatrix}$ b $\begin{pmatrix} 5 \\ -6 \end{pmatrix}$

2 a reflection in the y-axis
   b reflection in $y + 3x = 0$
3 Rotation about 0 through 60° anticlockwise.
4 Enlargement of S.F. $\frac{1}{3}$, C of E (2, −3)

Exercise 43A

2 176 large and 528 small.

3 ☐

4 b, c, d and e.

Exercise 44A

1 a 22 cm   b 28 m
2 a 84 cm   b 28 cm   c 8 cm   d **28 cm**   e 68 cm
   f 66.6 cm
3 132 cm (1.32 m)

Exercise 44B

1 a 12 cm^2 b 13.5 cm^2 c 15.75 cm^2 d 70 cm^2
   e 42.75 cm^2 f 151.29 cm^2
2 a 60 cm^2 b 300 cm^2 c 75 cm^2 d 43.75 cm^2
   e 264.48 cm^2 f 240.25 cm^2
3 a 120   b 378   c 31.35   d 2
4 13 700 cm^2 5 a 30 m^2 b £299.70   6 26 m^2
7 a 144 cm^2 b 256 cm^2 8 a 12 m   b 50 m
9 30.25 cm^2
10 a 50 000 cm^2 b 864 square inches   c 1260 mm^2

Exercise 44C

1 a 28   b 6   c 41.31   d 7.5   e 1.5
2 a 11.25 cm^2 b 32.76 m^2
3 a 88 cm^2 b 245 mm^2 c 63 in^2 d 19.32 m^2
4 a 10 m   b 16 cm
5 a 16 cm   b 8 cm   c 12 mm   d 12.24 m^2
   e 23.65 in^2
6 a 72.7 cm^2 b 12.7 cm^2 c 122.2 m^2 d 130.4 m^2
   e 33.6 cm^2 f 12.2 cm^2

Exercise 44D

1 a 120 cm^2 b 75.6 cm^2 c 451.1 in^2
   d 3380 cm^2 (0.338 m^2)   e 1978 mm^2 (19.78 cm^2)
2 a 2.4 cm   b 0.33 m   c 13.4 cm^2 d 4 m   e 2.4 yd
3 a 94.6   b 54.9   c 378
4 a 8 cm   b 4 cm   c 5.25 in   5 a 7 cm   b 1.80 cm

Exercise 44E

1 60 cm^2 2 224 cm^2 3 139 in^2 4 99 cm^2

Exercise 44F

1 a 5 cm, 31.4 cm   b 8.6 in, 4.3 in   c 7 mm, 22.0 mm
   d 16.2 cm, 8.1 cm   e 1.9 m, 1.0 m
2 72.26 mm   3 3.46 cm   4 10.68 m   5 11.0 m
6 a 45.5   b 59.1   c 25.5   7 26.5

Exercise 44G

1 a 158 cm^2 b 2730 in^2 c 145 mm^2 d 37.7 cm^2
   e 33.8 m^2 2 a 3 cm   b 7 in   c 6 cm   d 4 m
   e 10 ft   3 a 2.93 cm   b 6.91 in   c 5.86 cm
   d 3.91 m   e 9.77 ft   4 8.0 m   5 £7.35   6 56 cm

Exercise 44H

1 a 74.3 m   b 11 000 m^2
2 a (i) 104 cm   (ii) 745 cm^2
   b (i) 123 cm   (ii) 914 cm^2
   c (i) 85.7 cm   (ii) 442 cm^2
   d 99.4 cm   (ii) 547 cm^2
3 a 773 m^2 b 514 cm^2 c 1500 cm^2 d 373 m^2
4 a 9.82 cm^2, 6.28 cm^2, 3.53 cm^2 b 9.82 cm^2

c The area of the large semicircle equals the sum of
   the areas of the other two semicircles.
   d 25.6 cm^2
5 a πR^2 b πr^2 c $\pi R^2 - \pi r^2$ d 221 cm^2
6 88.0 m^2 7 a 151 in   b 1670 in^2

Exercise 44I

1 a 90 cm^3 b 51.3 cm^3 c 0.336 m^2 d 1450 cm^3
   e 1.2 m^3
2 3.33 cm   3 a 512 cm^3 b 1331 cm^3 c 68.9 cm^3
4 4.0 cm   5 a 27 in^3 b 3 in   6 a 96 cu ft   b 2.27 ft
7 a 26 m   b 28 m^2 c 56   d 42 m^3
8 a 788 cm^3 b 136 cm^3 c 657 cm^3 9 313 m^3
10 a 603 cm^3 b 3050 cm^3 c 1690 in^3 d 388 cm^3
11 a 1.43 cm   b 2.60 cm   c 1.72 in
12 15 800 m^3 13 79.7 cm^3 14 a 9 m^2 b 135 m^3

Exercise 44J

1 Perimeter   2 Area   3 Volume   4 None
5 Volume   6 None   7 Area
8 a 222 cm^2
   b The product of two lengths (r and h) gives **area** not
   volume.

Exercise 45A

1 b 37°, 53°
2 a 13 cm   b 2.67 mm   c 2.8 in
3 no   4 a $x = 7.67$ in   b $p = 8.94$ cm, $q = 7.35$ cm
   c $x = 10.3$ cm, $y = 11.1$ cm
   d $x = 13.4$ cm, $y = 12.7$ cm, $z = 12.0$ cm
5 13.4 cm   6 24.0 cm   7 129 ft   8 145.5 cm
9 134.5 cm   10 170 cm

Exercise 45B

1 6.09 cm   2 28.8°   3 6.36 cm
4 $x = 11.0$ cm, $y = 3.59$ cm
5 a 7.26 cm   b 26.9 cm^2 6 36.0°
7 a 5.49 cm   b 5.85 cm   8 31.9°   9 18.2 cm

Exercise 45C

1 a 4.84 cm   b 7.21 cm   2 a 7.28 cm   b 9.10 cm
3 a 22.2°   b 16.1°   4 a 33.6°   b 51.1°
5 $x = 18.0$ cm, $y = 5.27$ cm
6 $p = 13.8$ cm, $q = 2.16$ cm   7 11.3 cm, 33.7°
8 34.4°   9 21.8 cm

Exercise 45D

1 13.1°   2 7.84 cm   3 7.77 cm
4 $p = 8.38$ cm, $q = 5.50$ cm   5 1.44 m
6 a 86.4 cm   b 85.1 cm   7 48.0°   8 37.7°
9 a 118 cm   b 38.3°   10 a 4.85 m   b 68.0°

Exercise 45E

1 229 ft   2 15.6°   3 736 ft   4 449 mph   5 7.70 m
6 12.5°   7 a 87.5°   b 89.0°   8 5.00°   9 49.8 m
10 21.3 m   11 a 25.0 ft   b 50.2°

Exercise 45F

1 34.4 in   2 7.92 m   3 7.46 ft, 23.7 ft   4 2.54 m
5 5.51 in   6 56.3°   7 a 5.79 cm   b 46.4°
8 a 65.1 m^2 b 782 m^3 c 35.4°
9 6.65°   10 25.3°   11 a 9.95 m   b 37.5°   12 21.5°

Exercise 45F

1 34.4 in   2 7.92 m   3 7.46 ft, 23.7 ft   4 2.54 m
5 5.51 in   6 56.3°   7 a 5.79 cm   b 46.4°
8 a 65.1 m²   b 782 m³   c 35.4°
9 6.65°   10 25.3°   11 a 9.95 m   b 37.5°   12 21.5°

Exercise 46A

1 a −0.105   b −0.914   c 0.700   d −0.785
2 Any two of …, −241.2°, 61.2°, 118.8°, 421.2°, …
3 Any two of …, −51.3°, 128.7°, 308.7°, 488.7° …
4 Any two of …, −69.4°, 69.4°, 290.6°, 429.4° …

Exercise 46B

1 a 41.6 cm²   b 25.9 cm²   c 12 cm²
2 207 cm²   3 8 ft²

Exercise 46C

1 a 53°   b 3.19 cm   c 3.27 cm
2 a 21.1°   b 46.9°   c 5.67 in
3 129 m, 15.8 m   4 612 mph

Exercise 46D

1 a 70.2° or 109.8°
   b 68.7° or 29.1°, 7.23 cm or 3.77 cm
2 10.6 km/h or 18.3 km/h
3 6.04 (or 2.30) m

Exercise 46E

1 X = 75.2°, Y = 65.6°, Z = 39.2°
2 a 8.01 in   b 56.7°
3 1.17 miles
4 2.90 km

Exercise 46F

1 34.7°; 128 m   2 7.23 m
3 025.9°; 17.2 min   4 340 miles
5 a 17.9 miles   b 58.4°

Exercise 47A

1 a 5.24 cm   b 16.8 cm   c 8.38 cm   d 30.6 cm
   e 25.4 cm   f 119 cm
2 a 51.8 cm   b 8.25 cm   3 182°   4 105 cm
5 2.93 cm   6 271 cm

Exercise 47B

1 51.3 cm²   2 106 cm²   3 29.5 cm²   4 7.86 cm
5 6.77 in   6 70.2°   7 203°

Exercise 47C

1 a 0.189 cm²   b 16.9 cm²   c 9.51 cm²   d 228 cm²
2 170 cm²   3 39.3 cm²   4 18.6 cu ft   5 1460 cm³

Exercise 47D

1 20 cm²   2 309 cm²   3 104 cm²   4 66 cm²
5 65.4 cm² 6 a 6.93 cm   b 2490 cm³   c $2\pi r^2 + 6r^2$

Exercise 47E

1 28 cm²   2 950 cm²   3 7.75 cm²   4 8.5 cm²   5 2
6 4   7 $7\frac{1}{2}$ 8 3:4

Exercise 47F

1 a 192 in³   b 8 in   2 a 7 cm³   b 35 cm²
3 15 cm   4 a 3:2   b 9:4   c 42 cm
5 a 378 cm³   b £2.02 (£2.03)

Exercise 47G

1 a 905 cm³   b 230 cm³   c 589 in³   2 64 cm³
3 160 cm³   4 144 cm³
5 a 268 cm³   b 435 cm³   c 20.6 in³
6 a 603 cm³   b 222 cm³   c 120 in³   7 3.92 cm
8 9.23 cm   9 3.57 in   10 1230 cm³

Exercise 47H

1 a 150 cm²   b 216 cm²   c 394 cm²
2 a 180 cm²   b 113 cm²   c 194 in²
3 4 cm, 96 cm²   4 3.04 cm, 55.3 cm²
5 0.8 cm, 101 cm²   6 211 cm²   7 249 cm²   8 893 cm²

Exercise 47I

1 303 cm²   2 1100 cm²   3 a 283 cm   b 5650 cm²

Exercise 47J

1 a 314 cm²   b 804 cm²   c 452 in²
2 a 201 cm²   b 1260 in²   c 460 cm²
3 a 1.49 cm   b 1.50 in   4 a 8.41 cm³   b 20.3 cm³

Answers to Appendix

Exercise A1

1 a 1497   b 2103   c 15 252   d 1076   e 3269
2 a 454   b 518   c 463   d 5155   e 1197

Exercise A2

1 296	5 6576	9 15 102
2 1224	6 2461	10 60 516
3 1075	7 2006	
4 6468	8 38 409	

Exercise A3

1 14	5 57	9 33
2 69	6 28	10 17
3 68	7 16	
4 96	8 27	

Exercise A4

1 41.9	3 133.97	5 16.71
2 180.3	4 16.7	6 51.75

Exercise A5

1 169.4	5 72.31	9 0.0072	13 400
2 702.36	6 47.86	10 0.0015	14 8
3 1770	7 6 000 000	11 0.000 210	15 500
4 7.8	8 0.028	12 9	16 0.025

Exercise A6

1 $\dfrac{4 \times 5}{2} = 10$ 4 $\dfrac{5 \times 63}{3} = 105$

2 $\dfrac{6 \times 12}{9} = 8$ 5 $\dfrac{12 \times 4}{5 + 3} = 6$

3 $\dfrac{43 \times 9}{6} = 64.5$ 6 $\dfrac{40 \times 20}{40} = 20$

7 $\dfrac{40 + 20}{3 + 3} = 10$

8 $(30 - 20)(30 + 10) = 400$

9 $10 \times 10 + (8 - 2)^2 = 136$

10 $\dfrac{20 \times 20}{50 - 10} = 10$

Index